ENCYCLOPAEDIA OF
NANO BIOTECHNOLOGY

ENCYCLOPAEDIA OF
NANO BIOTECHNOLOGY

Robert William

ANMOL PUBLICATIONS PVT. LTD.
NEW DELHI - 110 002 (INDIA)

ANMOL PUBLICATIONS PVT. LTD.
H.O.: 4374/4B, Ansari Road, Darya Ganj,
New Delhi-110 002 (India)
Ph.: 23278000, 23261597

B.O.: No. 1015, Ist Main Road, BSK IIIrd Stage
IIIrd Phase, IIIrd Block,
Bangalore - 560 085 (India)
Visit us at: www.anmolpublications.com

Encyclopaedia of Nano Biotechnology

First Published, 2006

ISBN 81-261-2980-8

PRINTED IN INDIA

Printed at Mehra Offset Press, Delhi.

Contents

Preface

This book "Encyclopaedia of Nano Biotechnology" mainly deals with subject area where the science of Nanotechnology and biotechnology interact. This new science uses biological structures as building blocks of nanoscale devices, for example nanomotors. Nano Biotechnology is also called `wet dry' technology, representing biological and engineering components respectively as wet a dry and is the real future science. Consideration progress has been achieved experimentally, besides the number of theoretically proposed bio-nanodevices is huge. As part of the popular (bio) nanobot concept science includes study of molecular motors, biomaterials single molemlet manuplation technologies, biochip technologies and artificial cells. There are so many other applications of nano chemistry ranging from mundane to incredibly profound. Other applications include energy storage materials, photovoltaic cells, fuel cells, polymer nanocomposites, super-tough nanocoatings, landmine detectors, and thousands more. Although it is a field that has already seen its share of hype, there is a general consensus that over the long term, nano chemistry is likely to reconfigure everything. Nanostructured materials do not represent a new phenomenon. However, the ability to probe, manipulate, understand and engineer matter at atomic scales has only recently come within our grasp. Developments such as the invention of the Scanning Tunneling Microscope in 1981 have since made nanoscale science a reality. Today, the term nanotechnology is used to describe specific construction and manipulation performed on an atomic and molecular scale. Generally, systems with a maximum size of less than a hundred nanometers are classified as nanotechnology. Nanotechnology is a completely new production process in a molecular world, opening up possibilities far beyond what we have known before. Nature provides the inspiration for this new technology – from time immemorial6 , nanomachines have been at work in all organic cells, building our world of plants, animals and people from a store of chemical elements.

Experts world over believe nanotechnology will have a more far-reaching impact than the silicon integrated chip with many more applications than electronics. Nanotechnology is expected to revolutionize certain areas, such as semiconductors, pharmaceuticals and materials. The total societal impact of nanotechnology is expected to be much greater than that of the silicon integrated circuit because it is applicable in many more fields than electronics. Nanotechnology experts believe that because of nano chemistry, we would see more changes in the next 30 years than we saw in the entire last century.

This book tends to cover the aforementioned subject areas in their broadest possible terms. All useful supplementary research and reference tools have been included for readers' further investigations in the subject area of Nano chemistry. This book mainly focuses on the following areas: Nanotechnology; Nanoparticles, Nanomaterials and Nanostructures. This book seeks to explain this complex field of study. It also seeks to provide an understanding of what is happening now in the field of nano chemistry, as well as what to expect in future by generations to come.

—Editor

1

Introduction to Nano Biotechnology

Some pragmatic readers may be wondering what is so special about molecular nanotechnology, that existing or anticipated biotechnology could not accomplish just as well? After all, biotechnology is already an established medical capability. It has real applications and real products already on the market. Reflecting upon the future possibility of sophisticated mechanical medical nanorobots equipped with powerful nanocomputers, in 1989, one well-known cryobiologist mused that "what is not clear is just what need we would have for such devices." The simplest answer, suggested, is that each of the three contemporary branches of "nanotechnology" offers something of unique value to the practice of medicine.

Nanoscale materials technology has already found widespread use in medicine, including biocompatible materials and analytical techniques, surgical and dental practice, nerve cell research using intracellular electrodes, biostructures research and biomolecular research using near-field optical microscopy, scanning-probe microscopy and optical tweezers, and vaccine design, and also many 20th century bulk chemical and biochemical manufacturing techniques along with much of classical pharmacology.

As for "biotechnology," the original meaning of this word contemplated "the application of biological systems and organisms to technical and industrial processes". In recent times, the field has expanded to include genetic engineering and now takes as its ultimate goal no less than the engineering of all biological systems, even completely artificial organic living systems, using biological instrumentalities.

The third branch, molecular nanotechnology, takes as its purview the engineering of all complex mechanical systems constructed from the molecular level — potentially offering new tools for medical practice, the principal subject of this book. Observes G.M. Fahy, "the difference between nanotechnologists and biotechnologists is that the former do not restrict themselves to the biological limitations of the latter, and they are much more ambitious about the kinds of accomplishments that they want to achieve."

Doctors can utilize solutions to medical problems from all three approaches. As noted earlier, 80-90% of medical complaints resolve themselves via natural homeostatic processes, or without the necessary involvement of active biotechnological or molecular-nanotechnological agents. But by employing biotechnology, the range and efficacy of treatment options greatly increases. With molecular nanotechnology, the range, efficacy, comfort and speed of possible medical treatments again expands enormously. Molecular nanotechnology is essential when the damage to the human body is extremely subtle, highly selective, or time-critical (as in head traumas, burns, or fast-spreading diseases), or when the damage is very massive, overwhelming the body's natural defenses and repair mechanisms.

An earlier discussion from G.M. Fahy, makes these options more explicit. At every difficulty level, most classes of medical problems may be resolved with varying efficacy within the homeostatic/nanomaterials, biotechnological, or molecular-nanotechnological approaches. As the chosen technology becomes more precise, active, and controllable, the range of options broadens and the quality of the options improves. Thus the question is not whether molecular nanotechnology is required to accomplish a given medical objective. In many cases, it is not — though of course there are some things that only biotechnology and molecular nanotechnology can do, and some other things that only molecular nanotechnology can do. Rather, the important question is which approach offers a superior solution to a given medical problem, using any reasonable metric of treatment efficacy. For virtually every class of medical challenge, a mature molecular nanotechnology offers a wider and more effective range of treatment options than any other approach.

It is quite possible to imagine an advanced biotechnology that uses an engineered white cell, fibroblast, or macrophage chassis, energized by native oxygen and glucose and modified mitochondrial powerplants, driven by pseudopodia, cilia or flagella, communicating and navigating via biochemical signals, and even incorporating onboard digital biocomputers to make microscopic biorobots. Principal arguments favoring the biotechnology approach for medical purposes are: (1) that we are already somewhat familiar with such systems, after half a century of intensive molecular biology research; (2) that we have already "built" precursor systems, such as a whole living amoeba constructed from five distinct parts, bioengineered viruses and bacteria as DNA insertion devices, and natural replication stimulated in genetically engineered starter microbes; (3) that biocompatibility will not be a major issue since fibroblasts (which express no HLA Class II antigens, hence stimulate no rejection response; could be used as the starter material; (4) that both engineered viruses and bacteria are already in wide commercial and research use; and (5) the greater complexity of self-repair in mechanical systems, should it be needed. Many believe that the development pathway to early biorobots may be considerably shorter than for the mechanical nanorobots of the molecular nanotechnology approach, for which in 1998 not a single working prototype yet existed, even in research laboratories.

It is also possible to imagine a molecular nanotechnology that uses mechanical nanorobotic systems. Such systems will have many constitutional differences from biological-based systems. For instance, mechanical systems will transport parts, materials, energy and instructions via fixed channels, whereas most (but not all) biological systems operate by diffusion. Mechanical systems will have structures constrained by specific geometries, whereas biological systems have structures defined by patterns of containment and interconnection — the shape of a membrane compartment in a cell matters less than its continuity and the contents of the volume it defines. Mechanical systems will be deterministically manufactured by operations analogous to manual construction, whereas engineered ribosomes will self-assemble via diffusion and stochastic matching of complementary parts; in other words, biology uses recipes, while mechanical systems use blueprints. Cells grow, with their parts adapting to one another; mechanical nanorobots may be constructed from parts of fixed structure. Biology uses self-repair; mechanical systems generally do not — largely because self-repair (by component exchange) will not be needed in molecular mechanical systems, whose designs may be made more simple by relying upon high component redundancy. These and other differences imply a number of important advantages that mechanical-based medical systems will enjoy over biological-based medical systems, which, taken together, strongly suggest that predominantly mechanical nanosystems may be the approach of choice for a mature medical nanotechnology. The advantages of molecular nanotechnology (e.g. the mechanical tradition) are many:

1. *Speed of Medical Treatment.* Doctors may be surprised by the incredible quickness of nanorobotic action when compared to the speeds available from fibroblasts or leukocytes. Normal homeostatic processes such as dermal wound repair via natural fibroblasts may require weeks to run to completion. Typical fibroblast movements occur at 0.1-1 microns/sec, but mechanical nanomanipulators can operate at 1-10 cm/sec speeds or faster, a speed advantage of 4-5 orders of magnitude. Even the strongest biological fibers (e.g. intermediate filaments) have a failure strength 3 orders of magnitude below the strongest mechanical fibers. Biological cilia beat at ~30 Hz while mechanical nanocilia may cycle up to ~20 MHz, though practical power restrictions and other considerations may limit them to the ~10 KHz range for most of the time. Flexible mechanical surfaces can complete a morphing motion in ~0.1 millisec, compared to the ~100 millisec snapback time for pinched red cell membrane, again a thousandfold advantage in speed. Thus we expect that mechanical therapeutic systems can reach their targets up to ~10,000 times faster, all else equal, and treatments which require ~10^5 sec for a biological system may need only ~10^2 sec for a mechanical system; tachyiatria improves both patient and physician comfort. In either biological or mechanical systems, large numbers of devices of comparable physical size (e.g. ~microns) may be employed to do the work, so numbers alone cannot offset the mechanical speed advantage.

2. *Power Density and Transduction.* Biological cells typically employ power densities

of 10^3-10^4 W/m^3, with maximum densities of ~10^6 W/m^3 in honeybee flight muscle cells and bacterial flagellar motors. By contrast, nanomechanical power systems can produce power densities of 10^9-10^{12} W/m^3, an advantage of 10^3-10^8 for mechanical over biological systems. By 1998, conducting polymer-based actuators generated 20-100 times the force for a given cross-sectional area as mammalian skeletal muscle. Additionally, amoebic locomotion in motile cells requires diffusion-limited cytoskeletal disassembly and reassembly to achieve movement; mechanical motility systems may employ simple cable-pulling, winches, or ratchets, which are faster and more direct. Some biological energy transducers are reversible, but muscle contraction is irreversible — not only cannot muscle actively re-expand, but stretching it doesn't make it produce much useful chemical energy. In contrast, electric motors can be run backwards to generate electricity, forcing pistons makes them pump, and loudspeakers can be used as microphones.

3. *Superior Building Materials*. Typical biological materials have tensile failure strengths in the 10^6-10^7 N/m^2 range, with the strongest biological materials such as wet compact bone having a failure strength of ~10^8 N/m^2, all of which compare poorly to ~10^9 N/m^2 for good steel, ~10^{10} N/m^2 for sapphire, and ~10^{11} N/m^2 for diamond and carbon fullerenes, again showing a 10^3-10^5 advantage for mechanical systems that use nonbiological materials. Nonbiological materials can be much stiffer, permitting the application of higher forces with greater precision of movement, and they also tend to remain stable over a larger range of temperature, pressure, salinity and pH. Proteins are heat sensitive in part because much of the functionality of their structure is due to noncovalent bonds involved in folding, which are broken more easily at higher temperatures; in diamond, sapphire, and many other rigid materials, structural shape is covalently fixed, hence is far more temperature-stable. Most proteins tend to become dysfunctional at cryogenic temperatures, unlike diamond-based mechanical structures. Biomaterials are not ruled out for all nanomechanical systems, but represent only a small subset of the materials that can be used in nanorobots. Mechanical systems can employ a wider variety of atoms and molecular structures in their design and construction, with novel functional forms that might be difficult to implement in a biological system such as steam engines or nuclear power. As another example, an application requiring the most effective bulk thermal conduction possible should use diamond, the best conductor available, not some biomaterial with inferior thermal performance.

4. *Nondegradation of Treatment Agents*. Diagnostic and therapeutic agents constructed of biomaterials generally are biodegradable in vivo, although there is a major branch of pharmacology devoted to designing drugs that are moderately non-biodegradable — anti-sense DNA analogues with unusual backbone linkages and peptide nucleic acids (PNAs) are difficult to break down. However, suitably designed nanorobotic agents constructed of nonbiological materials are not biodegradable. An engineered fibroblast may not stimulate an immune response when transplanted into a foreign host, but its biomolecules are subject to chemical attack in vivo by free radicals, acids, and enzymes. Even "mirror" biomolecules or "Doppelganger proteins" comprised exclusively of unnatural

D-amino acids have a lifetime of only ~5 days inside the human body. Nonbiological materials such as diamond and sapphire are highly resistant to chemical breakdown or leukocytic degradation in vivo, and pathogenic biological entities cannot easily evolve useful attack strategies against these materials.

5. *Control of Nanomedical Treatment*. Present-day biotechnological entities are not programmable and cannot be switched on and off conditionally during task execution. A digital biocomputer, while possible in theory, represents a considerable conceptual departure from the usual biological paradigm. Even assuming that a digital biocomputer could be installed in a fibroblast, and that appropriate effector mechanisms could be attached, such a system would necessarily have slower clock cycles, less capacious memory per unit volume, and longer data access times, implying less diversity of action, poorer control, and less complex executable programs than would be available in nanoscale electromechanical computer systems. The mechanical approach emphasizes precise control of action, including control of physical placement, timing, strength, structure, and interactions with other (especially biological) entities. The biological approach emphasizes the use of poorly controlled natural structures, needlessly sacrificing huge blocks of the available functionality and design space.

6. *Nanodevice Versatility*. Mechanical systems can readily incorporate biological elements if necessary, but artificial biological systems can incorporate nonbiological materials such as carbon nanotubes or diamond/sapphire structural elements only with difficulty, in part because biology has a more limited repertoire of "effector" mechanisms. Artificial biological systems cannot easily incorporate nonbiological materials where desired because natural biological assembly methods make no provisions for these materials either in the coded instructions in DNA or in the attachment chemistries. Rebuilding or reconstructing the human body with nonbiological components, or augmentation of human body function with unnatural abilities or features, will be very difficult or impossible to achieve using purely biotechnological means.

7. *Avoiding Overspecialization*. M. Krummenacker notes that one of the most glaring shortcomings of bacteria and other naturally occurring molecular machinery — when viewed as systems subject to further engineering — is the rather limited range of molecular substrates they can utilize. For example, bacterial enzymes are highly specialized devices, with very narrow substrate specificities. Thousands of different enzymes are needed in each organism, and the substance classes capable of digestion are limited to some sugars, various amino acids including proteins that can be degraded by excreted proteases, lipids, and a few other smallish oxygen-functionalized carbon molecules such as glycerol and ethanol. Some bacteria can metabolize CO_2 and a handful of aromatic compounds, but there is a vast range of organic chemicals that most bacteria cannot degrade or manufacture. A much smaller set of substantially more general molecular tools can probably be designed using the mechanical approach; mechanosynthesis can fabricate and assemble, or disassemble, a far wider range of molecular structures than are available to the cellular machinery of life.

8. *Faster and More Precise Diagnosis*. The analytic function of medical diagnosis requires rapid communication between the injected devices and the attending physician. If limited to chemical messaging, biotechnology devices will require minutes or hours to complete each diagnostic loop. Nanomachines, with their more diverse set of input-output mechanisms, can outmessage the results of in vivo reconnaissance or testing literally in seconds. Such nanomachines can also run more tests of greater variety in less time. Mechanical nanoinstruments, including molecule-by-molecule disassemblers, will make comprehensive cell mapping and cell interaction analysis possible. Bacterial resistance can be assayed at the molecular level, allowing new treatment agents to more easily be composed, manufactured and immediately deployed.

9. *More Sensitive Response Threshold for High-Speed Action*. Unlike natural systems, an entire population of nanobiotic devices can be triggered globally by just a single local detection of the target antigen or pathogen. The natural immune system takes $>10^5$ sec to become fully engaged after exposure to a systemic pathogen or other antigen-presenting intruder. A biotechnologically enhanced immune system that can employ the fastest natural unit replication time ($\sim10^3$ sec for some bacteria) will require $\sim10^4$ sec for full deployment post-exposure. By contrast, a nanobiotic immune system can probably be fully engaged (though not finished) in at most two blood circulation times, or $\sim10^2$ sec.

10. *More Reliable Operation*. Engineered macrophages would probably individually operate less reliably than mechanical nanorobots. Many pathogens, such as *Listeria monocytogenes* and *Trypanosoma cruzi*, are known to be able to escape from phagocytic vacuoles into the cytoplasm; while biotech drugs or cell manufactured proteins could be developed to prevent this (e.g. cold therapy drugs are entry-point blockers), nanorobotic trapping mechanisms can be more secure. Proteins assembled by natural ribosomes typically incorporate one error per $\sim10^4$ amino acids placed; current gene and protein synthesizing machines utilizing biotechnological processes have similar error rates. A molecular nanotechnology approach will improve error rates by at least a millionfold. Mechanical systems can also incorporate sensors to determine if and when a particular task needs to be done, or when a task has been completed. Finally, it is unlikely that natural organisms will be able to infiltrate mechanical nanorobots or to co-opt their functions. By contrast, a biological-based robot could be diverted or defeated by microbes that can piggyback on its metabolism, interfere with its normal workings, or even incorporate the device wholesale into their own structures, causing the engineered biomachine to perform some new or different function than was originally intended. There are many examples of such co-option among natural biological systems, including the protozoan mixotrichs found in the termite gut that have assimilated bacteria into their bodies for use as motive engines, and the nudibranch mollusks (marine snails without shells) that steal nematocysts (stinging cells) away from coelenterates such as jellyfish (i.e. a Portuguese man-of-war) and incorporate the stingers as defensive armaments in their own skins, a process which S. Vogel has called "stealing loaded guns from the army."

11. *Verification of Progress and Treatment*. Using a variety of communication modalities, nanorobots can report back to the attending physician, with digital precision, a summary of diagnostically- or therapeutically-relevant data describing exactly what was found, and what was done, and what problems were encountered, in every cell visited. A biological-based approach relying upon chemical messaging is necessarily slow with limited signaling capacity. Also unlike mechanical nanorobots, biotechnological systems generally cannot monitor their own functions while working, and, except for a few highly specialized DNA proofreading systems, cannot directly inspect their work while it is in progress or after it is finished.

12. *Minimum Side Effects*. Almost all drugs have side effects, such as conventional cancer chemotherapy which causes hair loss and vomiting, although computer-designed drugs have high specificity and relatively few side effects. Carefully tailored cancer vaccines under development in the late 1990s were expected unavoidably to affect some healthy cells. Even well-targeted drugs are distributed to unintended tissues and organs in low concentrations, although some bacteria can target certain organs fairly reliably without being able to distinguish individual cells. By contrast, mechanical nanorobots may be targeted with virtually 100% accuracy to specific organs, tissues, or even individual cellular addresses within the human body. Such nanorobots should have few if any side effects, and will remain safe even in large dosages because their actions can be digitally self-regulated using rigorous control protocols that affirmatively prohibit device activation unless all necessary preconditions have been, and continuously remain, satisfied. G.M. Fahy has noted that these possibilities could transform "drugs" into "programmable machines with a range of sensory, decision-making, and effector capabilities [that] might avoid side effects and allergic reactions...attaining almost complete specificity of action....Designed smart pharmaceuticals might activate themselves only when, where, and if needed." Additionally, nanorobots may be programmed to excuse themselves from the site of action, or even from the body, after a treatment is completed; by contrast, spent biorobotic elements containing ingested foreign materials may have more limited post-treatment mobility, thus lingering at the worksite causing inflammation when naturally degraded or removed R. Bradbury notes that artificial eukaryotic biorobots may possess an apoptotic pathway which could be activated to permit clean and natural self-destruction that avoids inflammation in surrounding tissues. Artificial prokaryotic biorobots could also be designed for human biocompatibility; by replacing bacterial genes with generic human genes, such a device may be undetectable to the immune system.

13. *Reduced Replicator Danger*. Drexler points out that living systems are evolved systems, while nanomechanical replicators would be designed: "The former are shaped to serve the goal of their own survival and replication in a natural environment, whereas the latter will be shaped (whether well or poorly) to serve human goals, perhaps in an artificial environment." Genetic engineering involves not design of replicators from scratch, but tinkering with the molecular machinery of existing bioreplicators. Since bioreplicators were not designed, they are not necessarily structured

in ways that lend themselves to complete understanding, and processes based on diffusion and matching allow complex nonlocal interactions that can be hard to trace. Bioreplicators can be crippled, but having evolved in nature, they resemble systems that can survive in nature. Typically, they are able to exchange genetic information with wild organisms, raising the possibility of the introduction of new, unconstrained replicators in the natural environment. Having evolved to evolve, they have a capacity for further evolution — to serve their own survival, not human goals. For example, even when stripped of key pieces of DNA to interfere with its replication powers, a live attenuated AIDS vaccine can slowly recover its virulence and can attack immune cells. R. Bradbury suggests that artificial biorobots could incorporate multiple fail-safe mechanisms including required external essential nutrients or suicide suppressors, self-destruct triggers, countdown timers like telomeres, and engineering to reject foreign DNA (the basis for restriction enzymes), but it remains logically easier for a system with inchoate capacity to evolve to resume doing so, than for a system which has never had this capability to spontaneously develop it.

In contrast, nanomechanical replicators will be designed from scratch and thus will differ fundamentally from biological systems. The parts and structures of designed mechanical systems will be known, and the relationships among their parts will also be designed and fixed. More important, nanoreplicators will be fundamentally alien to the biosphere, unrelated to anything that has evolved to survive in nature. Certainly the capacity to fail can appear by accident, and emergent capabilities cannot be completely ruled out, but engineering experience shows that the ability to perform complex organized activities (such as replication in a natural environment) does not normally appear spontaneously. Also, and purely as a geometrical consideration, adding a new part inside a densely organized geometric structure (as would be found in a nanomachine) typically requires changes in the relative positions of many other parts, and hence corresponding adjustments in design. Adding a part inside a densely organized topological structure (as found in living systems) typically leaves topologies unchanged — room can be made by stretching and shifting other parts, with no change in their essential design, hence permitting easier modification to biological structures by exogenous agencies.

Note that mechanical medical nanodevices need not be capable of replication. There is no requirement for replication in vivo; such replication would be needlessly dangerous, and adding this capability would reduce effectiveness in carrying out the primary medical task. Analogously, viral vectors employed in genetic therapies are modified to be "incapable" of replication.

14. *Assured Patentability*. Microscopic biorobots, unavoidably derived from natural biological material, may someday be deemed unpatentable under a general prohibition on "genetic colonialism" or other emerging legal doctrines. In contrast, mechanical nanorobots, being fully-artificial and designed machines, should always be patentable provided they satisfy the customary legal criteria.

On the 25th anniversary of Diamond vs. Chakrabarty,* the US Supreme Court's landmark decision (June 16, 1980) that opened the floodgates to the patenting of living organisms, ETC Group releases a new report, "Nanotech's 'Second Nature' Patents." Since Chakrabarty, the biotech industry has worked hand-in-hand with governments to allow for the patenting of all biological products – the first monopoly grab over life. Chakrabarty set the stage for today's nanotechnology patents, where the reach of exclusive monopoly is not just on life – but the building blocks of life – nanotech's 'second nature' patents," explains Hope Shand, Research Director of ETC Group.

ETC Group's new report examines current trends in intellectual property and nanotechnology and the implications for the developing world. Nanotechnology refers to the manipulation of matter at the scale of atoms and molecules, where size is measured in billionths of meters. The world's largest transnationals, leading academic labs and nanotech start-ups are all racing to win monopoly control of tiny tech's colossal market. "Control and ownership of nanotech is a vital issue for all governments and civil society because nanomaterials and processes can be applied to virtually any manufactured good across all industry sectors," said Kathy Jo Wetter of ETC Group. "Patents are being granted that cut across multiple industry sectors – a single nano-scale innovation may span pharma, food, electronics and materials alike," continues Wetter. The US National Science Foundation predicts that nanotechnology will capture a $1 trillion dollar market within six or seven years.

ETC Group finds that breathtakingly broad nanotech patents have been granted that cut across multiple industry sectors and include sweeping claims on entire areas of the Periodic Table. Although industry analysts assert that nanotechnology is in its infancy, "patent thickets" on fundamental nano-scale materials, tools and processes are already creating thorny barriers for would-be innovators. Claims are often broad, overlapping and conflicting—a scenario ripe for massive patent litigation battles in the future. ETC Group's report provides case studies of patent activity involving four of nanotech's hottest and potentially most lucrative nanomaterials and one essential tool: carbon nanotubes; inorganic nanostructures; quantum dots; dendrimers; scanning probe microscopes.

G8: Downsizing Development? When the G8 Summit meets in Scotland next month, the leaders of the world's most powerful countries will unveil a "Pro-Poor Science" strategy to turn new technologies like nanotech into a silver bullet for social injustice. "Despite rosy predictions that nanotech will provide a technical fix for hunger, disease and the environment, the extraordinary pace of nanotech patenting suggests that developing nations will participate primarily via royalty payments," said Pat Mooney, Executive Director of ETC Group. "In a world dominated by proprietary science, researchers in the global South are likely to find that participation in the nanotech revolution is highly restricted by patent tollbooths, obliging them to pay royalties and licensing fees to gain access," said Mooney. "Ultimately, nanotech will profoundly affect the South's economy, regardless of its handling of intellectual property," explains

Silvia Ribeiro from ETC Group's Mexico City office. "Nano-scale technologies will revolutionize the way that new materials are designed and manufactured – changes that could turn commodity markets upside-down and make geography, raw materials, even labour, irrelevant. Nanotech underpins a new strategic platform for global control of materials, food, agriculture and health, and patent monopoly is a powerful tool for realizing that strategy," said Ribeiro.

Many South nations are still grappling with unresolved controversies over biotechnology, but by the end of this year, ready or not, even the world's "least developed" nations who are members of the World Trade Organization will be obligated by its Trade-Related Aspects of Intellectual Property (WTO-TRIPs) to evaluate and enforce nanotech patents.

Lessons learned from Diamond v. Chakrabarty: Despite all the hype about Mr. Chakrabarty's oileating microbe and how it would gobble up oil spills, the patented microorganism never worked. Instead of curing environmental ills, the biotech industry has introduced its own contamination problems – unwanted gene flow from genetically modified crops, a particularly serious problem for centres of genetic diversity in the developing world. Unlike 25 years ago, today's nanotech-related patents have not required major rule changes. As a result, many governments are unaware of the nanotech patent rush. ETC Group recommends that the World Intellectual Property Organization (WIPO) initiate a global suspension of patent approvals related to nanotechnology until South governments and countries-in-transition can undertake a full evaluation of their impacts, and until social movements can cooperate with WIPO, the Food and Agriculture Organization (FAO) and the United Nations Conference on Trade and Development (UNCTAD) to examine the impact of nanotech-related intellectual property on monopoly practices, technology transfer and trade. The full text of the 32-page report is available for downloading, free-of-charge, on the ETC Group.

The Action Group on Erosion, Technology and Concentration, formerly RAFI, is an international civil society organization headquartered in Canada. The ETC Group is dedicated to the advancement of cultural and ecological diversity and human rights. www.etcgroup.org. The ETC Group is also a member of the Community Biodiversity Development and Conservation Programme (CBDC). The CBDC is a collaborative experimental initiative involving civil society organizations and public research institutions in 14 countries. The CBDC is dedicated to the exploration of community-directed programmes to strengthen the conservation and enhancement of agricultural biodiversity.

Implications for the Global South is Nanotech "Pro-Poor Science" or is the G8 Downsizing Development? On the 25th anniversary of the US Supreme Court decision that opened the floodgates to the patenting of all life forms, ETC Group reports on current trends in intellectual property relating to nanoscale technologies. With nanotechnology, the reach of exclusive monopoly patents is not just on life, but all of nature. Accordingly, ETC Group refers to nanotech's "second nature" patents. Nanotech patent claims offer the South (and society) an advance look at who's likely to own nanotechnology and dominate 21st century commodity markets. Breathtakingly broad nanotech patents are being granted that span multiple industry sectors and include sweeping claims on entire classes of the Periodic Table. Yet, when the G8 nations meet this July they will be unveiling a "Pro-Poor Science" strategy to turn nanotechnology into a silver bullet antidote to social injustice. Is nanotech the solution – or just another big downer for development?

The race is on to win monopoly control of tiny tech's colossal market. The US National Science Foundation predicts that the immensely broad power and scope of nano-scale technologies will revolutionize manufacturing across all industry sectors – capturing a $1 trillion market within six or seven years. Although industry analysts assert that nanotech is in its infancy, "patent thickets" on fundamental nano-scale materials, building blocks and tools are already creating thorny barriers for would-be innovators. Industry analysts warn that, "IP roadblocks could severely retard the development of nanotechnology." Some insist that nano-scale technologies will address the most pressing needs of the South's marginalized peoples. But in a world dominated by proprietary science, it is the patent owners and those who can pay license fees who will determine access and price. Even as South governments are grappling with confusion and controversies over biotechnology, the World Trade Organization's Trade-Related Aspects of Intellectual Property (TRIPs) obligates even "least developed" countries to enforce nanotech patents by 2006.

The world's largest transnationals, leading academic labs and nanotech start-ups are all racing in the patent gold rush. Increasingly, universities are licensing on an exclusive basis. Nanotech's "second nature patents" are positioning multinational matter moguls to own and control novel materials, devices and their manufacturing processes.

While WIPO ponders a "development agenda" in Geneva, patent offices in Washington, Munich and Tokyo are deciding who will gain exclusive monopoly over a technology that will bring profound changes in demand for raw materials and manufacturing around the globe. Despite rosy predictions that nanotech will provide a technical fix for hunger, disease and environmental security in the South, the extraordinary pace of nanotech patenting suggests that developing nations will participate via royalty payments. Accordingly, national governments must protect their sovereignty and intergovernmental agencies must move quickly to prevent multi-sector monopolies, technology barriers and the formation of a powerful new oligopoly that seeks control over the economy's 'second nature.'

Control and ownership of nanotechnology is a vital issue for all governments and civil society because nanomaterials and processes can be applied to virtually any manufactured good across *all* industry sectors. Nanotechnology refers to the manipulation of matter at the scale of atoms and molecules, where size is measured in billionths of meters (one nanometer = one-billionth of a meter). Nanotech isn't a single technology – but a range of technologies converging at the nano-scale – including biotechnology, genomics, neurosciences, robotics and information technologies. For a brief introduction to nanotechnology, please see, *A Tiny Primer on Nano-Scale Technologies* [http:// www.etcgroup.org] Worldwide, industry and governments invested more than $10 billion in nanotech R&D last year, with two-thirds coming from corporate and private funds.3 There are an estimated 1200 nanotech start-up companies, half of which are US-based.4 Virtually all Fortune 500 companies invest in nanotech R&D. A recent survey conducted by MIT's *Technology Review* predicts that products involving nanotech will account for over $100 billion by 2008; the US government predicts that the nanotech market will explode to $1 trillion by 2012. Intellectual property (IP) will play a major role in deciding who will capture nanotech's trillion dollar market, who will have access to nano-scale technologies and at what price. At stake is control over innovations that span multiple industry sectors – from electronics, energy, mining and defense to new materials, pharmaceuticals and agriculture. As the *Wall St. Journal* put it, "companies that hold pioneering patents could potentially put up tolls on entire industries." Even industry insiders admit that current intellectual property trends related to nanotech are chaotic.

Many broad patents on nanotech-related materials, tools and processes have been granted too early and too often. In 2002, the US-based industry trade group, Nanotechnology Business Alliance, was already warning in testimony before the US Congress, "...several early nanotech patents are given such broad coverage, the industry is potentially in real danger of experiencing unnecessary legal slowdowns."

More recently, nanotech industry analysts observe that the "euphoria for patenting" in the US combined with the US Patent & Trademark Office's inability to handle a flood of patent applications has resulted in "the rejection of valid claims, the issuance of broad and over-lapping claims, and a fragmented and somewhat chaotic IP landscape." The writers warn, "These IP roadblocks could severely retard development of nanotechnology."

This report aims to provide information to civil society and South policymakers about current trends in nanotechnology and intellectual property. What is the current IP landscape? What is being patented and by whom? What are the implications for innovation and development in the South? How will intellectual property affect the South's participation in the nanotech revolution? "Like biotechnology in the early 1980s, the IP landgrab mentality pays off for early pioneers. Don't bet

Patents on nano-scale materials, tools and processes are a fair measure of the tsunami-like strength of this latest industrial revolution. Estimates vary on the number of nanotech patents issued since the early 1990s, but all agree that both companies and public sector entities are "rushing to the patent office in record numbers to patent nanotechnology inventions."

The lack of uniform definitions for nanotechnology means that identifying the number of nanotechrelated patents granted over the past decade is a very imprecise science. When the US PTO announced it had created a new classification for nanotechnology patents in October 2004 it defined nanotechnology patents narrowly: Classification 977 includes only those patents 1) whose subject matter is in the scale of approximately 1-100 nanometers in at least one dimension; and 2) that involve materials, structures, devices or systems that have novel properties and functions because of their nano-scale size. Nanotech patent searches often use broad search terms (for example, the prefix "nano"), which can result in exaggerated counts. There is wide consensus, however, that major patent offices worldwide are granting nanotech patents at an extraordinary pace. The chart below illustrates the overall trend, which is similar at the World Intellectual Property Organization's Patent Cooperation Treaty and at the US PTO. In response to the demand for nanotech patents, more than a dozen major law firms in the US have recently established nanotech patent law specialties. Researchers from the University of Arizona and the US National Science Foundation examined nanoscale science and engineering patents at the US Patent & Trademark Office from 1976-2003. 12 They found that 8,630 nanotech-related patents were issued by the US PTO in 2003 alone, an increase of 50% over the previous three years. The top 5 countries represented were: US (5,228 patents), Japan (926), Germany (684), Canada (244) and France (183). The top 5 entities winning nanotech-related patents included four multinational electronic firms and one university: IBM (198 patents), Micron Technologies (129), Advanced Micro Devices (128), Intel (90) and University of California (89). A new report by Lux Research, Inc. identifies far fewer nanotech patents granted by the US PTO. In April 2005 Lux announced that it had identified 3,818 nanotech-related patents issued between 1985-March 2005, with an additional 1,777 patent applications.

ETC Group takes a Closer Look at Nanotech Patents Granted by the US PTO: In October 2004, the US PTO announced it had created a new classification for nanotechnology patents – Class 977 – which would serve as a cross-reference to help examiners, among others, search prior art. Before Class 977 existed, examiners relied on keyword searches to find relevant information and related patents.15 As defined by the US PTO, nanotechnology patents in Class 977 must meet the following criteria:

- relate to research and technology development in the length scale of approximately 1-100 nm in at least one dimension
- provide a fundamental understanding of phenomena and materials at the nanoscale and create and use structures, devices, and systems that have size-dependent novel properties and functions.16 The US PTO is now reviewing patents issued before the classification was created and re-classifying those that meet the 977 criteria. According to Bruce Kisliuk, a patent examiner group director, the Patent Office hopes to be caught up

reviewing already-issued patents within a year.17 Until the review is complete, the PTO is referring to Class 977 as a "digest." Eventually, Class 977 will be broken down further into sub-classes. As of May 25, 2005, Class 977 included 726 patents issued by the US PTO, dating from May 24, 2005 back to 1980. In its incomplete state, Class 977 should be seen as a sample rather than a comprehensive picture of nano-patenting in the US. (At present, Class 977 contains only six nano-patents dating from the second half of 2004, for example.) Even though the classification is incomplete, the sample is worth looking at because it offers an advantage over more random patent searches (e.g., searching for keywords such as "nano" or "quantum"): we can be reasonably confident that the patents in Class 977 are bona fide nano-patents because they have been reviewed and each was determined to meet the classification criteria – relating to the nanometer scale (1-100 nm) in at least one dimension and exploiting size-dependent properties. A search through Class 977 reveals:

- The USPTO has issued nano-patents in the fields of electronics (including cameras, computers and other devices), instruments and tools (including scanning probe microscopes and sensors), pharmaceuticals (including drug delivery), food, agrochemicals, devices, materials (including fullerenes, nanotubes and quantum dots) and processes for creating and engineering nano-sized particles.
- More than 290 different primary patent examiners were assigned to evaluate these Class 977 patents. The extraordinarily large number of examiners raises suspicions that over-lapping and conflicting patents were granted by different examiners at about the same time; it also illustrates the multi-disciplinary and cross-sectoral nature of nano-scale science.
- Out of the 235 Class 977 patents issued in 2003 (the most complete year):
- 112 (48%) are assigned to US-based companies
- 29 (over 12%) are assigned to US universities
- 70 (30%) are assigned to companies based outside the US (of these, 26 are assigned to companies based in Japan, 10 in Germany, 9 in Australia – 8 of which are assigned to one company, Silverbrook Research. Companies from Canada, Venezuela, France, Korea, Belgium, Denmark, Italy, Ireland, Netherlands, Taiwan and Antilles make up the other non-US assignees.)
- 15 (6.5%) are assigned to non-US government entities or national research agencies
- The US government has rights to 31 patents (either because it provided

funding for research or through outright ownership), which represents 22% of the patents assigned to US entities

Class 977 patents issued by the USPTO range from mundane to alarming to otherworldly. One would expect only great things from the Delphi Oracle Corporation and it does not disappoint: Delphi owns patent 6,493,910, "a shoelace with enhanced knot retention and method of manufacture." (It's a nanopatent because the shoelace makes use of nano-scale fumed silica.) On the disturbing end of the spectrum, the US government won patent 5,805,657 back in 1998 for "nuclear fuel elements made from nanophase materials," which makes one wonder in the event of infringement – will there be anyone around to sue? Zyvex Corporation owns patent 6,510,359, a "method and system for self-replicating manufacturing stations." In case anyone devises one of these in the next 20 years, it looks like Zyvex will own it. They claim a "non-biological self replicating manufacturing system" comprising:

- a translating machine capable of translating at least a first surface in relation to at least a second surface, wherein said at least a first surface comprises at least one assembly station thereon and wherein said at least a second surface comprises parts arranged thereon for use in constructing at least one other assembly station; and
- a control system for controlling the operation of said translating machine and for controlling said at least one assembly station to cause said at least one assembly station to construct at least one other assembly station from said parts on said second surface.

Patent Paradox: For most nanotech start-up companies (especially those that operate without products or profits), intellectual property is the most essential asset. In the words of one industry executive, the strength of a nanotech company's patent portfolio is what separates the winners from the losers.19 Consider, for example: Nanogen's stock jumped over 50 percent on the day it announced it had won a patent on a technology to detect genetic variants; NVE Corporation's stock surged 41 percent when it announced its new patent on magnetic random access memory.

But the transaction costs of filing for and defending patents are enormous. Competing in the high-stakes patent game requires not only winning and defending patents, but also using them as bargaining chips to cross-license other proprietary technologies. Intellectual property is one of the biggest expenses for fledgling start-ups, and nanotech industry insiders predict that long and costly patent litigation battles could inflict mortal wounds. Consider, for example:

- In the US, depending on the complexity, claims, and length of the application, filing fees can go as high as $1000.21

- For nanotech start-up companies, patent fees are often a major cost of doing business – sometimes second only to payroll.22 Legal fees for winning a patent in the US can cost $25,000- $30,000, and up to $250,000 for international patent claims.23
- Outside the US, countries charge around $5,000 per year on each pending patent. The translation fees alone to win a patent in Japan are between $12,000 and $20,000.24
- The number of patent lawsuits filed in the US more than doubled over the decade of the 1990s. Based on a survey of intellectual property lawyers in 2000, the cost of defending a large (more than $25 million at risk) patent infringement suit ranges from $2 million to $4.5 million. For cases with less than $1 million at risk, the cost was $300,000 to $750,000, or about half the amount in dispute.
- Most shocking of all, it is estimated that only about two percent of all issued patents end up generating more revenue than the cost of obtaining the patent. Among this small group of "profitable" patents, only some will end up being worth the price of litigation.

Many intellectual property experts in the US are predicting that large-scale nanotech patent litigation is inevitable, and it's likely to be ugly. Because of the large number of over-lapping and conflicting patents being granted, nanotech companies must be prepared to vigorously defend their patents in court. It's conventional wisdom that, in most patent battles, it's the largest enterprises – not the most innovative – that will prevail. According to authors Josh Lerner and Adam Jaffe, "the firm with the best lawyers or the greatest capacity to withstand the risk of litigation wins the innovation wars – rather than the company with the brightest scientists or most original, valuable ideas."

"Nobody's making any money right now, so there's no particular reason to fight the battles yet. But a lot of people are coming out with patents, and a lot of claims are certainly overlapping, if not certainly *conflicting. The fights are going to be brutal."* – Matthew Nordan, vice-president of research, nanotech analyst firm, Lux Research Firms can use patents as "strategic weapons to strangle competition." If a larger firm believes that its dominance in a nanotechnology market is threatened, it "can slowly starve their start-up competitors into extinction by waging a protracted battle on the IP front."30 As with biotech in the 1980s and 1990s, nanotech start-ups holding key patents may also become attractive takeover targets because it's less costly for a multinational firm to acquire the company than to litigate in court. In part two of this report ETC Group provides case studies of nanotech and intellectual property, focusing on some of the key nanomaterials or tools that are the subject of multiple (and likely) over-lapping patent claims. The case studies examine: carbon nanotubes, Nanosys

Inc.'s patents on inorganic nanomaterials, quantum dots, dendrimers and scanning probe microscopes. What are the major trends in Nanotech-related IP?

Patents on nanotech's fundamental building blocks and tools. Stanford University law professor Mark Lemley asserts that nanotechnology "is the first new field in a century in which people started patenting the basic ideas at the outset." In contrast to most other major enabling technologies of the 20th century (such as computer hardware, software, the Internet, and even biotechnology), writes Lemley, the most basic ideas and fundamental building blocks in nanotechnology "are either already patented or may well end up being patented." ETC Group does not share Lemley's view that biotech's fundamental enabling tools were not patented early on – but we do share his view that nanotechnology's most basic ideas and building blocks are being patented early and often.

In the nanotech arena, it's not just the opportunity to patent the most basic enabling tools, but the ability to patent the nanomaterials themselves, the products they are used in and the methods of making them. At the US Patent & Trademark Office there are three primary types of patent claims:

1) composition of matter claims (that is, nanomaterials such as nanotubes, nanowires and nanoparticles),
2) device, apparatus or system claims (including, for example, tools used to characterize and control nanomaterials—or devices incorporating nanomaterials),
3) method claims (processes for synthesizing nanomaterials or constructing nano-scale devices).

The CEO of Nanosys, Inc., Larry Bock, explains the once-in-a-lifetime attraction of being able to get in on the ground floor by consolidating fundamental nanotech patents: "One of the key things that got me excited about this whole area three years ago is that from an IP standpoint you could control IP from things like composition of matter to the simplest functional devices to the end application all under one IP portfolio. That doesn't come around very often in one's career."

Nanomaterials are chemical elements or compounds less than 100 nm in size. Taking advantage of quantum physics, nanotech companies are engineering novel materials that may have entirely new properties never before identified in nature. The "raw materials" for creating nanomaterials and devices are the chemical elements of the Periodic Table – the building blocks of *everything* – both living and nonliving. Whereas biotechnology patents make claims on biological products and processes – nanotechnology patents may literally stake claim to chemical elements, as well as the compounds and the devices that incorporate them. With nano-scale technologies the issue is not just patents on life – but on all of nature. In short, atomic-level manufacturing provides new opportunities for sweeping monopoly control over both animate and inanimate matter. In essence, patenting at the nano-scale could mean monopolizing the basic elements that make life possible.

Proprietary Periodic Table: Exclusive monopoly patents on chemical elements are not new. Glenn Seaborg, the 1951 Nobel Prize-winning physicist, won US patent #3,156,523 for the chemical element *Americium* (element no. 95 on the periodic table) on November 10, 1964. Seaborg's patent is recognized for having the shortest patent claim on record: "What is claimed is Element 95." Seaborg's second patented element was Curium #96 – US patent # 3,161,462 granted on December 15, 1964. "It is true that one cannot patent an element found in its natural form; however, if you create a *purified form of it that has industrial uses—say, neon—you can certainly secure a patent."* – Lila Feisee, Biotechnology Industry Organization's Director for Government Relations and Intellectual Property

It's Elemental: When Harvard University's Charles Lieber obtained a key patent (US patent 5,897,945) on nano-scale metal oxide nanorods, he didn't claim nanorods composed of a single type of metal—but instead claimed a metal oxide selected from up to 33 chemical elements. Harvard's claims on nanorods include those comprised of titanium, zirconium, hafnium, vanadium, niobium, tantalum, chromium, molybdenum, tungsten, manganese, technetium, rhenium, iron, osmium, cobalt, nickel, copper, zinc, cadmium, scandium, yttrium, lanthanum, a lanthanide series element, boron, gallium, indium, thallium, germanium, tin, lead, magnesium, calcium, strontium, and barium. In a single patent, Lieber's claims extend to nearly one-third of the chemical elements in the Periodic Table – spanning 11 of the 18 Groups. Patent lawyers have identified Harvard's patent (licensed to Nanosys, Inc.) as one of the top 10 patents that could influence the development of nanotechnology.35 Similarly, a key patent on semiconductor nanocrystals (quantum dots) held by the University of California (licensed to Nanosys, Inc. and Quantum Dot Corp.) claims semiconductor nanoparticles from elements in Groups III-V of the Periodic Table. The claims in US patent number 5,505,928 extend to boron, aluminum, gallium, indium, nitrogen, phosphorus, arsenic, antimony as well as those compound semiconductors that result from combining elements in Groups III-V (such as gallium arsenide).

Cross-industry Patent Claims: Nanotechnology is not only cross-disciplinary, a single nano-scale innovation may have diverse applications that span multiple industry sectors. Mark Lemley of Stanford Law School observes, "a significant number of nanotechnology patentees will own rights not just in the industry in which they participate, but in other industries as well."36 When ETC Group examined the 700+ patents that the US Patent & Trademark Office had identified as nanotechnology patents as of May 25, 2005, it was not surprising to find that the patents had originally been assigned to *all* of the major patent classes – including electricity; human necessities; chemistry/metallurgy; performing operations and transporting;

mechanical engineering (lighting, heating, weapons, blasting); physics; fixed construction; textiles and paper.

Masters of the Industrial Universe? The crucial aspect to understand about nano IP is not simply that the patents span a broad range of fields, but that a single invention can be relevant for widely divergent applications. It doesn't come as a big surprise to find an invention that is applicable to related fields (e.g., pharmaceuticals and foods), but consider the following examples from US PTO's Class 977:

- US5,874,029 – University of Kansas, 23 February 1999: Methods for particle micronization and nanonization by recrystallization from organic solutions sprayed into a compressed antisolvent: The invention can be used in the pharmaceutical, food, chemical, electronics, catalyst, polymer, pesticide, explosives, and coating industries, all of which have a need for smalldiameter particles.
- US6,667,099 – Creavis Gesellschaft für Technologie und Innovation mbH, 23 December 2003: Meso-and nanotubes: The invention relates to mesotubes and nanotubes (hollow fibers) having an inner diameter of 10 nm-50 µm and to a method for the production thereof ...*The hollow fibers* are used in separation technology, catalysis, micro-electronics, medical technology, material technology or in the clothing industry.
- US6,641,773 – The USA as represented by the Secretary of the Army, 11 November, 2004: Electro spinning of submicron diameter polymer filaments: An electro spinning process yields uniform, nanometer diameter polymer filaments...*The filament is particularly useful for weaving body* armor, for chemical/biological protective clothing, as a biomedical tissue growth support, for fabricating micro sieves and for microelectronics fabrication. The reason that the same invention can be used inside the human body, in clothing and in computers, as in the third example above, is that at the molecular level biological and non-biological material can be integrated – whether this is a seamless integration is a matter yet to be determined by toxicological research.

Second Nature Patents: While biotech's raw materials are biological, nano-scale technologies involve the manipulation of both living and non-living materials, sometimes in combination. When this is the case, the discipline is known as *nanobiotechnology*. A nanostructured material used inside the body as a bone replacement is one example of nanobiotechnology, but so is a hybrid organism created from living and non-living materials, such as the nano-scale silicon and muscle-tissue hybrid announced by researchers in early 2005.37 Closely related to and sometimes overlapping nanobiotech is the new field of "synthetic biology" in which living systems are built to order and then programmed to perform specific tasks. These, too, often combine biological and non-biological parts. Patents on the products of nanobiotechnology provide the opportunity to monopolize the basic elements that are the building blocks of the entire natural world, bringing a whole new dimension to the notion of "life patenting." The table below provides examples of the possible range of nanobiotechnology and synthetic biology patents recently issued by the US PTO. It includes, for example: hybrid devices combining a nanomaterial and muscle tissue, which generate electrical power and which the inventor has described as "absolutely alive"38 (Montemagno); membranes made from biological and non-biological materials to be used in electricity production or water purification (Montemagno); a method for controlling the properties of semiconductor nanoparticles by creating them with the help of biological material (Belcher); synthetic DNA base pairs that do not occur in nature (Benner); a method for genetically modifying cells by pricking them with carbon nanotube "needles" and injecting foreign DNA (McKnight); a gene switch that uses "switching agents" to control gene expression by turning them on or off.

Role of Public Sector Universities in Nanotech IP: One of the unique features of nanotechnology, according to Stanford Law School professor, Mark Lemley, is that universities and public research foundations hold "a grossly disproportionate share of nanotech patents" that he believes are critically important to downstream nanotech products. In 2004 a patent attorney specializing in nanotechnology identified 10 key patents that he believed could have the greatest impact on the development of nanotechnology. Seven of the 10 patents are owned by universities.

Because they conduct basic research, it's not surprising that universities are the early-stage engines for nanotechnology. But unlike early-stage researchers 25 years ago, the new generation of US public researchers has become "extremely aggressive patenters" largely because of the Bayh-Dole Act of 1980 – US legislation designed to encourage technology transfer by permitting universities to patent their federally funded research projects. Before 1980, universities worldwide were granted about 250 US patents per year. By 2003, the number of university-owned patents increased almost 16-fold, to 3,933.41 Because university labs aren't in the business of commercializing products, they try to re-coup their research costs by patenting their employees' early scientific innovations – in the hope of earning royalties or licensing fees. Exclusive licensing is generally the more lucrative deal – and therefore the most appealing to technology transfer offices. In general, universities are acting more and more like businesses. Not only are universities patenting nanotech early and often, they are more frequently licensing their inventions on an exclusive basis. US policymakers who favor Bayh-Dole would argue that universities are benefiting society by transferring science and technology to the private sector for commercialization. But in many cases, consumers end up paying twice – once by paying taxes to support governmentfinanced research, and again when they purchase a new, proprietary technology developed with taxpayer funds. With the emphasis on winning exclusive monopoly patents, the traditional academic culture of open communication and exchange is also undermined and eroded.

From 2003 to early 2005 the *Nanotechnology Law & Business Journal* identified 55 publicly announced nanotech patent license agreements – 20 of which involved a university or public research entity as the licensor. Of the 20 license agreements involving university or research entities as licensor, all but one was granted on exclusive terms (and its terms were not disclosed).

The largest single holder of nanotech patents in the world is a Chinese researcher, Yang Mengjun, who is taking ancient Chinese medicinal herbs, reducing them to nano-scale formulations, and claiming exclusive monopoly over the herbs or the process used to nano-size them. He holds over 900 patents on nanoscale versions of traditional Chinese medicinal plants.43 Similar patents are being granted in the US and Europe. For example, the Pacific Corporation (Korea) has won a European patent on nanoscale ginseng for use in cosmetic products.44 The Pacific Corporation claims that an emulsion of ginseng at the nano-scale (reduced to small particles between 50-500nm) allows it to penetrate the skin, exerting an antiaging effect.

Patent claims on nano-scale formulations of traditional herbal plants are providing insidious pathways to monopolize traditional resources and knowledge – one more reason why the Convention on Biological Diversity and FAO should address the implications of nanotechnology. Diamond vs. Chakrabarty Remembered June 16, 2005 marks the 25th anniversary of Diamond vs. Chakrabarty, the landmark US Supreme Court decision that opened the floodgates to the patenting of all life forms. The anniversary offers a timely opportunity to examine current trends in intellectual property relating to nano-scale technologies – the world's newest technological wave.

In 1971, Ananda Chakrabarty, an employee of General Electric, applied for a patent on a genetically modified, oil-eating microbe. His patent application was rejected by the US Patent & Trademark Office (US PTO) on the grounds that animate life forms were not patentable. When Chakrabarty won his case on appeal, the PTO Commissioner, Sidney Diamond, took the case to the US Supreme Court. On June 16, 1980 by a narrow 5-4 margin, the US Supreme Court ruled that Chakrabarty's oil-eating microbe was not a product of nature; living organisms could be seen as human made inventions and are therefore patentable subject matter. An ironic footnote to the saga is that the "invention" didn't work. The monumental importance of the Chakrabarty decision did not register with the Court – or the public – at the time. (Some environmentalists were eager to embrace life patenting if it meant microbes could devour oil spills.) In 1980 the Supreme Court specifically noted that the Chakrabarty decision was a narrow case that would not affect the "future of scientific research."45 The Court got it wrong. According to lawyer and activist Andrew Kimbrell, "The complete failure by the Court to correctly assess the impacts of the Chakrabarty decision may go down as among the biggest judicial miscalculations in the Court's long history."

As a result of Chakrabarty, the slippery slope of IP on living organisms became a patent landslide, and a bonanza for the biotech industry. Over the course of a single decade, the US government re-interpreted intellectual property laws to allow for exclusive monopoly control over *all* biological products and processes. After Chakrabarty, the once unthinkable patenting of genes, plants, animals, microorganisms and human genetic material would become common practice in the US – positioning industry and the US government to set the precedent for IP regimes worldwide via the World Trade Organization and through bilateral and regional trade agreements.

Chakrabarty "Life Patenting" Timeline

1980 – Diamond v. Chakrabarty – US Supreme Court case establishes precedent for the patenting of living organisms.

1980 – US Bayh-Dole Act allows businesses, universities and non-profit organizations to retain title to patents that result from federally-funded research, and to exclusively license them.

1984 – University of California wins US patent on a cell line developed from cancerous tissue of John Moore, a leukemia patient whose cancerous spleen cells were patented and commercialized without his knowledge.

1985 – US PTO rules that genetically engineered plants, seeds and plant tissue are patentable subject matter.

1986 – Uruguay Round of General Agreement on Tariffs and Trade (GATT) begins with Trade-Related Intellectual Property (TRIPs) on the table.

1987 – US PTO rules that genetically modified animals are patentable subject matter.

1988 – US PTO issues first patent on a living animal – a transgenic mouse. Harvard licenses the "OncoMouse" to Dupont.

1993 – Patent claim by the US Secretary of Commerce on the cell line of a 26-year old Guaymi indigenous woman from Panama. Following worldwide controversy, the US government abandoned its patent claim on the Guaymi cell line in November 1993.

1994 – EPO issues species-wide patent on all genetically modified soybeans to Agracetus (later acquired by Monsanto).

1995 – WTO comes into force at the conclusion of the Uruguay Round of GATT.

1995 – Civil society organizations and social movements in Europe defeat European Patent Directive that aims to harmonize patenting of genetic material within the EU.

1995 – US Supreme Court rules in Asgrow vs. Winterboer that farmers no longer have the right to harvest and re-sell proprietary seed for reproductive purposes (that is, proprietary seed protected by Breeders' Rights – plant variety protection – may be saved only for the purpose of re-planting the farmers' own acreage).

1995 – US PTO issues patent to US National Institutes of Health for an unmodified human cell line from an indigenous person from Papua New Guinea. Due to the international controversy, the US government is forced to disclaim the patent in December 1996.

1998 – European Parliament gives final approval to a controversial biotechnology "patent directive" that aims to harmonize national legislation on the patenting of genetic material within the EU. The Directive creates, for the first time in European history, an explicit legal right to patent higher organisms such as plants and animals.

2002 – Canadian Supreme Court rules against patenting of higher life forms– rejects patenting of genetically modified mouse.

2004 – Canadian Supreme Court (Monsanto v. Schmeiser) affirmed Monsanto's right to prosecute farmers who are found to have proprietary GM crops growing on their land – whether they wanted them or not.

2004 – US Patent & Trademark Office establishes nanotechnology patent class.

Lessons learned from Chakrabarty:

- Historic decisions allowing exclusive monopoly control of all biological products and processes involved no public input or wider societal debate; these decisions were made by a handful of individuals in the courts and patent offices – not by the US Congress. In essence, it was the courts and not citizens who gave biotech the green light in the US. Similarly, at the international level, intellectual property rules have been crafted by and for a narrow group of corporate interests.
- Following Chakrabarty, the US government's aggressive life patenting policies set the bar for the rest of the world – especially at the World Trade Organization.
- The slope is slippery indeed. The history of patent monopoly demonstrates that patent holders typically seek wider patentability, more expansive scope of patent claims, longer patent terms and greater harmonization of patent rules worldwide.
- For many developing nations the rationale for accepting stronger IP

regimes has been the argument that their economies would prosper from increased technology transfers and foreign direct investment. In the case of biotechnology, however, the vast majority of key enabling technologies are proprietary products and processes, tightly concentrated in the hands of multinational gene giants. Under these conditions, stronger levels of IP obligate developing countries to make a massive transfer of resources to the North, in order to acquire licenses for proprietary technologies. A new study by the World Bank concludes that the effects of stronger intellectual property regimes in creating greater trade flows to developing countries are "theoretically ambiguous." The authors conclude, however, that stronger levels of intellectual property in developing countries are not a factor in spurring high-technology trade flows.

The rallying cry "no patents on life" has become a line in a technological and legal sandstorm. Although the notion of intellectual monopolies can be traced back to early Greece, patents did not come into their own until Britain's Industrial Revolution when the inventors of textile machinery demanded "protection." Recognizing that patents would make technology accessible only to well-heeled manufacturers, smaller enterprises protested. The response: "Don't worry. We only seek to patent the machines we invented."

In the 1920s and 30s, when rose and chrysanthemum breeders demanded intellectual property for their flowers, they argued that it was unfair to grant patents to machine inventors but to deny equal rights to ornamental inventors. Although some were repelled by the idea that living things could be patented, the flower companies replied, "Don't worry. These patents protect only decorative plants – not food crops."

In the 1960s, when plant breeders called upon governments to grant them intellectual property over food crops, they said it was unfair to recognize the minor contributions of ornamental breeders without recognizing the contributions of the breeders of crop varieties. The companies chided their critics by saying, "Don't be alarmed. We just want breeders' rights to protect plant varieties; we're not patenting plants, animals or human genetic material, and we would never stop farmers from saving seed."

In 1980, the Gene Giants won patents on genetically modified microbes. A few years later they applied for patents on plants and animals. When civil society protested, industry responded, "Why all the fuss? If you allow the patenting of micro-organisms, why not plants and lab rats?"

In the 1990s, corporations and governments began to patent genes, snippets of DNA, and entire human cell lines. When indigenous peoples protested, patent offices responded, "Don't worry. Human cell lines are just microorganisms."

Meanwhile, patents made it illegal for farmers to save and re-use proprietary seed.

The seed/biotech industry denounced the 12,000-year old right of farmers to save harvested seed as patent infringement.

With the advent of nano-scale technologies, corporations are patenting essential building blocks of all living and non-living things. Industry is redefining life to create hybrid organisms that will take on machine functions. When we tell them they have gone too far, they will reply, "Don't worry. We're all just machines."

Impact of Nanotech IP in the Global South

Over the past two decades the role of intellectual property in all areas of science and technology has exploded globally—primarily due to rules prescribed by the World Trade Organization's Trade-Related Aspects of Intellectual Property (TRIPs) and by bilateral/regional trade agreements. The TRIPs agreement obligates all WTO member countries to adopt and enforce minimum standards of intellectual property. WTO has 150 members, and claims that it accounts for over 97% of all world trade.50 In 1996 the World Intellectual Property Organization (WIPO) and WTO established a collaborative relationship in order to implement the TRIPs Agreement. (WIPO has 182 member states, which represents over 90% of the world's countries.) Among other IP rules, WTO members must allow patenting in *all* fields of technology. The initial grace periods and flexibility allowed by TRIPs for developing country members have nearly expired. By 2006, the so-called "least developed countries" are required to adopt the WTO/TRIPs standards. Over the past decade, civil society, social movements, the UN Human Rights Commission and some governments have warned of the inequities of IP for the global South. Recently, even at WIPO – the UN body whose mission is to promote and protect intellectual property – the uneven IP playing field and the negative impacts of TRIPs have become undeniable and untenable for many developing nations. In September 2004 the "Geneva Declaration on the Future of the World Intellectual Property Organization" warned that current IP regimes are having negative impacts in the developing world, resulting in lack of access to essential medicines, anticompetitive practices that hinder innovation and the misappropriation of social and public goods. At WIPO's General Assembly meeting (September 27-October 5, 2004), Brazil and Argentina, supported by 14 developing country co-sponsors, proposed that WIPO adopt a "development agenda," stating that Intellectual property protection cannot be seen as an end in itself, nor can the harmonization of intellectual property laws leading to higher protection standards in all countries, irrespective of their levels of development. The role of intellectual property and its impact on development must be carefully assessed on a case-by-case basis. IP protection is a policy instrument the operation of which may, in actual practice, produce benefits as well as costs, which may vary in accordance with a country's level of development. Action is therefore needed to ensure, in all countries, that the costs do not outweigh the benefits of IP protection. WIPO's General Assembly adopted the decision to welcome a development agenda. But the US, UK and other industrialized nations are balking at the decision to give development concerns a higher profile within WIPO, acknowledging only that WIPO should give greater technical assistance to developing countries.52 Reports will be prepared and the issue will be considered at WIPO's General Assembly meeting in September 2005.53 Meanwhile, developing nations are now facing a new technology wave – and the requirement to accommodate nanotechnology-related inventions – even while still grappling with unresolved controversies over biotechnology and information technologies. By next year, ready or not, most of the world's developing nations will be obligated to evaluate and enforce nanotech patents. ETC Group is not suggesting that nanotech, unencumbered by patents, will provide solutions for the South's most pressing needs. On the contrary, ETC Group believes that a technological fix can never right social wrongs. However, much ink has been spilled of late on the benefits nanotech will bring to developing nations while ignoring the realities of technology transfer and intellectual property.54 Multinational corporations, universities and nanotech start-ups (primarily in the OECD countries) have already secured numerous patents on essential nanotech tools, materials and processes. To the extent that these are "foundational" patents – that is, seminal breakthrough inventions upon which later innovations are built, researchers in the developing world could be shut-out. Nanotech patent thickets are already causing concern in the US and Europe. Researchers in the global South are likely to find that participation in the proprietary "nanotech revolution" is highly restricted by patent tollbooths, obliging them to pay royalties and licensing fees to gain access. Ultimately, nanotech will profoundly impact the South's economy, regardless of its level of direct participation or its handling of intellectual property. Nano-scale manufacturing platforms could make geography, raw materials, as well as labour, irrelevant. By employing nanotech to build from the bottom-up rather than processing down, the quantity of raw materials required could be sharply reduced. In short, nano-scale technologies are poised to become the strategic platform for global control of materials, food, agriculture and health in the immediate years ahead. Intellectual property monopoly is a crucial and powerful tool for realizing that strategy. As a starting place for further debate, ETC Group offers the following recommendations related to nanotech IP.

Six "Pro-South Science" Policy Proposals:

1. Poor applause: When the G8 meets in July to discuss its dubious "Pro-Poor Science" strategy, leaders of rich nations shouldn't ignore the restrictions implicit in intellectual property that make it difficult or impossible for the South to develop its own independent technology solutions and to have access to the useful technologies of others. A truly "Pro-South" science policy would establish a global ten-year sunset clause on all monopoly patents.

2. Patent pause: WIPO should initiate a global suspension of patent approvals related to any applications that meet the US PTO's Class 977 standard (the criteria for nanotechnology patents) until further social review, including wide public debate, is undertaken on the impacts of nanotech IP.

3. Technology flaws: In close cooperation with social movements, including trade unions, the Food and Agriculture Organization (FAO) and the United Nations Conference on Trade and Development (UNCTAD) should cooperate with WIPO in producing a study on the impact of nanotech-related intellectual property on monopoly practices, technology transfer and trade. The UNCTAD Commission on Science and Technology for Development, also in conjunction with social movements, should examine the implications for technology transfer and the needs and interests of developing countries.

4. TRIPs clause: South governments and countries-in-transition should suspend any Class 977 equivalent patent grants or applications pending a full evaluation of their impacts. In particular, governments should determine whether or not such patents compromise access to the basic elements of nature or contravene national legislation or international agreements, such as WTO TRIPs, concerning intellectual property over living material.

5. Ordre public laws: National governments and relevant international organizations such as WIPO and UNCTAD should examine the social and *ordre public* significance of Class 977 equivalent patents that could compromise access to the fundamental components of nature.

6. Diversity laws: With input from indigenous peoples and peasant farmers' organizations, the Convention on Biological Diversity (CBD), the Commission on Sustainable Development (CSD) and FAO should consider the impact of intellectual property on fundamental elements of nature with respect to biodiversity and national sovereignty over genetic resources (especially in such fields as synthetic biology or nanobiotechnology). In a wider context, either the CBD or the CSD should adopt a permanent agenda item to monitor developments in nano-scale technologies.

Gas Sensors

MS Modeling's $DMol^3$ has been used to investigate the nitrogen dioxide sensing mechanism of tin dioxide nanoribbons. Understanding the sensing mechanism will enable the efficient design and manufacture of nanoscale chemical sensors—an important commercial application of nanotechnology.

Practical challenges with carbon nanotubes involving cost of synthesis, control of chirality and diameter, separation from bundles, and attachment of functional groups have prompted researchers to explore other types of one-dimensional nanostructures. Two of the most promising candidates in this regard are nanowires and nanoribbons. As the name suggests, nanoribbons are solid objects (unlike nanotubes, which are hollow) with a near-uniform rectangular cross-section.

So far, nanoribbons have primarily been synthesized from the oxides of metals and semiconductors. In particular, SnO_2 and ZnO nanoribbons have been materials systems of great current interest because of potential applications as catalysts, in optoelectronic devices, and as chemical sensors for pollutant gas species and biomolecules. Although they grow to tens of microns long, the nanoribbons are remarkably single-crystalline and essentially free of dislocations. Thus they provide an ideal model for the systematic study of electrical, thermal, optical, and transport processes in one-dimensional semiconducting nanostructures, and their response to various external process conditions.

Recent experiments with SnO_2 nanoribbons indicate that these are highly effective in detecting even very small amounts of harmful gases like NO_2. Upon adsorption of these gases, the electrical conductance of the sample decreases by more than an order of magnitude. More interestingly, it is possible to get rid of the adsorbates by shining UV light, and the electrical conductance is completely restored to its original value. Such single-crystalline sensing elements have several advantages over conventional thin-film oxide sensors: low operating temperatures, no ill-defined coarse grain boundaries, and high active surface-to-volume ratio.

To be able to fully commercialize their potential, it is important to better understand the sensing mechanism of such systems. With this goal in mind, researchers at the Brookhaven National Laboratory, Lawrence Berkeley National Laboratory, and Accelrys used MS Modeling's DMol[3] to investigate the nitrogen dioxide sensing mechanism of SnO_2 nanoribbons. Reporting in *Nano Letters*, the researchers examined the NO_2-sensing mechanism of SnO_2 nanoribbons with exposed (1 0'1) and (0 1 0) surfaces.

The density functional theory (DFT) calculations revealed that:

- The most stable adsorbed species involved an unexpected NO_3 group doubly bonded to Sn centers
- An orders-of-magnitude drop in electrical conductance can be explained by significant electron transfer to the adatoms
- Computed binding energies were consistent with adsorbate stability up to 700 K, with X-ray absorption spectroscopy indicating predominantly NO_3 species on the nanoribbon surface.

The ability of the nanoribbons to sense O_2 and CO was also investigated. In the case of O_2, the response of the nanoribbon was highly sensitive to the concentration of O-vacancies on the surface. Thus, in the absence of any surface vacancies, the calculation predicted negligible charge transfer. However, when surface vacancies were present, an O_2 molecule can adsorb as a peroxide bridge, and withdraw a significant amount of electronic charge from the nanoribbon surface, thereby decreasing its electrical conductance. In the case of CO adsorption, there was a net electron transfer from the CO to the nanoribbon surface. Thus the calculation predicted an increase in nanoribbon electrical conductance upon CO-adsorption, in agreement with experimental results.

Researchers from the Department of Energy's Sandia National Laboratories and the University of New Mexico (UNM) have developed a new way of mimicking photosynthetic proteins to manipulate platinum at the nanoscale. The method has the potential of changing the metal's properties and benefiting emerging technologies. While we are in the early stages of research, we see the possibility of manipulating the nanoscale structure of platinum so that we can have control over the size, porosity, composition, surface species, solubility, stability and other functional properties of these metal nanostructures," says John Shelnutt, the Sandia scientist leading the research effort. "Such control means that the redesigned platinum could be used in many new applications, including catalysis, sensors and optoelectronic and magnetic devices." He adds that while research groups have reported a few platinum nanostructures — including nanoparticles, nanowires, nanosheets and others — the addition of new types of nanostructures is "highly desirable and potentially technologically important." Working with Shelnutt in the research are Frank van Swol from Sandia, UNM graduate student Yujiang Song and Eulalia Pereira from the University of Porto in Portugal. The new method of manipulating platinum was detailed in a paper in the *Journal of American Chemical Society* published in December. The idea for the technique is similar to photosynthesis, in which plants use the energy from sunlight to produce sugar. But instead of manufacturing sugar, the new method changes a platinum ion to the neutral metal atoms. The photosynthetic protein mimicks this repeatedly, allowing metal to be deposited as desired at the nanoscale.

The method involves putting porphyrins — the active part of photosynthetic proteins — along with the platinum salt in an aqueous solution of ascorbic acid at room temperature. The porphyrins are placed in specific locations in the solution where it is intended that metal should be deposited. For example, the porphyrins may be confined to micelles, or liposomes. Micelles are spherical assemblies of detergent molecules in which the heads are exposed to the water and the tails stick together in the interior. Liposomes are similar structures, but they are larger and have water on the inside and outside separated by a closed membrane — sort of like a cell. The membrane is composed of two layers of detergent molecules, with the heads on the inner and outer surface facing the water and the tails forming the interior of the membrane. When light is shined on the porphyrins located in these detergent structures, the porphyrins excite, becoming catalysts for platinum reduction and deposition. As this occurs, the metal grows onto the surfaces of the surfactant structures as a thin sheet or in other ways. In the case of micelles, the platinum grows into balls that look like the common toy "Koosh" ball. The ball size can be controlled by the amount of porphyrins and platinum in the solution, the amount of light illuminating the solution, and the amount of time the light is on.

For the metals platinum and palladium that form these nanostructures, it is enough for the porphyrin molecule to grow only a small metal "seed" particle composed of about 500 atoms. When it reaches this size, the seed starts to catalyze its own rapid growth (by oxidation of ascorbic acid), budding off arms in all directions and creating the Koosh-ball-like nanostructures. The porphyrin provides a convenient method of making these seeds at the location and time desired, leading to a uniform and selectable nanostructure size The platinum nanostructures take on a different form when they are prepared under different conditions. When the porphyrin is in a micelle, the platinum nanostructures produced look like Koosh balls. When the porphyrin is in the bilayer membrane of a liposome, the platinum grows in 2-nanometer thick sheet or platinum lace on the outer surface of the membrane, giving circular sheets — sort of like two-dimensional Koosh balls. Under solution conditions for which the liposomes aggregate, growth can occur along the interfaces of the liposomes to give platinum foamlike materials and foam nanoballs. The type of nanostructure is mainly determined by the type of surfactant assembly upon which the platinum grows and the extent of growth from the individual seed nanoparticles. Since the porphyrin remains attached to the platinum nanostructure and active in the presence of light, it can also perform other functions besides growing itself. For example, when illuminated with light, the platinum nanostructure evolves hydrogen from water. This reaction is similar to one of interest to car manufacturers looking for new ways to build automobiles powered by hydrogen fuel cells. Shelnutt says that in addition to structuring the platinum, the process also happens very fast. A few minutes in light will create many seeds, which then grow into the mature nanostructures in tens of minutes. And the process is easy to do.

Scientific Application Semiconductor Superlattices

A brief review of semiconductor superlattices is presented—their definition, basics of theory, types of materials and methods used for their preparation, the most important and unusual electrical and optical properties, which have already found their practical applications.

When alternating layers of two semiconductors with different band gaps are deposited and thickness of these layers lies between 1-10 nm (which corresponds approximately to 10–100 atomic layers), what we obtain is semiconductor superlattice. The mostly known example of such a structure is GaAs/AlAs system depicted schematically together with its corresponding potential profile. Differences in band gaps of materials lead to discontinuities in valence and conduction band and a set of square potential wells (low band gap semiconductor) separated by potential barriers (high band gap semiconductor) forms. Let us consider one of these wells from the quantum mechanical point of view. If we assume that an electron mean free path within a well is substantially longer than well width and the barrier width is infinite, what we obtain is a classical problem of a particle confined in a square well of finite depth. Unfortunately, in the real-life systems we cannot obtain perfect separation of adjacent wells and this results in coupling of their wave functions. Thus, instead of discrete levels, this yields us subbands. An example of theoretical analysis of the GaAs/AlAs system shown, where an equal width, a, of well and barrier layer is assumed. Two main features common for other superlattice structures are easily seen when a is decreased. First, an energy distance between ground state and subsequent bound states increases, which happens because the well width is getting smaller. Second, the discrete levels split into the subbands as a result of thinner barrier layers and poorer separation of neighbouring wells. At this point, it is also important to notice, that the energy discontinuity, i.e. well depth, cannot be simply derived from such material constants as an energy gap, the conduction and valence band structure etc. in each semiconductor because of the lack of knowledge of the common energy reference. This parameter must be either calculated or measured.

Historically, the crystalline superlattices where the first to produce, starting from pioneering work of Esaki and Tsu (1970). What is crucial in fabrication of such structures is nearly perfect match in their lattice constants. Otherwise the density of defects becomes so large that it results in strong electron scattering and obscures quantum phenomena. This is why a molecular beam epitaxy technique is utilised in their deposition and the number of materials, which can be combined to create superlattices, is restricted. The best known and studied are GaAs/AlAs and GaAs/AsxGa1-xAs, InAs/GaSb are also worth mentioning. If interface defects cause such difficulties, why not to approach the problem from the opposite side—let us build the superlattice from the materials, where lattice constants no longer exist, i.e. amorphous semiconductors. This idea, apparently irrational, proved absolutely sensible in 1983, when Abeles and Tiedje reported in their paper fabrication of a-Si1-xNx:H/a-Si:H and a-Ge:H/a-Si:H systems. They stated that the strict requirements concerning lattice matching in crystalline superlattices can be relaxed due to amorphous structure by its very nature and a hydrogen content which plays an important role by passivating coordination defects. Since then, many other materials were combined to produce superlattices : a-Si:H/a-C:H, a-Si:H/a-SiN:H, a-Si:H/a-SiNx, a-Si:H/a-SiOx, a-Si:H/a-Ge:H, CdSe/SeTe, CdSe/Se, a-Si3N4:H/a-Si:H, a-C:H superlattices where the band gap is tailored by changing deposition conditions. An impressive, comparing to the crystalline case, number of types of amorphous superlattices results from the fact that in this case we have much more freedom in choosing components of our systems as we are not limited by the lattice matching condition.

It is also important to emphasise that the techniques utilised to produce amorphous films -plasma deposition, thermal evaporation, sputtering—are relatively cheaper than molecular beam epitaxy and easier to implement in industrial applications. On the other hand, in case of amorphous multilayers it is more difficult to obtain abrupt interfaces between subsequent layers. Different methods are employed : pulsed application of microwave plasma during continuos r.f. glow discharge in Si2H6/N2, by which N2 is decomposed and incorporated into the growing film, nitridation of a-Si:H to attain a-Si3N4:H, lowering the growth rate in order to avoid fabrication of transition layers when the deposition conditions change, quick exchange of source gases. However, sometimes smooth transition from one layer to another is intentionally introduced, which may lead to sinusoidal potential profile, instead of a step-like,.

For the sake of convenience, electrical properties, particularly conduction phenomena, may be considered in terms of what is happening when we study currents flowing in-plane (parallel to the layers) or perpendicular to them. As it has been stated above, quantum confinement causes shift of the first bound state allowed for charge carriers. This must of course result in the increase of an activation energy of conductivity, E_a. In general, the smaller well width, the higher E_a, thus in-plane electrical conductivity decreases as compared to the unlayered material. These effects are presented in a number of research papers, typical behaviour for diamond-like carbon films (DLC) superlattices, is shown in Figs. 20.3. and 20.4. However, the opposite situation may occur as well. This is because in amorphous semiconductors, unlike crystalline ones, there is a finite density of states within a mobility gap. When two such materials are put into contact, charge carriers may be transferred through an interface, building up accumulation and depletion layers, and as a result, the conductivity in one of the materials increases to such a degree that obscures quantum effects. A completely different picture appears when conductivity in the direction perpendicular to the layers is considered. If there is no electric field, the current does not flow of course. When the small field is applied, a tunneling current can flow through the barriers. As the field is increased, an alignment between the energy subbands in adjacent wells is getting worse so observed maximum is a compromise between a driving force (electric field) and subbands mismatch. When the field is increased further, the subbands misalignment is so high, that charge transport becomes very inefficient and the current—field characteristics exhibits minimum.

If the field becomes higher, we can have the situation, that the lower subband in one quantum well is aligned with the upper one in the adjacent well. Transport is efficient again. The most important feature of this characteristics is a region of negative differential resistance where an increase of the electric field leads to a decrease of the electric current. More detailed theoretical considerations of this matter can be found elsewhere. This kind of behaviour has been observed in both crystalline and amorphous multilayers; however, in the crystalline case it is more distinct and resembling model behaviour. An example of characteristics for DLC superlattice is shown, where current-voltage profile together with its first and second differentials is presented. Bumps observed on the current-voltage curve are typical of resonant tunneling. The start of the negative differential resistance region at 3.1 V corresponds to the loss of subbands alignment. It should be noted that the conductivity in the direction perpendicular to the superlattice layers is by its nature very similar to what Esaki found in *p-n* germanium junctions and what is now applied in construction of tunnel diodes. Typical *blue shift* effect is shown in Figs. 20.7 and 20.8. The solid lines represents fit to a theoretical model of a one-dimensional quantum well with no coupling between energy levels of adjacent wells.

We have discussed electrical and optical phenomena which characterise superlattice structures which found their explanation on the basis of quantum size effect. However, alternative interpretations of experimental data do exist as well. Arsenault et al. claim that bumps observed in the current-voltage characteristics for perpendicular conductivity in amorphous multilayers, usually attributed to resonant tunneling can be explained in a different way. They suggest that at least one of these bumps is simply a transition between two transport mechanisms: space charge limited current (low electric field) and multiple hopping (high field). The most convincing justification of their hypothesis is that they observed the same bumps for both quantum well structures and single barriers. Quantum mechanical explanation of the *blue shift* has been criticised in a number of research papers as well. In our discussion it has been assumed that described multilayers were periodic system, characterized, from the geometrical point of view, only by one well and one barrier width. However it does not have to be the only approach. Another and very original view is presented in some researches. Superlattices of these researches consist of the layers of a-Ge and a-Si deposited using dual-target magnetron sputtering. The most interesting thing is their topology generated by Cantor sets. The modulation algorithm according to the so called "Cantor middle—third" set together with corresponding multilayer structure is presented.

Opportunities in Nano-Enabled Drug Delivery Systems

Nanomedicine is widely expected to become a multi-billion dollar business over the next decade. Although some of the applications for nanomedicine are quite futuristic, others are on the verge of generating significant revenues. One of these is nano-enabled drug delivery systems which promise to increase the bioavailability of drugs and reduce the likelihood of toxicity. Researches pinpoint several areas where nano-engineering of drug delivery systems is leading to greater control over active compounds, resulting in the physician's ability to send the smallest dose possible in order to achieve the desired result in the patient. This chapter reviews the main opportunities that are emerging for nano-enabled drug delivery systems. In most cases, these are based on delivering drugs in nanoparticulate form. Most drugs perform better as nanoparticles, because they can be targeted better and there are fewer side effects. In addition, by using smaller amounts of drugs, treatment costs may be reduced. For example, according to researches, patients show a strong preference for nanoparticle inhalers as an alternative to the widely used injectable methods, which may permit far lower doses of expensive protein-based drugs like insulin.

It is expected that the dosing benefits of nano-enabled drug delivery systems will be extended to compounds used in treating both infectious disease and cancer, and has identified six types of drug delivery systems in which nanotechnology is likely to have an impact.

Although injectable delivery systems have some inherent discomfort for the patient and other types of delivery may be preferred by both the patient and the physician, nanotechnology may be applied to produce improvements and additional revenue streams for the drug company. A case in point is Johnson & Johnson, which, at the beginning of 2005, announced that its Élan NanoCrystal technology would be used in a Phase III clinical trial for an injectable formulation of paliperidone palmitate, an anti-schizophrenia drug. This is a new "nano formulation" of an older drug which overcomes the original's insolubility, by reducing the particle size to under 200 nm. This new formulation enables the product life cycle to be extended by a significant amount.

Implantable drug delivery systems are often preferable to the use of injectable drugs, because the latter frequently display first-order kinetics (e.g. initial concentration is elevated, which then drops exponentially over time). This may cause difficulties with toxicity, if and when the peak concentration remains above the therapeutic range. Conversely, drug efficacy often diminishes as the drug concentration falls below the targeted range. By contrast, implantable "time release" systems may help minimize peak plasma levels and reduce the risk of adverse reactions; allow for more predictable and extended duration of action; reduce the frequency of re-dosing; and improve patient acceptance and compliance. To these advantages nanotechnology can add the benefits listed at the beginning of this White Paper. Implantable drug delivery systems are expected to be used for the delivery of proteins, hormones, pain medications, and other pharmaceutical compounds. An example of how nanotechnology is already making its impact felt in this part of the drug delivery business is provided by pSivida's BioSilicon product. BioSilicon is a novel nanostructured material that effectively stores an active compound in nanosized pockets that controllably release minute amounts of drug as the silicon dissolves. pSivida is currently exploring biodegradable implantable methods for tissue engineering and ophthalmic delivery. Nano-implants will also be used in the not-too-distant future for treating cancer. Among the first nanoscale devices to show promise in anti-cancer therapeutics and drug delivery are structures called "nanoshells," which it is believed may afford a degree of control never before seen in implantable drug delivery products. Nanoshells typically have a silicon core that is sealed in an outer metallic core. By manipulating the ratio of wall to core, the shells can be precisely tuned to scatter or absorb very specific wavelengths of light. For example, Gold encased nanoshells have been used to convert light into heat, enabling the destruction of tumors by selective binding to malignant cells. A physician can use infrared rays to pass harmlessly through soft tissue, while initiating a lethal application of heat when the nanoshells are excited. Some researchers are experimenting with temperature-sensitive drug delivery control methods, using nanoshells that release their payload only when illuminated with the proper infrared wavelength.

The vast majority of consumers prefer an oral drug delivery system to implantables or injectables. With this in mind, various development groups are working to enhance traditional oral delivery systems with nanoengineered improvements. There are some areas where nano-enhanced drugs could make a big difference in increasing oral bioavailability and reducing undesirable side effects. By increasing bioavailability, nanoparticles can increase the yield in drug development and more importantly may help treat previously untreatable conditions. Because of the blood brain barrier (BBB) many new chemical entities aimed at treating brain disorders have proved not to be clinically useful. By contrast, nanoparticles have been demonstrated to cross the BBB with little difficulty. Companies such as Germany's NanoPharm have developed systems capable of reaching the brain for anesthesia (Dalargin, an analgesic), cancer drugs, and various therapeutics. The company claims several advantages over existing systems, including:

(1) no requirement to open the BBB,

(2) the ability to use potentially any drug, whether hydrophilic or hydrophobic, and

(3) the drug does not need to be modified.

Meanwhile, researchers at the University of Texas at Austin have described a means of using nanospheres for oral drug delivery. These nanosphere carriers are derived from hydrogels, which are highly stable organic compounds that swell when their environment becomes more acidic. They have been successfully formulated into controlled-release tablets and capsules, which release active compounds when the hydrogel body swells.

NanoMarkets' research indicates that nanomaterials provide a unique opportunity for rapid topical delivery of active compounds. Given their very small size, nanoparticles are able to enter human tissues and cells quickly. Companies such as Novavax have already developed regulated topical systems that take advantage of the unique properties of micellar nanoparticles. It has two hormone replacement therapies dubbed Estrasorb (which received FDA approval in October 2003) and Androsorb (which successfully completed Phase I human trials in 2003). The number of FDA-approved polymers available for use on-skin is increasing rapidly. With this in mind, the industry will be presented with opportunities to create new transdermal platform designs that improve on-skin properties and diffusion of active molecules from patches. This interesting trend is expected to result in smaller and less invasive patches that increase the universe of available drug candidates. In some cases, electronics are even being integrated into patch-like platforms involving wound care, monitoring, and diagnostic methods. As such, patch-like platforms will play an important role in healthcare to treat, measure, diagnose, and generally improve quality of life.

Colloidal dispersions have been demonstrated to remove potentially lethal compounds from the bloodstream, including high concentrations of lipophilic therapeutics, illegal drugs, and chemical and biological agents. A team of scientists at the University of Florida and Clarkson University in Potsdam, N.Y., has demonstrated favorable results to this end, using biocompatible microemulsions. These oil-in-water systems have a rapid and efficient absorption capacity for many target molecules that are frequently overdosed, whether this be intentional or accidental. The microemulsions use a polymeric surfactant, in combination with an ionic co-surfactant. Other oil compositions have been made with varying absorption characteristics for different toxins, in all cases avoiding hemolysis and unwanted blood coagulation. The mechanism by which nanoparticles trap toxins involves aromatic or cyclodextrin receptors with electron deficiencies, which form complexes with selected toxins, sometimes referred to as "benzenoid moieties. The analysis above establishes that the impact of nanotechnology on the drug delivery business is likely to be very broad in three senses:

- Nanotechnological improvements promise to impact almost every common mode of drug delivery oral, implantable, injectable, etc.
- Such improvements will themselves cover a lot of ground, including lower drug toxicity, lowering cost of treatments, improved bioavailability and extension of the economic life of proprietary drugs.
- Nano enabled drug delivery will provide improvements in the treatment of some of the most widespread diseases, such as cancer and diabetes.

This is an impressive list. Also impressive is the fact that many of the categories of nano-enabled drug delivery systems are already close to or at the point of marketing. It is believed that not only will the nano-enabled drug delivery market be one of the first true nanomedicine markets to evolve, but as it does so, the revenues from nano-enabled drug delivery systems will be quite large.

The pharmaceutical industry has to discover and develop innovative medicines for a wide range of diseases in a marketplace that is (1) likely to experience growing regulatory and pricing pressures and (2) that is increasingly targeted towards diseases that have been traditionally resistant to pharmacotherapy (e.g., solid tumors). This chapter profiles a number of tools that are emerging from nanotechnology that are believed will be important aids in the process of drug discovery in this environment. Today, nearly all pharmaceutical companies follow common technology processes for discovering drugs. These include cloning and expressing human receptors and enzymes in formats that allow high-throughput, automated screening and the application of combinatorial chemistry. Thus, random screening can now be achieved with libraries sufficiently large and diverse to have a relatively high probability of finding a novel molecule. These libraries are possible because they can be generated by the techniques of combinatorial chemistry (combichem). Backing this are important developments in the application of genetics and genomics to understand associations between diseases and gene products. Importantly, bioinformatics are beginning to identify putative targets for a number of diseases.

An important consequence of today's approach is that more lead molecules are being discovered for diverse targets, giving the medicinal chemist more scope to find a candidate molecule. Yet the probability of success of launching a candidate molecule into the market remains basically unchanged at around 8 percent, and the perception is that the time to market is not reducing as might be expected with greater automation. Numerous problem areas remain:

- Current methodologies for compound synthesis.
- Storage and retrieval are inefficient and inadequate.
- Better information is needed to improve target selection and validation.
- More predictive screens are essential.
- More emphasis must be placed on acquiring good ADMET data earlier in the discovery process.
- Novel screening methodologies are needed that dramatically increase throughputs and reduce reagent consumption.

It is becoming clear that many of the important solutions will be nanotechnological in nature. This is because the size range that holds the most interest in drug discovery is more or less the same as that which defines nanotechnology from 100 nm down to the atomic level (approximately 0.2 nm). Nanotechnology cannot provide the ultimate solutions to all these problems. However, it is becoming clear that many of the tools developed to pursue nanotechnology and nanoscience will have important roles in drug discovery.

While AFM microscopes are invaluable for imaging objects at the nanoscale, until recently they haven't been able to see how components of a cell react in biological processes, such as their response to a specific chemical or compound. New imaging techniques involve attaching antibodies specific to individual proteins to the tip of an atomic force microscopes' probe. When an antibody reacts with the protein to which it is targeted, it creates a variance in the microscope's reading compared to a reading with a bare tip, showing the presence of a protein or other specific material in the region being scanned. This technique leads to a greater understanding of the chemical dynamics involved in how cells react to stimuli, and could prove particularly significant for drug discovery. However, although NanoMarkets believes that AFM is a useful tool to improve the understanding of molecular interactions, our research also indicates that the widespread use of this technology in bioanalytical applications has been hindered by the limited throughput of these techniques and the high experimental burden imposed by complicated and expensive instrumentation. Highly parallel micrometer and submicrometer cantilever arrays, which are currently being developed, might increase the throughput of AFM-based force spectroscopy.

NSOMs allow the study of optical properties of the sample surface with a resolution better than the wavelength of the light. By scanning the optical probe at very small distances from the sample (a few nanometers) "evanescent waves" from the surface are detected by the probe. The use of evanescent waves allows bypassing of the wavelength limitation of traditional optical techniques (the traditional limit of resolution being half the wavelength in use). NSOM can be used to image biological samples and could ultimately be a significant contributor to drug discovery. However, there is an obstacle to such deployment at the present time manufacturing the probes reliably, the requirement for a very small hole through which the light passes being problematic.

SPR is a phenomenon that occurs when light is reflected off thin metal films and a small amount interacts with electrons in the film, reducing the light intensity. The refractive index of the materials sandwiching the film dictates the angle at which the light reduction (essentially a shadow) occurs. Using an SPR-based approach, the interaction of biomolecules can be detected in real time, offering applications largely for observing biological systems in action on a very small scale, but also with potential for biosensors. Surface plasmons were the subject of some interesting general research in late 2002 and early 2003. There is still a lot to be learned in this field, but NanoMarkets believes that there will be important improvements for imaging systems (and thus for drug discovery) as this learning process proceeds.

Current technologies used for proteomic studies are based on a variety of separation techniques followed by identification of the separated proteins and proteolytic peptides using mass spectrometry (MS). One popular technique is high-resolution two-dimensional (2D) gel electrophoresis, which is capable of resolving 2500 or more distinct protein spots from complex samples. The in-gel protein spots are then identified using sensitive MS and sequence database searching. An alternative approach to 2D gel electrophoresis is chromatographic separation of peptides with electrospray ionization (ESI)-MS or tandem MS (MS/MS) detection. The need for relatively large sample volumes is a major challenge for conventional ESI-MS analysis. There are now systems that address this challenge by applying nanotechnology to the electrospray nozzles. Thus, Advion's ESI Chip contains an array of nanoelectrospray nozzles, each one-fifth the diameter of a human hair, etched in a silicon wafer.

Nanolithography is the art and science of etching, writing, or printing at the nanoscopic level. DPN is a scanning probe nanopatterning technique in which an AFM tip is used to deliver molecules to a surface via a solvent meniscus, which naturally forms in the ambient atmosphere. This direct-write technique offers high-resolution patterning capabilities for a number of molecular and biomolecular "inks" on a variety of substrates, such as metals, semiconductors, and monolayer functionalized surfaces. DPN is becoming a significant tool for the scientist interested in fabricating and studying soft- and hard-matter on the nanoscale. DPN enables precise multiple patterns with near-perfect registration. It's both a fabrication and imaging tool, as the patterned areas can be imaged with clean or ink-coated tips. The ability to achieve precise alignment of multiple patterns is an additional advantage earned by using an AFM tip to write, as well as read, nanoscopic features on a surface. NanoMarkets believes that together, all these attributes make DPN a valuable tool for studying fundamental issues on colloid chemistry, surface science, and nanobiotechnology.

The central paradigm in proteomics studies has been to identify differential protein levels in healthy and diseased cells, characterize these proteins and determine the protein's role in biochemical pathways. These proteins can then serve as diagnostic markers and potential drug targets. Traditionally, 2D gel electrophoresis has been the workhorse in proteomics on the front end of mass spectroscopy. This technology, however, has been described as a bottleneck in high-throughput proteomic studies. As a result there has been a significant push to develop alternative technologies that serve to alleviate this rate failing step in protein separation prior to further analysis. At the forefront of these emerging and growing technologies are protein and DNA microarrays that allow the highly specific capture and analysis of a large number of proteins expressed in various cell types exposed to given perturbations in a high throughput manner.

Microarrays are generally defined as a substrate and surface chemistry onto which a biomolecule or capture agent has been immobilized for the purpose of expression analysis and functional characteristic. The potential to decrease the time required for the critical steps or protein separation and characterization are serving as a considerable driver in the protein microarray market. Microarray technologies can be considered platforms for nanoscale bioanalysis, and these products have already proven their value in the marketplace. However, currently available microarray technologies suffer from certain limitations that prohibit the exploitation of the full range of drug discovery applications. These limitations are already being addressed at the nano level with "nanoarrays," which are ultra-miniaturized versions of the traditional microarray that can measure interactions between individual molecules down to resolutions of as little as one nanometer. Nanoarrays are being touted as the next evolutionary step in the miniaturization of bioaffinity tests for proteins, nucleic acids, and receptor-ligand pairs. It is believed that nanometer-scale resolution capabilities of nanoarray technology offer many advantages in the emerging field of functional proteomics. Unlike nucleic acids, which can easily be multiplied by amplification methods, individual proteins cannot be easily increased in quantity. Using nanoarray technology, very small quantities of individual proteins can be effectively screened against a large set of drug targets. In addition, nanoarrays can be incorporated as sensors in ways that would be impossible with larger microarrays. In addition, unlike traditional microarrays, a nanoarray can be used "in solution." Microarrays, once exposed to the target protein, must be removed from solution and rinsed before they are scanned to see whether or not there have been any "hits." Not only does this take time, but the rinsing process can also easily result in the degradation of the array itself.

QDots are nanometer-sized semiconductor crystals, or electrostatically confined electrons. While other bio-imaging tools such as fluorescent organic dyes have been available for many years. QDots hold the promise of a true revolution in bio-imaging quality and sensitivity. Since 1914, researchers studying everything from blood tests to the effects of cancer drugs on the workings of a cell have used fluorescent dyes to tag cells. The dyes, however, can be problematic. Each dye molecule requires a laser of the same wavelength to cause it to illuminate. The dyes are also imprecise and have a tendency to bleed together. They can only be lit up for very short periods of time, typically just a few seconds after a light source is applied. QDots take advantage of the quantum confinement effect predicted by quantum mechanical theory to fluoresce extremely brightly when excited by a light source such as a laser. And they don't have any of the disadvantages of conventional technologies. By varying the size of the crystals one can cause a rainbow of colors to fluoresce. In addition, QDots stay lit for much longer periods of time than conventional dyes often hours or days. QDots enable the tagging of a variety of different biological components like proteins or different strands of DNA with specific colors. QDots are designed to bond with and illuminate individual biological targets of choice whether genes, nucleic acids, proteins, small molecules, cancer cells, or even entire blood vessels. It may be noted that some of the world's largest pharmaceutical and biotech companies, such as GlaxoSmithKline, Pfizer, and AstraZeneca, and leading biotech firm Genentech are applying QDots in high-content drug screening and have completed initial drug screens using QDots as the biological read-out. Genentech is applying QDots to the detection of breast cancer. This chapter demonstrates how nanotechnology offers several improvements in drug discovery technology. These include:

- Improved imaging (including real-time imaging) that enables researchers to better understand how biological processes are affected by a new compound.
- The use of dip pen technology to improve fundamental research techniques for drug discovery applications.
- Major improvements in today's microarray and fluorescent dye technologies.

As the drug industry strives to meet the increasingly difficult task of producing new drugs and especially new blockbuster drugs, it is increasingly seeking new drug discovery technologies that can improve R&D success rates and time to market. It is therefore believed that these improvements are likely to lead to substantial new business revenues over the next few years.

REFERENCES

Averback, R.S., J. Bernholc, and D.L. Nelson. 1991. *MRS Symposium Proceedings* Vol. 206.

Baird, D. & Shew, A.: 2004, 'Probing the History of Scanning Tunneling Microscopy', in: D. Baird, A. Nordmann & J. Schummer (eds.), *Discovering the Nanoscale*, Amsterdam: IOS Press, pp. 145-156.

Baird, D.; Nordmann, A. & Schummer, J. (eds.): 2004, *Discovering the Nanoscale*, Amsterdam: IOS Press.

Baker, R.T.K. Synthesis, properties and applications of graphite nanofibers. In *R&D status and trends*, ed. Siegel et al.

Bate, R., Frazier, G., Frensley, W., Reed., M., "An Overview of Nanoelectronics," *Texas Instruments Technical Journal*, July-August 1989, pp. 13-20.

Bensaude-Vincent, B.: 2001, 'The Construction of a Discipline: Materials Science in the United States', *Historical Studies in the Physical and Biological Sciences*, 31, 223-248.

Berube, D.M.: 2004, 'The Rhetoric of Nanotechnology', in: D. Baird, A. Nordmann & J. Schummer (eds.), *Discovering the Nanoscale*, Amsterdam: IOS Press, pp. 173-192.

Bowes, C.L., A. Malek, and G.A. Ozin. 1996. *Chem. Vap. Deposition* 2:97.

Braun, T.; Schubert, A. & Zsindely, S.: 1997, 'Nanoscience and nanotechnology on the balance', *Scientometrics*, 38, 321-325.

Brinker, C.J. 1996. *Curr. Opin. Solid State Mater. Sci.* 1:798.

Brown, N.: 2004, 'Needle on the Real: Technoscience and Poetry at the Limits of Fabrication', in: N.K. Hayles (ed.), *Nanoculture: Implications of the New Technoscience*, Bristol, UK: Intellect Books, pp. 173-190.

Bueno, O.: 2004, 'The Drexler-Smalley Debated on Nanotechnology: Incommensurability at Work?', *Hyle: International Journal for Philosophy of Chemistry*, 10(2), 83-98.

Bueno, O.: 2004, 'Von Neumann, Self-Reproduction and the Constitution of Nanophenomena', in: D. Baird, A. Nordmann & J. Schummer (eds.), *Discovering the Nanoscale*, Amsterdam: IOS Press, pp. 101-115.

Carroll, J.S.: 2001, 'Social Science Research Methods for Assessing Societal Implications of Nanotechnology', in: M.C. Roco and W.S. Bainbridge (eds.), *Societal Implications of Nanoscience and Nanotechnology*, Dordrecht: Kluwer, pp. 188-192.

Chen, C-Y., S.L. Burkett, H.-X. Li, and M.E. Davis. 1993. *Microporous Mater.* 2:27.

Chianelli, R.R. 1998. Synthesis, fundamental properties and applications of nanocrystals, sheets, and fullerenes based on layered transition metal chalcogenides. In *R&D status and trends*, ed. Siegel et al.

Chianelli, R.R., M. Daage, and M.J. Ledoux. 1994. *Advances in Catalysis* 40:177.

Claeson, T., and Likharev, K.," Single Electronics," *Scientific American*, June 1992, pp. 80-85.

Coenen, C.: 2004, 'Nanofuturismus: Anmerkungen zu seiner Relevanz, Analyse und Bewertung', *Technikfolgenabschätzung ' Theorie und Praxis*, 13 (2), 78-85.

Crandall, B.C. and Lewis, J., *Nanotechnology: Research and Perspectives*, MIT Press, Cambridge, 1992.

Crow, M. & Sarewitz, D.: 2001, 'Nanotechnology and Societal Transformation', in: M.C. Roco and W.S. Bainbridge (eds.), *Societal Implications of Nanoscience and Nanotechnology*, Dordrecht: Kluwer, pp. 45-54.

Decker, M.; Fiedeler, U.; Fleischer, Th.: 2004, 'Ich sehe was, was Du nicht siehst... zur Definition von Nanotechnologie', *Technikfolgenabschätzung ' Theorie und Praxis*, 13 (2), 10-16.

Dresselhaus, M., and G. Dresselhaus. 1995. *Ann. Rev. Mat. Sci.* 25:487.

Dresselhaus, M.S. 1998. Carbon-based nanostructures. In *R&D status and trends*, ed. Siegel et al.

Dresselhaus, M.S., G. Dresselhaus, and P. Eklund. 1996. *Science of fullerenes and carbon nanotubes*. San Diego: Academic Press.

Drexler, K.E.: 2004, 'Nanotechnology: From Feynman to Funding', *Bulletin of Science, Technology & Society*, 24, 21-27.

Dupuy, J.-P.: 2004, 'Complexity and Uncertainty: A Prudential Approach To Nanotechnology', in: European Commission (Community Health and Consumer Protection): *Nanotechnologies: A Preliminary Risk Analysis on the Basis of a Workshop, Brussels, 1-2 March 2004*, pp. 71-93 (www.europa.eu.int/comm/health/ ph_risk/documents/ev_20040301_en.pdf).

Dupuy, J.-P.: 2004, 'Pour une évaluation normative du programme nanotechnologique', *Annales des Mines*, (February 2004), 27-32.

Einsiedel, E.F.; Goldenberg, L.: 2004, 'Dwarfing the Social? Nanotechnology Lessons from the Biotechnology Front', *Bulletin of Science, Technology & Society*, 24, 28-33.

ETC: 2003a, *The Big Down: Atom Tech ' Technologies Converging at the Atomic Scale*. Winnipeg, Canada: Action Group on Erosion, Technology and Concentration [www.etcgroup.org/documents/TheBigDown.pdf].

ETC: 2003b, 'No Small Matter II: The Case for a Global Moratorium ' Size Matters!' *Occasional Paper Series* 7(1) [www.etcgroup.org/documents/Occ.Paper_Nanosafety.pdf].

Etzkowitz, H., 2001, 'Nano-Science and Society: Finding a Social Basis for Science Policy', in: M.C. Roco and W.S. Bainbridge (eds.), *Societal Implications of Nanoscience and Nanotechnology*, Dordrecht: Kluwer, pp. 121-128.

European Comission: *European Workshop on Social and Economic Research on Nanotechnologies and Nanosciences, Brussels, 14-15 April 2004* [www.stage-research.net/STAGE/PAGES/Nano.html].

European Commission (Community Health and Consumer Protection): 2004, *Nanotechnologies: A Preliminary Risk Analysis on theBasis of a Workshop, Brussels, 1-2 March 2004* [www.europa.eu.int/comm/health/ ph_risk/documents/ev_20040301_en.pdf].

European Commission: 2004, *Converging Technologies: Shaping the Future of European Societies (Report of the High Level Expert Group "Foresighting the New Technology Wave")*, Brussels: European Commission Research [http://europa.eu.int/comm/research/conferences/2004/ntw/index_en.html].

Feynman, R., "There's Plenty of Room at the Bottom: An invitation to Enter a New Field of Physics," Talk at the Annual Meeting of the American Physical Society, 29 December 1959.

Fielder, F.A.; Reynolds, G.H.: 1994, 'Legal Problems of Nanotechnology: An Overview', *Southern California Interdisciplinary Law Journal*, 3, 593-629.

Fleischer, Th.; Decker, M. & Fiedeler, U. (eds.): 2004, *Große Aufmerksamkeit für kleine Welten ' Nanotechnologie und ihre Folgen*, special issue of *Technikfolgenabschätzung ' Theorie und Praxis*, 13 (2), 5-85 [www.itas.fzk.de/tatup/042/inhalt.htm].

Fogelberg, H. & Glimell, H.: 2003, 'Molecular Matters: In Search of the Real Stuff', in: H. Fogelberg & H. Glimell, *Bringing Visibility to the Invisible: Towards A Social Understanding of Nanotechnology*, Göteborg: Göteborg University, pp. 5-32 [www.sts.gu.se/publications/STS_report_6.pdf].

Fogelberg, H. & Glimell, H.: 2003, *Bringing Visibility to the Invisible: Towards A Social Understanding of Nanotechnology*, Göteborg: Göteborg University [www.sts.gu.se/publications/STS_report_6.pdf].

Fogelberg, H.: 2003, 'The Grand Politics of Technoscience: Contextualizing Nanotechnology', in: H. Fogelberg & H. Glimell, *Bringing Visibility to the Invisible: Towards A Social Understanding of Nanotechnology*, Göteborg: Göteborg University, pp. 33-54 [www.sts.gu.se/publications/STS_report_6.pdf].

Frazier, G., "An Ideology For Nanoelectronics," in *Concurrent Computations: Algorithms, Architecture, and Technology*, Plenum Press, New York, 1988.

Glimell, H.: 2001, 'Challenging Limits' Excerpts from an Emerging Ethnography of Nano Physicists', in: H. Glimell & O. Johlin (eds.), *The Social Production of Technology: On the Everyday Life with Things*, Göteburg, SE: BAS Publisher, chapter 7, pp. 111-131 (reprinted in: H. Fogelberg & H. Glimell, *Bringing Visibility to the Invisible: Towards A Social Understanding of Nanotechnology*, Göteborg: Göteborg University, pp. 115-139.

Glimell, H.: 2001, 'Dynamics of the Emerging Field of Nanoscience', in: M.C. Roco and W.S. Bainbridge (eds.), *Societal Implications of Nanoscience and Nanotechnology*, Dordrecht: Kluwer, pp. 156-160.

Glimell, H.: 2003, 'A Nano Narrative: The Mircopolitics of a New Generic Enabling Technology', in: H. Fogelberg & H. Glimell, *Bringing Visibility to the Invisible: Towards A Social Understanding of Nanotechnology*, Göteborg: Göteborg University, pp. 55-77 [www.sts.gu.se/publications/STS_report_6.pdf].

Glimell, H.: 2003, 'Dynamics of the Emerging Field of Nanoscience', in: H. Fogelberg & H. Glimell, *Bringing Visibility to the Invisible: Towards A Social Understanding of Nanotechnology*, Göteborg: Göteborg University, pp. 79-85 [www.sts.gu.se/publications/STS_report_6.pdf].

Glimell, H.: 2004, 'Grand Visions and Lilliput Politics: Staging the Exploration of the 'Endless Frontier'', in: D. Baird, A. Nordmann & J. Schummer (eds.), *Discovering the Nanoscale*, Amsterdam: IOS Press, pp. 231-246.

Goldhaber-Gordon, D., Montemerlo, M.S., Love, J.C., Opiteck, G.J., and Ellenbogen, J.C., "Overview of Nanoelectronic Devices," submitted to the *Proceedings of the IEEE*, February 1997. For more information, please send e-mail to nanotech@mitre.org.

Gorman, M.E.; Groves, J.F. & Shrager, J.: 2004, 'Societal Dimensions of Nanotechnology as a Trading Zone: Results from a Pilot Project', in: D. Baird, A. Nordmann & J. Schummer (eds.), *Discovering the Nanoscale*, Amsterdam: IOS Press, pp. 63-73.

Grinbaum, A. & Dupuy, J.-P.: 2004, 'Living with Uncertainty: Toward the Ongoing Normative Assessment of Nanotechnology', *Techne: Research in Philosophy and Technology*, 8(3) (forthcoming).

Grinbaum, A.: 2004, 'La condition de l'homme moderne et les nanotechnologies', in: G. Nivat (ed.), *Les limites de l´humain. 39èmes Rencontres Internationales de Genève*, L´Age d´Homme, Genève, p. 141.

Grunwald, A.: 2004, 'Ethische Aspekte der Nanotechnologie. Eine Felderkundung', *Technikfolgenabschätzung ' Theorie und Praxis*, 13 (2), 71-78.

Gupta, V.K. & Pangannaya, N.B.: 2000, 'Carbon nanotubes: bibliometric analysis of patents', *World Patent Information*, 22, 185-189.

Hansson, S.O.: 2004, 'Great Uncertainty about Small Things', *Techne: Research in Philosophy and Technology*, 8(3) (forthcoming).

Haruta, M. 1997. *Catalyst surveys of Japan* 1:61 and references therein.

Hayles, N.K. (ed.): 2004, *Nanoculture: Implications of the New Technoscience*, Bristol, UK: Intellect Books.

Hayles, N.K.: 2004, 'Connecting the Quantum Dots: Nanotechscience and Culture', in: N.K. Hayles (ed.), *Nanoculture: Implications of the New Technoscience*, Bristol, UK: Intellect Books, pp. 11-26.

Hemerén, G.: 2004, 'Nano Ethics Primer', in: European Commission (Community Health and Consumer Protection): *Nanotechnologies: A Preliminary Risk Analysis on theBasis of a Workshop, Brussels, 1-2 March 2004* (www.europa.eu.int/comm/health/ ph_risk/documents/ ev_20040301_en.pdf), pp. 95-102.

Hennig, J.: 2004, 'Changes in the Design of Scanning Tunneling Microscopic Images from 1980 to 1990', *Techne: Research in Philosophy and Technology*, 8(3) (forthcoming).

Hessenbruch, A.: 2004, 'Nanotechnology and the Negotiation of Novelty', in: D. Baird, A. Nordmann & J. Schummer (eds.), *Discovering the Nanoscale*, Amsterdam: IOS Press, pp. 135-144.

Hullmann, A. & Meyer, M.: 2003, 'Publications and Patents in Nanotechnology. An overview of previous studies and the state of the art', *Scientometrics*, 58 (3) 507-527.

Huo, Q., D.I. Margolese, U. Ciesla, P. Feng, T.E. Gier, P. Sieger, R. Leon, P.M. Petroff, F. Schuth, and G.D. Stucky. 1994. *Nature* 368:317.

Jena, P., S.N. Khanna, and B.K. Rao. 1996. In *Science and technology of atomically engineered materials*, ed. P. Jena. River Edge, NJ: World Scientific.

Johansson, M.: 2003, 'Plenty of room at the bottom: Towards an anthropology of nanoscience', *Anthropology Today*, 19 (No. 6), 3-6.

Johnson, A.: 2004, 'The End of Pure Science: Science Policy from Bayh-Dole to the NNI', in: D. Baird, A. Nordmann & J. Schummer (eds.), *Discovering the Nanoscale*, Amsterdam: IOS Press, pp. 217-230.

Jones, C.W., and W.J. Koros. 1994. *Carbon* 32:1419.

Jounet, C., W.K. Maser, P. Bernier, A. Loiseau, M. Lamy de la Chapelle, S. Lefrant, P. Deniard, R. Lee, and J.E. Fischer. 1997. *Nature* 388:756.

K.E. Drexler, *Nanosystems: Molecular Machinery, Manufacturing, and Computation,* Wiley, New York, 1992.

Karger, J., and D.M. Ruthven. 1992. *Diffusion in zeolites*. New York: J. Wiley.

Keiper, A.: 2003, 'The Nanotechnology Revolution', *The New Atlantis*, 2, 17-34.

Khushf, G.: 2004, 'A Hierarchical Architecture for Nano-scale Science and Technology: Taking Stock of the Claims About Science Made By Advocates of NBIC Convergence', in: D. Baird, A. Nordmann & J. Schummer (eds.), *Discovering the Nanoscale*, Amsterdam: IOS Press, pp. 21-33.

Khushf, G.: 2004, 'Systems Theory and the Ethics of Human Enhancement: A Framework for NBIC Convergence', *Annals of the New York Academy of Sciences*, 1013, 124-149.

Krätschmer, W., L.D. Lamb, K. Fostiropoulos, and D.R. Huffman. 1990. *Nature* 347:354.

Kresge, C.T., M.E. Leonowicz, W.J. Roth, J.C. Vartuli, and J.S. Beck. 1992. *Nature* 359:710.

Kupperman, A., S. Nadimi, S. Oliver, G. Ozin, J. Garcés, and M. Olken. 1993. *Nature* 365:239.

Kuusi, O.; Meyer, M.: 2002, 'Technological generalizations and leitbilder ' the anticipation of technological opportunities', *Technological Forecasting & Social Change*, 69, 625-639.

Landon, B.: 2004, 'Less is More: Much Less is Much More: The Insistent Allure of Nanotechnology Narratives in Science Fiction', in: N.K. Hayles (ed.), *Nanoculture: Implications of the New Technoscience*, Bristol, UK: Intellect Books, pp. 131-146.

Laszlo, P.: 2004, 'Is There Life After Partington?', *Hyle: International Journal for Philosophy of Chemistry*, 10(2), 169-178.

Lenhard, J.: 2004, 'Nanoscience and the Janus-Faced Character of Simulations', in: D. Baird, A. Nordmann & J. Schummer (eds.), *Discovering the Nanoscale*, Amsterdam: IOS Press, pp. 93-100.

Lent, C.S., Tougaw, P.D., Porod, W., Bernstein, G.H., "Quantum Cellular Automata," *Nanotechnology*, Vol. 4, p. 49, 1993.

Lewak, S.E.: 2004, 'What's the Buzz? Tell Me What's A-Happening: Wonder, Nanotechnology, and Alice's Adventures in Wonderland', in: N.K. Hayles (ed.), *Nanoculture: Implications of the New Technoscience*, Bristol, UK: Intellect Books, pp. 201-.

Lide, D.R., ed. 1993-1994. *CRC Handbook of Chemistry and Physics*, 74th ed.

Lin-Easton, P.C.: 2001, 'It's Time for Environmentalists to Think Small ' Real Small: A Call for the Involvement of Environmental Lawyers in Developing Precautionary Policies Molecular Nanotechnology', *Georgetown International Law Review*, 14, 106-134.

López, J.: 2004, 'Bridging the Gaps: Science Fiction in Nanotechnology', *Hyle: International Journal for Philosophy of Chemistry*, 10(2), 129-152.

Lösch, A.: 2004, 'Nanomedicine and Space: Discursive Orders of Mediating Innovations', in: D. Baird, A. Nordmann & J. Schummer (eds.), *Discovering the Nanoscale*, Amsterdam: IOS Press, pp. 193-202.

Marshall, K.: 2004, 'Atomizing the Risk Technology', in: N.K. Hayles (ed.), *Nanoculture: Implications of the New Technoscience*, Bristol, UK: Intellect Books, pp. 147-160.

Martin, T.P., N. Malinowski, U. Zimmerman, U. Naher, and H. Schaber. 1993. *J. Chem. Phys.* 99:4210.

Martin, T.P., U. Naher, H. Schaber, U. Zimmerman. 1993. *Phys. Rev. Lett.* 70:3079.

Mayer, S.: 2002, 'From genetic modification to nanotechnology: the dangers of 'sound science'', in: T. Gilland (ed.), *Science: Can We Trust the Experts?*, London: Hodder and Stoughton, pp. 1-15.

Mehta, M.D.: 2002, 'Nanoscience and Nanotechnology: Assessing the Nature of Innovation in These Fields', *Bulletin of Science, Technology & Society*, 22(4), 269-273.

Mehta, M.D.: 2004, 'From Biotechnology to Nanotechnology: What Can We Learn from Earlier Technologies?, *Bulletin of Science, Technology & Society*, 24, 34-39.

Meyer, M. & Kuusi, O.: 2004, 'Nanotechnology: Generalizations in an Interdisciplinary Field of Science and Technology', *Hyle: International Journal for Philosophy of Chemistry*, 10(2), 153-168.

Meyer, M.: 2000a, 'Does science push technology? Patents citing scientific literature', *Research Policy*, 29, 409-434.

Meyer, M.: 2000b, 'Patent citations in a novel field of technology: What can they tell about interactions of emerging communities of science and technology?', *Scientometrics*, 48, 151-178.

Meyer, M.: 2001a, 'Patent citations in a novel field of technology: An exploration of nano-science and nano-technology', *Scientometrics*, 51, 163-183.

Meyer, M.: 2001b, 'Socio-economic Research on Nanoscale Science and Technology: A European Overview and Illustration', in: M.C. Roco and W.S. Bainbridge (eds.), *Societal Implications of Nanoscience and Nanotechnology*, Dordrecht: Kluwer, pp. 217-241.

Meyer, M.; Persson, O.: 1998, 'Nanotechnology ' interdisciplinarity, patterns of collaboration and differences in application', *Scientometrics*, 47, 195'205.

Milburn, C.: 2002, 'Nanotechnology in the age of post-human engineering: science fiction as science', *Configurations*, 10, 261-295 [reprinted in: N.K. Hayles (ed.), *Nanoculture: Implications of the New Technoscience*, Bristol, UK: Intellect Books, 2004, pp. 109-130].

Milburn, C.: 2004, 'Nano/Splatter: Disintegrating the Postbiological Body', *New Literary History*, 35 (in print).

Mnyusiwalla, A.; Abdallah, S.D.; Singer, P.A.: 2003, 'Mind the gap: science and ethics in nanotechnology', *Nanotechnology*, 14, R9-R13.

Mody, C.C.M.: 2004, 'How Probe Microscopists Became Nanotechnologists', in: D. Baird, A. Nordmann & J. Schummer (eds.), *Discovering the Nanoscale*, Amsterdam: IOS Press, pp. 119-133.

Mody, C.C.M.: 2004, 'Instruments in Training: The Growth of American Probe Microscopy in the 1980s', in: D. Kaiser (ed.), *Pedagogy and the Practice of Science: Producing Physical Scientists, 1800-2000*, Cambridge, MA: MIT Press (forthcoming).

Mody, C.C.M.: 2004, 'Small, but Determined: Technological Determinism in Nanoscience', *Hyle: International Journal for Philosophy of Chemistry*, 10(2), 99-128.

Mody, C.C.M.: 2004, *Crafting the Tools of Knowledge: The Invention, Spread, and Commercialization of Probe Microscopy, 1960-2000*, Ph.D. dissertation, Cornell University.

Montemerlo, M.S., Love, J.C., Opiteck, G.J., Goldhaber, D. J., and Ellenbogen, J.C., "Technologies and Designs for Electronic Nanocomputers," MITRE Technical Report 96W0000044, The MITRE Corporation, McLean, VA, July 1996. For more information, send e-mail to nanotech@mitre.org.

Moor, J.H. & Weckert, J.: 2004, 'Nanoethics: Assessing the Nanoscale From an Ethical Point of View', in: D. Baird, A. Nordmann & J. Schummer (eds.), *Discovering the Nanoscale*, Amsterdam: IOS Press, pp. 301-310.

Munn Sanchez, E.: 2004, 'The Expert's Role in Nanoscience and Technology', in: D. Baird, A. Nordmann & J. Schummer (eds.), *Discovering the Nanoscale*, Amsterdam: IOS Press, pp. 257-266.

Nordmann, A.: 2003, 'Shaping the World Atom by Atom: Eine nanowissenschaftliche WeltBildanalyse', in: A. Grunwald (ed.), *Technikgestaltung zwischen Wunsch und Wirklichkeit*, Berlin: Springer, pp. 191-199.

Nordmann, A.: 2004, 'Molecular Disjunctions: Staking Claims at the Nanoscale', in: D. Baird, A. Nordmann & J. Schummer (eds.), *Discovering the Nanoscale*, Amsterdam: IOS Press, pp. 51-62.

Nordmann, A.: 2004, 'Nanotechnology: Convergence and Integration', Presentation at *EuroNanoForum, Trieste, December 10, 2003* (Proceedings in preparation).

Nordmann, A.: 2004, 'Nanotechnology's WorldView: New Space for Old Cosmologies', *IEEE Technology and Society Magazine*, 23 (forthcoming).

Nordmann, A.: 2004, 'Social Imagination for Nanotechnology', in: European Commission (Community Health and Consumer Protection): *Nanotechnologies: A Preliminary Risk Analysis on the Basis of a Workshop, Brussels, 1-2 March 2004*, pp. 111-113 [www.europa.eu.int/comm/health/ph_risk/documents/ev_20040301_en.pdf].

Nordmann, A.: 2004, 'Was ist TechnoWissenschaft? ' Zum Wandel der Wissenschaftskultur am Beispiel von Nanoforschung und Bionik', in: T. Rossmann & C. Tropea (eds.), *Bionik ' Neue Forschungsergebnisse aus Natur-, Ingenieur- und Geisteswissenschaften*, Berlin: Springer, 2004 (forthcoming)

Paschen, H.; Coenen, C.; Fleischer, T.; Grünwald, R.; Oertel, D. & Revermann, C.: 2004, *Nanotechnologie: Forschung, Entwicklung, Anwendung*, Berlin: Springer (*Nanotechnologie, TAB-Arbeitsbericht 92*, Berlin: Büro für Technikfolgen-Abschätzung beim Deutschen Bundestag).

Pitt, J.C.: 2004, 'The Epistemology of the Very Small', in: D. Baird, A. Nordmann & J. Schummer (eds.), *Discovering the Nanoscale*, Amsterdam: IOS Press, pp. 157-163.

Pressman, J.: 2004, 'Nano Narrative: A Parable from Electronic Literature', in: N.K. Hayles (ed.), *Nanoculture: Implications of the New Technoscience*, Bristol, UK: Intellect Books, pp. 191-200.

Prigogine, I., and S. Rice. 1988. *Advances in chemical physics*, Vol. 70, Parts 1 & 2. New York: J. Wiley.

R. T. Bate, "Nanoelectronics," *Nanotechnology*, Vol. 1, pp. 1-7, 1990.

Rademann, K., B. Kaiser, U. Even, F. Hensel. 1987. *Phys. Rev. Lett.* 59:2319.

Rao, C.N.R., B.C. Satishkumar, and A. Govindaraj. 1997. *Chem. Commun.* 1581.

Rao, M.B., and S. Sircar. 1993. *Gas Separation and Purification* 7:279.

Reed, M.A., "Quantum Dots," *Scientific American*, January 1993, pp. 118-123.

Roberts, J.A.: 2004, 'Deciding the Future of Nanotechnologies: Legal Perspectives on Issues of Democracy and Technology', in: D. Baird, A. Nordmann & J. Schummer (eds.), *Discovering the Nanoscale*, Amsterdam: IOS Press, pp. 247-255.

Robinson, C.: 2004, 'Images in NanoScience/Technology', in: D. Baird, A. Nordmann & J. Schummer (eds.), *Discovering the Nanoscale*, Amsterdam: IOS Press, pp. 165-169.

Robison, W.L.: 2004, 'Nano-Ethics', in: D. Baird, A. Nordmann & J. Schummer (eds.), *Discovering the Nanoscale*, Amsterdam: IOS Press, pp. 285-300.

Roco, M.C. & Bainbridge, W.S. (eds.): 2001, *Societal implications of nanoscience and nanotechnology*, (Proceedings of a workshop organized by the National Science Foundation, September 28-29, 2000), Kluwer: Dordrecht [available online at http://itri.loyola.edu/nano/societalimpact/nanosi.pdf]

Roco, M.C. & Tomellini, R. (eds.): 2002, *Nanotechnology: Revolutionary Opportunities and Societal Implications, Workshop, Lecce (Italy), 31 January - 1 February 2002*, Luxemburg: Office for Official Publications of the European Communities, [200 pp.]

Roco, M.C.; Bainbridge, W.S. (eds.): 2002, *Converging Technologies for Improving Human Performance: Nanotechnology, Biotechnology, Information Technology and the Cognitive Science*, Arlington, VA: National Science Foundation.

Rohlfing, E.A., D.M. Cox, and A. Kaldor. 1984. *J. Chem. Phys.* 81:3846.

Rosen, A. 1998. A periodic table in three dimensions: A sightseeing tour in the nanometer world. In *Advances in quantum chemistry*. In press.

Ruthven, D.M., S. Farooq, K.S. Knaebel. 1994. *Pressure swing adsorption*. New York: VCH Publishers.

Sarewitz, D.; Woodhouse, E.: 2003, 'Small is Powerful', in: A. Lightman, D. Sarewitz & Chr. Desser, (eds.), *Living with the Genie: Essays on Technology and the Quest for Human Mastery*, Washington, DC: Island Press, pp. 63-83.

Sayari, A. 1996. *Chem. Mater.* 8:1840.

Schiemann, G.: 2004, 'Dissolution of the Nature-Technology Dichotomy? Perspectives on Nanotechnology from an Everyday Understanding of Nature', in: D. Baird, A. Nordmann & J. Schummer (eds.), *Discovering the Nanoscale*, Amsterdam: IOS Press, pp. 209-213.

Schmidt, J.C.: 2004, 'Unbounded Technologies: Working Through Technological Reductionism of Nanotechnology', in: D. Baird, A. Nordmann & J. Schummer (eds.), *Discovering the Nanoscale*, Amsterdam: IOS Press, pp. 35-50.

Schuler, E.: 2004, 'Perception of Risks and Nanotechnology', in: D. Baird, A. Nordmann & J. Schummer (eds.), *Discovering the Nanoscale*, Amsterdam: IOS Press, pp. 279-284.

Schummer, J.: 2001, 'Ethics of Chemical Synthesis', *Hyle: International Journal for Philosophy of Chemistry*, 7, 103-124 [online: www.hyle.org/journal/issues/7/schummer.htm].

Schummer, J.: 2003, 'Aesthetics of Chemical Products: Materials, Molecules, and Molecular Models', *Hyle: International Journal for Philosophy of Chemistry*, 9, 73-104 [online: www.hyle.org/journal/issues/9-1/schummer.htm].

Schummer, J.: 2003, 'The Notion of Nature in Chemistry', *Studies in History and Philosophy of Science*, 34 (2003), 705-736.

Schummer, J.: 2004, 'Interdisciplinary Issues in Nanoscale Research', in: D. Baird, A. Nordmann & J. Schummer (eds.), *Discovering the Nanoscale*, Amsterdam: IOS Press, pp. 9-20.

Schummer, J.: 2004, 'Multidisciplinarity, Interdisciplinarity, and Patterns of Research Collaboration in Nanoscience and Nanotechnology', *Scientometrics*, 59, 425-465.

Schummer, J.: 2004, 'Naturverhältnisse in der modernen Wirkstoff-Forschung', in: K. Kornwachs (ed.), *Technik ' System ' Verantwortung*, Münster: LIT, pp. 629-638.

Schummer, J.: 2004, 'Why do Chemists Perform Experiments?', in: D. Sobczynska, P. Zeidler, E. Zielonacka-Lis (eds.), *Chemistry in the Philosophical Melting Pot*, Peter Lang (Frankfurt/M.), 2004, pp. 395-410.

Schummer, J.: 2005, 'Reading Nano: The Public Interest in Nanotechnology as Reflected in Book Purchase Patterns", *Public Understanding of Science*, 14 (2) (forthcoming).

Schummer, J.; Baird, D. (eds.): 2004-5, *Nanotech Challenges*, joint special issue of *Hyle: International Journal for Philosophy of Chemistry & Techne: Research in Philosophy and Technology* (forthcoming).

Schwarz, A.E.: 2004, 'Shrinking the 'Ecological Footprint' with NanoTechnoScience?', in: D. Baird, A. Nordmann & J. Schummer (eds.), *Discovering the Nanoscale*, Amsterdam: IOS Press, pp. 203-208.

Schwarz, R.B. 1998. Storage of hydrogen in powders with nanosized crystalline domains. In *R&D status and trends*, ed. Siegel et al.

Siegel, R.W. 1991. *Ann. Rev. Mater. Sci.* 21:559.

Siegel, R.W., E. Hu, and M.C. Roco, eds. 1998. *R&D status and trends in nanoparticles, nanostructured materials, and nanodevices in the United States*. Baltimore: Loyola College, International Technology Research Institute. NTIS #PB98-117914.

Soffer, A., J.E. Koresh, S. Saggy. 1987. United States Patent 4,685,940.

Stix Gary., "Toward 'Point One'," *Scientific American*, February 1995, pp. 90-95.

Stuckless, J.T, D.E. Starr, D.J. Bald, C.T. Campbell. 1997. *J. Chem. Phys*. 107:5547.

Suchman, M.: 2001, 'Envisioning Life on the Nano-Frontier', in: M.C. Roco and W.S. Bainbridge (eds.), *Societal Implications of Nanoscience and Nanotechnology*, Dordrecht: Kluwer, pp. 211-216.

Suchman, M.: 2002, 'Social Science and Nanotechnologies', M. Roco & R. Tomellini (eds.), *Nanotechnology ' Revolutionary Opportunities and Societal Implications*, Luxembourg: European Communities, pp. 95-99.

Sweeney, A.E.; Seal, S. & Vaidyanathan, P.: 2003, 'The promises and perils of nanoscience and nanotechnology: Exploring emerging social and ethical issues', *Bulletin of Science, Technology & Society*, 23 (4), 236-245.

Swiss Re: 2004, *Nanotechnology: Small Matter, Many Unknowns*, Zurich: Swiss Re [www.swissre.com].

Tanev, P.T., M. Chibwe, and T.J. Pinnavaia. 1994. *Nature* 368:321.

Taylor, K.J., C.L. Pettiette-Hall, O. Cheshnovsky, and R.J. Smalley. 1992. *J. Chem. Phys*. 96:3319.

ten Wolde, A. (ed.): 1998, *Nanotechnology: Towards a Molecular Construction Kit*, The Hague: The Netherlands Study Centre for Technology Trends (STT).

Tenner, E.: 2001, 'Nanotechnology and Unintended Consequences', in: M.C. Roco and W.S. Bainbridge (eds.), *Societal Implications of Nanoscience and Nanotechnology*, Dordrecht: Kluwer, pp. 241-246.

The Royal Society & The Royal Academy of Engineering: 2004, *Nanoscience and nanotechnologies: opportunities and uncertainties*, London [www.nanotec.org.uk/finalReport.htm].

Toumey, C.: 2004, 'Anticipating Public Reactions to Nanotechnology', *Techne: Research in Philosophy and Technology*, 8(3) (forthcoming).

Trudeau, M.L., and J.Y. Ying. 1996. *Nanostructured Materials* 7:245.

Tschope, A.S., W. Liu, M. Flyzani-Stephanopoulos, and J.Y. Ying 1995. *J. Catal*. 157:42.

Turton, R., *The Quantum Dot*, Oxford University Press, 1995.

Verbeek, A. & Callaert, J.: 2002, *Detailed analysis of the science-technology interaction in the field of Nanotechnology*, Luxemburg: Office for Official Publications of the European Communities (Linking Science to Technology. Bibliographic References in Patents, Vol. 9) [79 pp.].

Vermaas, P.E.: 2004, 'Nanoscale Technology: A Two-Sided Challenge for Interpretations of Quantum Mechanics', in: D. Baird, A. Nordmann & J. Schummer (eds.), *Discovering the Nanoscale*, Amsterdam: IOS Press, pp. 77-91.

Wald, C.A.: 2004, 'Working Boundaries on the NANO Exhibition', in: N.K. Hayles (ed.), *Nanoculture: Implications of the New Technoscience*, Bristol, UK: Intellect Books, pp. 83-108.

Weckert, J.: 2001, 'The Control of Scientific Research: The Case of Nanotechnology', *Australian Journal of Professional and Applied Ethics*, 3, 29-44.

Weil, V.: 2001, 'Ethical Issues in Nanotechnology', in: M.C. Roco and W.S. Bainbridge (eds.), *Societal Implications of Nanoscience and Nanotechnology*, Dordrecht: Kluwer, pp. 193-198.

2

Biotechnology, Self-Assembly & Nano World

Nanoworld: An Overview

Nanotechnology will be a bottom-up technology, building upward from the molecular scale. It will bring a revolution in human abilities like that brought by agriculture or power machinery. It can even be used to reverse many of the changes brought by agriculture or power machinery. But we humans are huge creations with no direct experience of the molecular world, and this can make nanotechnology hard to visualize, hence hard to understand. Scientists working with moleculas face this problem today. They can often calculate how molecules will behave, but to understand this behavior, they need more than heaps of numbers: they need pictures, movies, and interactive simulations, and so they are producing them at an ever-increasing pace. Molecules are objects that exert forces on one another. If your hands were small enough, you could grab them, squeeze them, and bash them together. Understanding the molecular world is much like understanding any other physical world: it is a matter of understanding size, shape, strength, force, motion, and the like—a matter of understanding the differences between sand, water, and rock, or between steel and soap bubbles. Today's visualization tools give a taste of what will become possible with tomorrow's faster computers and better "virtual realities," simulated environments that let you tour a world that "exists" only as a model inside the computer. Before discussing nanotechnology and how it relates to the technologies of today, let's try to get a more concrete understanding of the molecular world by describing a simulation embedded in a scenario. In this scenario, events and technologies described as dating from 1990 or before are historically accurate; those with later dates are either projections or mere scenario elements. The descriptive details in the simulation are written to fit designs and calculations based on standard scientific data, so the science isn't fiction.

We saw Joel Gregory manipulating molecules in the virtual reality of a simulated world using video goggles, tactile gloves, and a supercomputer. The early twenty-first century should be able to do even better. Imagine, then, that today you were to take a really long nap, oversleep, and wake up decades later in a nanotechnological world. In the twenty-first century, even more than in the twentieth, it's easy to make things work without understanding them, but to a newcomer much of the technology seems like magic, which is dissatisfying. After a few days, you want to understand what nanotechnology is, on a gut level. Back in the late twentieth century, most teaching used dry words and simple pictures, but now—for a topic like this—it's easier to explore a simulated world. And so you decide to explore a simulation of the molecular world. Looking through the brochure, you read many tedious facts about the simulation: how accurate it is in describing sizes, forces, motions, and the like; how similar it is to working tools used by both engineering students and professionals; how you can buy one for your very own home, and so forth. It explains how you can tour the human body, see state-of-the-art nanotechnology in action, climb a bacterium, etc. For starters, you decide to take an introductory tour: simulations of real twentieth-century objects alongside quaint twentieth-century concepts of nanotechnology. After paying a small fee and memorizing a few key phrases (any variation of "Get me out of here!" will do the most important job), you pull on a powersuit, pocket a Talking Tourguide, step into the simulation chamber, and strap the video goggles over your eyes. Looking through the goggles, you seem to be in a room with a table you know isn't really there and walls that seem too far away to fit in the simulation chamber. But trickery with a treadmill floor makes the walk to the walls seem far enough, and when you walk back and thump the table, it feels solid because the powersuit stops your hand sharply at just the right place. You can even feel the texture of the carvings on the table leg, because the suit's gloves press against your fingertips in the right patterns as you move. The simulation isn't perfect, but it's easy to ignore the defects. On the table is (or seems to be) an old 1990s silicon computer chip. When you pick it up, as the beginners' instructions suggest, it looks like

You feel as though you're falling toward the chip's surface, shrinking rapidly. In a moment, it looks roughly like With your thumb still there holding it. The world grows blurrier, then everything seems to go wrong as you approach the molecular level. First, your vision blurs to uselessness—there is light, but it becomes a featureless fog. Your skin is tickled by small impacts, then battered by what feel like hard-thrown marbles. Your arms and legs feel as though they are caught in turbulence, pulling to and fro, harder and harder. The ground hits your feet, you stumble and stick to the ground like a fly on flypaper, battered so hard that it almost hurts. You asked for realism, and only the built-in safety limits in the suit keep the simulated thermal motions of air molecules and of your own arms from beating you senseless. "Stop!" gives you a rest from the suit's yanking and thumping, and "Standard settings!" makes the world around you become more reasonable. The simulation changes, introducing the standard cheats. Your simulated eyes are now smaller than a light wave, making focus impossible, but the goggles snap your vision into sharpness and show the atoms around you as small spheres. (Real nanomachines are as blind as you were a moment ago, and can't cheat.) You are on the surface of the computer chip, between a cell and two blocky nanocomputers like the ones. The simulation keeps bombarding you with air molecules, but the standard settings leave out the sensation of being pelted with marbles. A moment ago you were stuck tight to the ground by molecular stickiness, but the standard settings give your muscles the effective strength of steel—at least in simulation—by making everything around you much softer and weaker. The tourguide says that the only unreal features of the simulation have to do with you—not just your ability to see and to ignore thermal shaking and bombardment, but also your sheer existence at a size too small for anything so complex as a human being. It also explains why you can see things move, something about slowing down everything around you by a factor of 10 for every factor of 10 enlargement, and by another factor to allow for your being made stronger and hence faster. And so, with your greater strength and some adjustments to make your arms, legs, and torso less sticky, you can stand, see, feel, and take stock of the situation.

The ground underfoot, like everything around you, is pebbly with atom-sized bumps the size of your fingertips. Objects look like bunches of transparent grapes or fused marbles in a variety of pretty but imaginary colors. The simulation displays a view of atoms and molecules much like those used by chemists in the 1980s, but with a sharper 3-D image and a better way to move them and to feel the forces they exert. Actually, the whole simulation setup is nothing but an improved version of systems built in the late 1980s—the computer is faster, but it is calculating the same things. The video goggles are better and the whole-body powersuit is a major change, but even in the 1980s there were 3-D displays for molecules and crude devices that gave a sense of touching them. The gloves on this suit give the sensation of touching whatever the computer simulates. When you run a fingertip over the side of the smaller nanocomputer, it feels odd, hard to describe. It is as if the surface were magnetic–it pulls on your fingertip if you move close enough. But the result isn't a sharp click of contact, because the surface isn't hard like a magnet, but strangely soft. Touching the surface is like touching a film of fog that grades smoothly into foam rubber, then hard rubber, then steel, all within the thickness of a sheet of corrugated cardboard.

Moving sideways, your fingertip feels no texture, no friction, just smooth bumps more slippery than oil, and a tendency to get pulled into hollows. Pulling free of the surface takes a firm tug. The simulation makes your atom-sized fingertips feel the same forces that an atom would. It is strange how slippery the surface is—and it can't have been lubricated, since even a single oil molecule would be a lump the size of your thumb. This slipperiness makes it obvious how nano-scale bearings can work, how the parts of molecular machines can slide smoothly. But on top of this, there is a tingling feeling in your fingers, like the sensation of touching a working loudspeaker. When you put your ear against the wall of the nanocomputer, you flinch back: for a moment, you heard a sound like the hiss of a twentieth century television tuned to a channel with no broadcast, with nothing but snow and static—but loud, painfully loud. All the atoms in the surface are vibrating at high frequencies, too fast to see. This is thermal vibration, and it's obvious why it's also called thermal noise.

Individual molecules still move too quickly to see. So, to add one more cheat to the simulation, you issue the command "Whoa!", and everything around seems to slow down by a factor of ten. On the surface, you now can see thermal vibrations that had been too quick to follow. All around, air molecules become easier to watch. They whiz about as thick as raindrops in a storm, but they are the size of marbles and bounce in all directions. They're also sticky in a magnetlike way, and some are skidding around on the wall of the nanocomputer. When you grab one, it slips away. Most are like two fused spheres, but you spot one that is perfectly round—it is an argon atom, and these are fairly rare. With a firm grip on all sides to keep it from shooting away like a watermelon seed, you pinch it between your steel-strong fingers. It compresses by about 10 percent before the resistance is more than you can overcome. It springs back perfectly and instantly when you relax, then bounces free of your grip. Atoms have an unfamiliar perfection about them, resilient and unchanging, and they surround you in thick swarms. At the base of the wall is a churning blob that can only be a droplet of water. Scooping up a handful for a closer look yields a swarm of molecules, hundreds, all tumbling and bumbling over one another, but clinging in a coherent mass. As you watch, though, one breaks free of the liquid and flies off into the freer chaos of the surrounding air: the water is evaporating. Some slide up your arm and lodge in the armpit, but eventually skitter away. Getting rid of all the water molecules takes too much scraping, so you command "Clean me!" to dry off.

Beside you, the smaller nanocomputer is a block twice your height, but it's easy to climb up onto it as the tourguide suggests. Gravity is less important on a small-scale: even a fly can defy gravity to walk on a ceiling, and an ant can lift what would be a truck to us. At a simulated size of fifty nanometers, gravity counts for nothing. Materials keep their strength, and are just as hard to bend or break, but the weight of an object becomes negligible. Even without the strength-enhancement that lets you overcome molecular stickiness, you could lift an object with 40 million times your mass–like a person of normal size lifting a box containing a half-dozen fully loaded oil tankers. To simulate this weak gravity, the powersuit cradles your body's weight, making you feel as if you were floating. This is almost like a vacation in an orbital theme park, walking with stickyboots on walls, ceilings, and whatnot, but with no need for antinausea medication. This looks like a cluster of grapes and is about the same size. It even feels a bit like a bunch of grapes, soft and loose. The parts don't fly free like a gas or tumble and wander like a liquid, but they do quiver like gelatin and sometimes flop or twist. It is solid enough, but the folded structure is not as strong as your steel fingers. In the 1990s, people began to build molecular machinery out of proteins, copying biology. It worked, but it's easy to see why they moved on to better materials. From a simulated pocket, you pull out a simulated magnifying glass and look at the simulated protein. This shows a pair of bonded atoms on the surface at 10 times magnification. The atoms are almost transparent, but even a close look doesn't reveal a nucleus inside, because it's too small to see. It would take 1,000 times magnification to be able to see it, even with the head start of being able to see atoms with your naked eye. How could people ever confuse big, plump atoms with tiny specks like nuclei? Remembering how your steel-strong fingers couldn't press more than a fraction of the way toward the nucleus of an argon atom from the air, it's clear why nuclear fusion is so difficult. In fact, the tourguide said that it would take a real-world projectile over a hundred times faster than a high-powered rifle bullet to penetrate into the atomic core and let two nuclei fuse. Try as you might, there just isn't anything you could find in the molecular world that could reach into the middle of an atom to meddle with its nucleus. You can't touch it and you can't see it, so you stop squinting though the magnifying glass. Nuclei just aren't of much interest in nanotechnology.

Taking the advice of the tourguide, you grab two molecular knobs on the protein and pull. It resists for a moment, but then a loop comes free, letting other loops flop around more, and the whole structure seems to melt into a writhing coil. After a bit of pulling and wrestling, the protein's structure becomes obvious: It is a long chain–longer than you are tall, if you could get it straight—and each segment of the chain has one of several kinds of knobs sticking off to the side. With the multicolored, glassy-bead portrayal of atoms, the protein chain resembles a flamboyant necklace. This may be decorative, but how does it all go back together? The chain flops and twists and thrashes, and you pull and push and twist, but the original tight, solid packing is lost. There are more ways to go wrong in folding up the chain than there are in solving Rubik's Cube, and now that the folded structure is gone, it isn't even clear what the result should look like. How did those twentieth-century researchers ever solve the notorious "protein folding problem"? It's a matter of record that they started building protein objects in the late 1980s. This protein molecule won't go back together, so you try to break it. A firm grip and a powerful yank straightens a section a bit, but the chain holds together and snaps back. Though unfolding it was easy, even muscles with the strength of steel—the strength of Superman—can't break the chain itself. Chemical bonds are amazingly strong, so it's time to cheat again. When you say, "Flimsy world–one second!" while pulling, your hands easily move apart, splitting the chain in two before its strength returns to normal. You've forced a chemical change, but there must be easier ways since chemists do their work without tiny superhands. While you compare the broken ends, they thrash around and bump together. The third time this happens, the chain rejoins, as strong as before. This is like having snap-together parts, but the snaps are far stronger than welded steel. Modern assembler chemistry usually uses other approaches, but seeing this happen makes the idea of molecular assembly more understandable: Put the right pieces together in the right positions, and they snap together to make a bigger structure. Remembering the "Whoa!" command, you decide to go back to the properly scaled speed for your size and strength. Saying "Standard settings!," you see the thrashing of the protein chain speed up to hard-to-follow blur.

At your feet is a ribbed, ringed cylindrical object about the size of a soup can—not a messy, loosely folded strand like the protein (before it fell apart), but a solid piece of modern nanotechnology. It's a gear like the one. Picking it up, you can immediately feel how different it is from a protein. In the gear, everything is held in place by bonds as strong as those that strung together the beads of the protein chain. It can't unfold, and you'd have to cheat again to break its perfect symmetry. Like those in the wall of the nanocomputer, its solidly attached atoms vibrate only slightly. There's another gear nearby, so you fit them together and make the atomic teeth mesh, with bumps on one fitting into hollows on the other. They stick together, and the soft, slick atomic surfaces let them roll smoothly. Underfoot is the nanocomputer itself, a huge mechanism built in the same rigid style. Climbing down from it, you can see through the transparent layers of the wall to watch the inner works. An electric motor an arm-span wide spins inside, turning a crank that drives a set of oscillating rods, which in turn drive smaller rods. This doesn't look like a computer; it looks more like an engineer's fantasy from the nineteenth century. But then, it is an antique design–the tourguide said that the original proposal was a piece of exploratory engineering dating from the mid-1980s, a mechanical design that was superseded by improved electronic designs before anyone had the tools to build even a prototype. This simulation is based on a version built by a hobbyist many years later.

The mechanical nanocomputer may be crude, but it does work, and it's a lot smaller and more efficient than the electronic computers of the early 1990s. It's even somewhat faster. The rods slide back and forth in a blur of motion, blocking and unblocking each other in changing patterns, weaving patterns of logic. This nanocomputer is a stripped-down model with almost no memory, useless by itself. Looking beyond it, you see the other block–which contains a machine powerful enough to compete with most computers built in 1990. This computer is a millionth of a meter on a side, but from where you stand, it looks like a blocky building looming over ten stories tall. The tourguide says that it contains over 100 billion atoms and stores as much data as a room full of books. You can see some of the storage system inside: row upon row of racks containing spools of molecular tape somewhat like the protein chain, but with simple bumps representing the 1s and 0s of computer data. These nanocomputers seem big and crude, but the ground you're now standing on is also a computer–a single chip roughly as powerful as the smaller, stripped-down nanocomputer at your side. As you gaze out over the chip, you get a better sense for just how crude things were a few decades ago. At your feet, on the smallest scale, the chip is an irregular mess. Although the wall of the nanocomputer is pebbly with atomic-scale bumps, the bumps are as regular as tile.

The chip's surface, though, is a jumble of lumps and mounds. This pattern spreads for dozens of paces in all directions, ending in an irregular cliff marking the edge of a single transistor. Beyond, you can see other ridges and plateaus stretching off to the horizon. These form grand, regular patterns, the circuits of the computer. The horizon–the edge of the chip–is so distant that walking there from the center would (as the tourguide warns) take days. And these vast pieces of landscaping were considered twentieth-century miracles of miniaturization?

Even back then, research in molecular biology had revealed the existence of smaller, more perfect machines such as the protein molecules in cells. A simulated human cell–put here because earlier visitors wanted to see the size comparisons–its on the chip next to the smaller nanocomputer. The tourguide points out that the simulation cheats a bit at this point, making the cell act as though it were in a watery environment instead of air. The cell dwarfs the nanocomputer, sprawling across the chip surface and rearing into the sky like a small mountain. Walking the nature trail around its edge would lead across many transistor-plateaus and take about an hour. A glance is enough to show how different it is from a nanocomputer or a gear: it looks organic, it bulges and curves like a blob of liver, but its surface is shaggy with waving molecular chains. Walking up to its edge, you can see that the membrane wrapping the cell is fluid (cell walls are for stiff things like plants), and the membrane molecules are in constant motion. On an impulse, you thrust your arm through the membrane and poke around inside. You can feel many proteins bumping and tumbling around in the cell's interior fluid, and a crisscrossing network of protein cables and beams. Somewhere inside are the molecular machines that made all these proteins, but such bits of machinery are embedded in a roiling, organic mass. When you pull your arm out, the membrane flows closed behind. The fluid, dynamic structure of the cell is largely self healing. That's what let scientists perform experimental surgery on cells with the old, crude tools of the 20th century: they didn't need to stitch up the holes they made when they poked around inside. Even a single human cell is huge and complex. No real thinking being could be as small as you are in the simulation: A simple computer without any memory is twice your height, and the larger nanocomputer, the size of an apartment complex, is no smarter than one of the submoronic computers of 1990. Not even a bendable finger could be as small as your simulated fingers: in the simulation, your fingers are only one atom wide, leaving no room for the slimmest possible tendon, to say nothing of nerves. For a last look at the organic world, you gaze out past the horizon and see the image of your own, full-sized thumb holding the chip on which you stand. The bulge of your thumb rises ten times higher than Mount Everest. Above, filling the sky, is a face looming like the Earth seen from orbit, gazing down. It is your own face, with cheeks the size of continents. The eyes are motionless. Thinking of the tourguide's data, you remember: The simulation uses the standard mechanical scaling rules, so being 40 million times smaller has made you 40 million times faster. To let you pull free of surfaces, it increased your strength by more than a factor of 100, which increased your speed by more than a factor of 10. So one second in the ordinary world corresponds to over 400 million here in the simulation. It would take years to see that huge face in the sky complete a single eyeblink. Enough. At the command "Get me out!", the molecular world vanishes, and your feeling of weight returns as the suit goes slack. You strip off the video goggles—and hugely, slowly, blink.

Self-Assembly Processes for Nano Biotechnology

Researchers at the University of Massachusetts Amherst have developed a series of novel techniques in nanotechnology that hold promise for applications ranging from highly targeted pharmaceutical therapies, to development of nutrition-enhanced foods known as "nutraceuticals," to nanoscopic sensors that might one day advance medical imaging and diagnostics. The research, published in the Jan. 10 issue of Science, was funded by the U.S. Department of Energy and the National Science Foundation. The team included faculty members Thomas Russell and Todd Emrick of the department of polymer science and engineering, and Anthony Dinsmore of the department of physics, and graduate students Yao Lin and Habib Skaff, both of polymer science and engineering. "Our findings open new avenues to revolutionize technology by the controlled fabrication of nanoscopic materials having unique optical, magnetic and electronic properties," said Russell.

The study details three major findings:

- A novel method to create robust capsules from nanometer-sized particles
- A new technique to make nanoscopic particles water-soluble
- Functionalizing regions of the capsules with tailored properties, such as luminescence.

Emrick's research explores the behavior of nanoparticles to which ligands organic molecules and polymers have been attached. Russell is an expert in the surface and interfacial properties of polymers, and polymer-based nanostructures. Dinsmore specializes in colloidal assemblies and interface physics. "This is a productive collaboration in that we really have all the bases covered in terms of synthesis, understanding of interfacial activity and mediation, and the physics issues including surface tension and particle interactions," Emrick said. The study details a new method for assembling nanoparticles into robust, three-dimensional structures by encapsulating and stabilizing water droplets. Nanoparticles suspended in oil will self-assemble around a droplet of water, fully coating it with a shell. Although scientists have long known that particles tend to assemble at fluid interfaces, "the idea of using liquid interfaces as scaffolds is exciting and tremendously useful since researchers can tailor or modify the nanoparticles from both sides of the interface," explained Dinsmore. "We have much more surface area to work with for adding or removing specific particles." "Nanoparticles have exciting properties due to their small size, and they can be prepared in various shapes and sizes. What's really key is that you attach ligands that extend from the nanoparticles like hairs, in order to preserve the nanoscopic integrity of the particles and prevent them from clustering," Emrick said. "Changing the nature of these organic ligands can really modify the behavior of the particles. You can endow the nanoparticles, and thus the capsules that they form upon interfacial assembly, with a wide range of properties based on which ligands are attached." The effect of the ligands on the interactions of nanoparticles with the surrounding environment is crucial in medical applications. "These organic molecules will dictate the solubility, miscibility, and charge transport properties of the particles," Emrick said.

UMass researchers also developed a method to take these nanoparticles, which are oil-soluble, and make them water-soluble, simply by shining light on them. "Developing nanoparticles that are water-soluble has significant implications for medicine in the biosensors area," Emrick said. "Using luminescent material, as we did, could lead to advances in very sophisticated medical-imaging techniques as the fluorescent nature of these particles allows them to be viewed and tracked over time." Finally, the UMass team discovered that when nanoparticles of different sizes compete for assembly at the interface, the bigger ones win, and segregate or cluster into patches on the droplet surface. "This opens a range of possibilities for developing nanoscopic capsules that have certain properties in specific areas," said Dinsmore. "You could build in an area with permeability, magnetism, or conductivity, so that one area would be functionally distinct." In just a few decades physicians could be sending tiny machines into our bodies to diagnose and cure disease. These nanodevices will be able to repair tissues, clean blood vessels and airways, transform our physiological capabilities, and even potentially counteract the aging process. It is the year 2031, and the age of advanced nanomedicine has arrived.

A young man arrives at his physician's office with a mild fever, nasal congestion, discomfort and a cough. The physician pulls from her pocket a lightweight, handheld device resembling a pocket calculator. She unsnaps from it a cordless, self-sterilizing, pencil-sized probe and inserts it into the patient's mouth as if it were a tongue depressor. On the tip of the probe are billions of molecular assay receptors, mounted on hundreds of self-guiding retractile stalks. Each receptor is sensitive to the chemical signature of a specific kind of bacterium or virus. "Ahhh," says the patient, and a few seconds later a three-dimensional, color-coded map of his throat appears on the display panel of the device. Beneath the map scroll columns of data, revealing the unique molecular signature of a known and unwelcome bacterial pathogen. With the diagnosis complete, the infectious microorganism can be exterminated. No need for antihistamines, cough drops and a week-long course of antibiotics. The physician keeps several generic classes of nanorobots in her office for just such a circumstance. Using a desktop appliance in her office, she programs billions of nanorobots to find, recognize and destroy the particular microbial strain. The nanomachines are suspended in a carrier fluid that the patient inhales into his lungs, after which the mobile devices march down the patient's throat, propelled on tiny legs. Following a search pattern, the nanorobots ingest and destroy the harmful bacteria they encounter using mechanical and chemical phagocytosis. The patient feels nothing: nanorobots are the size of bacteria, which constantly crawl on and inside the body without ever being noticed. After several minutes, the physician activates an acoustic homing beacon to guide the nanorobots back into the patient's mouth, where she retrieves them through a collection port on the tip of the homing device. A further survey with the original diagnostic probe reveals no evidence of the pathogen. This is a wonderful vision of medicine in the future, but how do we get there from here? Nanotechnology ("nano" from the Greek word for dwarf) involves engineering and manufacturing at the molecular, or nanometer, level. One nanometer is one billionth of a meter, about the width of six bonded carbon atoms. How does one build at this scale? Scientists at New York University have adopted the self-assembly approach, and have succeeded in producing complementary strands of DNA that can zip themselves into complex structures. They have built cubes, octahedra and other solids made of just a few thousand nucleotides each, by the billions per batch. Meanwhile at Cornell University, researchers have genetically engineered a natural biomotor normally found in the enzyme ATPase to allow it to incorporate nonbiological parts such as a silicon nitride bar 100 nanometers in length, making the first artificial hybrid nanomotor. In a microscopic video presentation, dozens of bars attached to artificial molecular motors in a large precise array could be seen spinning at 200 revolutions per minute, like a field of tiny propellers. Another way to build at the molecular level is by positional assembly—picking and placing molecular parts exactly where you want them. A device capable of positional assembly would work much like the robot arms that manufacture cars on automobile assembly lines, or like the ribosome that assembles amino acids one by one into proteins in our cells. The US engineering firm Zyvex Corp. is aiming to be the first to create an artificial molecular assembler using positional assembly to manufacture atomically precise structures. Elsewhere, researchers at Harvard University have constructed the first general-purpose "nanotweezers," using a pair of electrically controlled carbon nanotubes These have successfully grasped 300-nanometer clusters of polystyrene spheres and extracted a single semiconductor wire 20 nanometers wide from a mass of entangled wires. The scientists hope eventually to produce nanotweezers small enough to grab individual large molecules.

Such nanoresearch will eventually lead to applications affecting almost every aspect of modern life, including the home, transport, computers, energy and the environment. Some of the most important and life-changing applications, however, are likely to be seen in medicine. We may see the first nanomedical materials and devices in use within the next few years. Scientists at the University of Michigan's Center for Biologic Nanotechnology are currently pursuing the use of dendrimers as a safer and more effective genetic therapy agent. Dendrimers are branching synthetic molecules that can be grown nanometer by nanometer to reach the desired size. They take on a spherical shape, and have sufficiently large openings and cavities to carry small molecules—dendrimers could therefore be used to sneak DNA into cells without triggering an immune response. At the same center, researchers have reported using dendrimer "nanodecoys" to trap and deactivate influenza virus particles. Relatively simple nanodevices could soon offer cures for major conditions such as diabetes. Researchers at Ohio State University and the University of Illinois have managed to construct silicon-based microcapsules highly perforated with "nanopores," each as small as 20 nanometers in diameter. These pores are large enough to allow small molecules such as oxygen, glucose and insulin to pass through, but are small enough to impede the passage of much larger immune system molecules such as immunoglobulins. Microcapsules containing islet cells could be implanted beneath the skin of some diabetes patients, temporarily restoring the body's delicate glucose control feedback loop without the need for powerful immuno-suppressants that can leave the patient at serious risk for infection. The same method of encapsulation could be used to treat other enzyme- or hormone-deficiency diseases. Encapsulated neurons could also be implanted in the brain and then electrically stimulated to release neurotransmitters, possibly as part of a future treatment for Alzheimer's or Parkinson's disease.

Artificial "biobots" could be in our bodies within five to ten years. Advances in genetic engineering are likely to allow us to construct an artificial microbe—a basic cellular chassis—to perform certain functions. These biobots could be designed to produce vitamins, hormones, enzymes or cytokines in which the host body was deficient, or they could be programed to selectively absorb and break down poisons and toxins. A new company called EngeneOS, Inc., founded in late 2000, has already announced plans to develop artificial Engineered Genomic Operating Systems using the techniques of molecular biology. These systems will comprise a library of component device modules and proprietary modular components. This will allow the engineering and construction of programmable biobots with novel form and function. The greatest power of nanomedicine will emerge in the longer term, perhaps 10-20 years from now, when we learn how to design and construct complete artificial nanorobots using strong diamond-like materials, nanometer-scale parts, and onboard subsystems including sensors, motors, manipulators, power plants, and molecular computers. One example is an artificial mechanical cell called a respirocyte, which could be used to keep a patient's tissues safely oxygenated for up to about four hours (at maximum dosage) if their heart has stopped beating. These cells (side panel, 4) could also enable a healthy person to sprint at top speed for at least 15 minutes without breathing, or to sit underwater at the bottom of a swimming pool for hours. Still entirely theoretical, the respirocyte is a micron-wide spherical nanorobot made of 18 billion atoms precisely arranged in a diamondoid structure to form a tiny 1,000-atmosphere pressure tank. Several billion molecules of oxygen and carbon dioxide can be absorbed into, or released from, this tank using computer-controlled molecular pumps powered by serum glucose and oxygen. External gas concentration sensors would allow respirocytes to mimic the action of the natural hemoglobin-filled red blood cells, with oxygen released and carbon dioxide absorbed in the tissues, and vice versa in the lungs. Each respirocyte would be able to hold 200 times more gas per unit volume than a natural red cell, so a few cubic centimeters injected into the human bloodstream would exactly replace the gas carrying capacity of the patient's entire 5.4 liters of blood.

Other proposed medical nanorobots offer equally astonishing performance improvements over nature. For instance, micronsize artificial mechanical platelets could allow complete hemostasis (control of bleeding by clot formation) in just one second, even for moderately large wounds, a response time 100-1,000 times faster than the natural system. This would be achieved by rapidly unfurling a compactly stowed onboard biodegradable mesh under the control of an onboard computer. These "clottocytes" would be about 10,000 times more effective as clotting agents than an equal volume of natural platelets. Nanorobotic phagocytes called microbivores could patrol the bloodstream, seeking out and digesting unwanted pathogens including bacteria, viruses or fungi. Each nanorobot could completely destroy one pathogen in just 30 seconds—about 100 times faster than natural leukocytes or macrophages—releasing a harmless effluent of amino acids, mononucleotides, fatty acids and sugars. No matter that a bacterium has acquired multiple drug resistance to antibiotics or to any other traditional treatment. The microbivore will eat it anyway, achieving complete clearance of even the most severe septicemic infections in minutes to hours, as compared to weeks or even months for antibiotic-assisted natural phagocytic defenses, without increasing the risk of sepsis or septic shock. Related nanorobots could be programed to recognize and digest cancer cells, or to clear circulatory obstructions within minutes in order to rescue stroke patients from ischemic damage. More sophisticated medical nanorobots will be able to intervene at the cellular level, performing surgery within cells. Physician-controlled nanorobots could extract existing chromosomes from a diseased cell and insert newly manufactured ones in their place, a process called chromosome replacement therapy. This would allow a permanent cure of any pre-existing genetic disease, and permit cancerous cells to be reprogramed to a healthy state.

In recent years, many gerontologists have begun to think of aging as a genetic disease that might be cured. Evidence from the Human Genome Project suggests that just a few hundred genes at most may be directly involved in aging. If we can completely understand these few genes and how they work, we may be able to alter them to eliminate this unwanted syndrome. Using cytosurgical nanorobots, corrected genes could be installed in every one of the 10 trillion tissue cells in our bodies. We would then no longer naturally age, and our bodies would again repair themselves as well as they did when we were children. Although nanotechnology is in its infancy, researchers are steadily making major breakthroughs. If we can learn to harness and precisely control the ability to manipulate molecules, then many aspects of our lives will change forever. In particular, the ability to carry out medical procedures at the molecular level will revolutionize medical practice. The next few decades will be very interesting indeed.

A new era on medicine are expected to happen in the coming years. Due to the advances in the field of nanotechnology, nanodevice manufacturing has been growing gradually. From such achievements in nanotechnology, and recent results in biotechnology and genetics, the first operating biological nanorobots are expected to appear in the coming five years, and more complex diamondoid based nanorobots will become available in about ten years. In terms of time it means a very near better future with significant improvements in medicine. In this work we present a practical approach taken on developing nanorobots for medicine in the sense of using computational nanomechatronics techniques as ancillary tools for investigating manufacturing design, nanosystems integration, sensing and actuation for medicine applications. Thus, the work describes pathways that could enable design testability, but also help scientists and profit corporations in providing the helpful information needed to test and design integrated devises and solutions towards manufacturing biomedical nanorobots. The use of robots in surgery has provided additional tools for surgeons enabling minimally invasive intervention or even long distance teleoperated surgeries. Indeed we may trust on human creativeness and technical capabilities that can ever be improved in terms of technical achievements. In recent years the medicine has enabled significant wellness for the life quality and longevity of the world population. And for the coming years, we may be prepared to experiment even more benefits, as results from advances that are being pursued step by step in new fields of science, such as nanobiotechnlogy. With the expected miniaturization of devices provided by several works on nanoelectromechanical systems (NEMS), nanomanufacturing has actually become a reality. Hence, with the NEMS recent advances on building nanodevices, and the development of interdisciplinary works, altogether may be translated in few years through the development of integrated nanomachines, also known as nanorobots. With the use of techniques that are advancing rapidly, such as nano-transducers, and biomolecular computing, nanorobots are expected to be able to operate in a well defined set of behaviors performing pre-programed tasks. Thus, in the coming few years, nanorobots being teleoperated to perform surgery, or even nanorobots continually supervising the human body in order to assist organs that may require some kind of repair, is one of the most expected revolutionary tools for biomedical engineering problems.

The development of nanorobots is an emerging field with many aspects under investigation. Simulation is an essential tool for exploring alternatives in the organization, configuration, motion planning, and control of nanomachines exploring the human body. Nanorobot applications could be focused mainly on two major areas, as follows: nanorobots for surgical interventions, as well as their utilization for patients that need constant monitoring. The nanorobots require specific controls, sensors and actuators, basically in accordance with each kind of biomedical problem. Advanced simulations can include various levels of detail, giving a trade-off between physical accuracy and the ability to control large numbers of nanorobots over relevant time scales with reasonable computational effort. Another advantage is that simulation can be done in advance of direct experimentation. It is most efficient to develop the control technology in tandem with the fabrication technologies, so that when we are able to build these devices, we will already have a good background in how to control them.

Computational mechatronic approaches were proposed as a suitable way to enable the fast development of nanorobots operating in a fluid environment relevant for medical applications. Unlike the case of larger robots, the dominant forces in this environment arise from viscosity of low Reynolds number fluid flow and Brownian motion and such parameters are been implemented throughout a set of different investigations. Practical and innovative paradigms have been developed based on the Nanorobot Control Design (NCD) simulator that allows fast design testability comparing various control algorithms for nanorobots and their application for different tasks. Also such information generated by the NCD can be useful as parameters for building nanodevices, such as transducers and actuators.

In future decades the principal focus in medicine will shift from medical science to medical engineering, where the design of medically-active microscopic machines will be the consequent result of techniques provided from human molecular structural knowledge gained in the 20th and early 21st centuries. For the feasibility of such achievements in nanomedicine, two primary capabilities for fabrication must be fulfilled: fabrication and assembly of nanoscale parts. Through the use of different approaches such as biotechnology, supramolecular chemistry, and scanning probes, both capabilities had been demonstrated to a limited degree as early as 1998. Despite quantum effects which impart a relative uncertainty to electron positions, the quantum probability function of electrons in atoms tends to drop off exponentially with distance outside the atom. Even in most liquids at their boiling points, each molecule is free to move only ~0.07 nm from its average position. Developments in the field of biomolecular computing have demonstrated positively the feasibility of processing logic tasks by biocomputers, a promising first step toward building future nanoprocessors with increasing complexity. There has been progress in building biosensors and nanokinetic devices, which also may be required to enable nanorobotic operations and locomotion. Classical objections related to the feasibility of nanotechnology, such as quantum mechanics, thermal motions and friction, have been considered and resolved and discussions of techniques for manufacturing nanodevices are appearing in the literature with increasing frequency. One important challenge that has become evident as a vital problem in nanotechnology industrial applications is the automation of atomic-scale manipulation. The starting point of nanotechnology to achieve the main goal of building systems at the nanoscale is the development of control automation for molecular machine systems, which could enable the massively parallel manufacture of nanodevice building blocks. Governments all around the world are directing significant resources toward the fast development of nanotechnology. At least 30 countries have initiated activities in this field, and beyond that, with government investments to nanotechnology of US$ 800 million in Japan and US $ 774 million in USA. The U.S. National Science Foundation has launched a program in "Scientific Visualization" in part to harness supercomputers in picturing the nanoworld. A US $ 1 trillion market consisting of devices and systems with some kind of embedded nanotechnology is projected by 2015. More specifically, the firm 'display search' predicts rapid market growth from US $ 84 million today to $ 1.6 billion in 2007. A first series of commercial nanoproducts has been announced as foreseeable by 2007. To reach the goal of building organic electronics, firms are forming collaborations and alliances that bring together new nanoproducts through the joint efforts of companies such as IBM, Motorola, Philips Electronics, Xerox/PARC, Hewlett-Packard, Dow Chemical, Bell Laboratories, and Intel Corp., among others. For such goals, new methodologies and theories to explore the nanoworld are the key technology.

A useful starting point for achieving the main goal of building nanoscale devices is the development of generalized automation control for molecular machine systems which could enable a manufacturing schedule for positional nanoassembly manipulation. In our work we consider more specialized scheduling problems with a focus on nanomedicine: describing in a detailed fashion the nanorobot control designs and the surrounding virtual workspaces modelling that are required for the main kinematics aspects in the physically-based simulations. Here the biomolecular assembly manipulation could be automatically performed by smart agents, which are given a set of possible tasks for biomedical engineering problems, embedded in a complex 3D environment. Virtual Reality could be considered as a suitable technique for nanorobot design and for the use of macro and microrobotics concepts given certain theoretical and practical aspects that focus on its domain of application. The collective nanorobotics approach proposed is one possible method to perform a massively-parallel positional nanoassembly manipulation . We constructed and demonstrated the applicability of multi-robot teams in timely sequenced set of works with practical applications that could enable the establishment of generalized control guidelines for nanorobotics. The use of multirobot teams working cooperatively to achieve a single global task applied to nanotechnology is a field of research that is very new and challenging. Research on collective robotics suggests that we should consider emulating the methods of the social insects to build decentralized and distributed systems that are capable of accomplishing tasks through the interaction of agents with the same structures and pre-programed actions and goals. Thus, a careful decomposition of the main problem task into subtasks with action based on local sensor-based perception could generate multirobot coherent behaviors. Several techniques was applied for such aims, as Neural Networks algorithms, Evolutionary computation, chemical based sensors and actuators, and even temperature time-gradients, just to quote a few. Among other interesting nanorobot applications, we could foresee their use to process specific chemical reactions in the human body as ancillary devices for injured organs. Nanorobots equipped with nanosensors could be used to detect glucose demand in diabetes patients. Moreover, nanorobots could also be applied in chemotherapy to combat cancer through superior chemical dosage administration, and a similar approach could be taken to enable nanorobots to deliver anti-HIV drugs. Successful nanorobotic systems must be able to respond efficiently in real time to changing aspects of microenvironments not previously examined from a control perspective. Unlike some prior simulators for simple robots, in the present work we have developed nanorobots that are not restricted to a fixed grid nor behave as simple cellular automata with very simple environments. Also by contrast, most CAD approaches provide only animation or visualization tools, while the NCD is a physically-based simulator. The set of experiments we have carried out include the main physical properties existent in the environment where the nanorobots are being currently projected to be operating in the coming years. Hence, the majority of results from our investigation should be useful as well on analysing integrated manufacturing capabilities. In fact, including aspects of the physical environment in conjunction with graphical visualization provide a feasible approach for automation and control design.

Furthermore, the architecture that was developed intended to enable the incorporation and evaluation of several control methods and distinct nanorobot shapes analyses. The automation, control, and manufacturing of nanorobots is a challenging and very new field. Realizing revolutionary applications of nanorobots to health or environmental problems raises new control challenges. The design and the development of complex nanomechatronic systems with high performance should be addressed via simulation to help pave the way for future applications of nanorobots in biomedical engineering problems.

Nanoscale Chemistry: Ingredients & Tools

Chemistry is one of the best developed, all encompassing and well defined fields of science today. Since ancient times, mankind has been using chemical techniques to manipulate matter on the atomic and molecular levels. As chemists improve our ability to organize atoms into atomically precise structures known as 'molecules,' fields including medicine, materials science, computer science and engineering reap the benefits. It may seem that there is little difference between the achievements of chemistry and the goals of nanotechnology, that is, to understand and manipulate molecular scale phenomena. However, nanotechnology takes a novel and revolutionary approch to this goal.

While chemistry deals with molecules in a statistical sense, nanotechnology deals with them as discrete entities, each requiring special attention. As Richard Feynman accurately *predicted in 1959*, 'The problems of chemistry and biology can be greatly helped if our ability to see what we are doing, and to do things on an atomic level, is ultimately developed—a development which I think cannot be avoided.' Indeed, *Scanning Probe Microscopy* has allowed sub-angstrom resolution of surfaces, leading to radical advancements in chemistry, biology and other fields.

For example, consider chemical synthesis. In Feynman's 1959 lecture, he explained, 'The chemist does a mysterious thing when he wants to make a molecule. He sees that it has got that ring, so he mixes this and that, and he shakes it, and he fiddles around. And, at the end of a difficult process, he usually does succeed in synthesizing what he wants.' Despite enormous progress in chemical methods since that time, the mixing of reagents described by Feynman is still the primary method used by synthetic chemists today. Due to the random nature of the molecular collisions in a solution, synthesis of complex (atomically precise) molecules suffers from the statistical law of entropy. Thus no reaction gives 100% yield. Furthermore, since most syntheses require multiple reactions, valuable starting materials are lost in unwanted side reactions and purification processes.

A biomimetic solution to this problem involves the idea of a molecular assembly line. Biological systems often carry out multistep reactions by passing a substrate from one enzyme to the next, the method used by Henry Ford to mass produce Model-

T automobiles with nearly 100% yield. Molecular Manufacturing is a sub-field of nanotechnology that seeks to realize synthetic methods such as this one.

Nanoscale Reaction Vessels

Vesicles are abundant in biological organisms, where they often serve the function of transporting various molecules. They are bubble-like structures composed of a lipid bilayer membrane, which can be created by means of biomolecular *self-assembly*. Vesicles of diameters ranging from 25 nm to 20 microns can be fabricated in a laboratory with some self-assembly technique. For instance, 25 nm vesicles can be made by sonicating a bilayer membrane at high frequencies, while 20 micron vesicles are produced when a low frequency alternating current is applied to the solution.

Zare et. al. has demonstrated the possiblility of carrying out reactions in vesicles containing attoliter to zeptoliter volumes. Each vesicle can contain as little as one molecule of a reagent. In this case, the reagents were deposited by electroporation or by electrofusion, methods involving application of an electrical pulse to propel charged particles through the solution.

It may seem that that shrinking reaction vessels and using molecular assembly lines could be considered a simple extension of current chemical techniques. However, the discrete nature of nanoscale chemistry requires an adaption of the fundamental principles defined by traditional chemistry. For instance, in a previous article entitled '*quantum biology*' it was noted that our statistical method for measuring temperature breaks down at the nanoscale. The current definition of temperature is especially meaningless in systems that are not at equilibrium, such as a biological cell. Indeed, the term 'equilibrium' itself loses meaning (breaks down) at the nanoscale due to its statistical definition. In order to study and design nanoscale systems, one cannot rely only on macroscale statistics that have proven invaluable to nearly all fields of science. Since temperature is one of the 7 fundamental units upon which other units rely, it is becoming necessary for nano-scientists to tweak this and other units to better reflect nanoscale phenomena.

Nanoscale Temperature

Temperature is defined as 'the average kinetic energy of all particles in the system.' According to the theory, there should be a standard deviation to go along with this average, but I have yet to see a thermometer that gives standard deviation. When one considers temperature in mixtures of different kinds of molecules or systems that are not at equilibrium, at the nanoscale there should be localized regions of various temperatures contributing to this average.

Temperature distribution is frequently calculated in non-equilibrium molecular dynamics simulations. In these theoretical models, the kinetic energy of each particle at various points in time is taken into account in order to provide a better understanding of molecular scale events. One might then define local temperature as—the average kinetic energy of a specified set of atoms within the system. While this is still a statistical method, it allows discrete statements to be made about localized parts of a system. Take, for intance, the *myosin* molecule in the previous article. If one were to specify as the set of atoms those included in the beta sheet adjacent to the ATP binding site, then, considering the 6 trillion myosin molecules [*myology*] in a muscle cell may even provide a sufficient sample size for spectroscopic measurements.

While local temperature is relatively easy to describe theoretically, measurements are more difficult to make. However, novel methods are being developed that have the potential to verify theoretical models experimentally. For example, local temperatures and temperature gradients can now be directly measured by *Scanning Thermal Microscopy (SThM)*. While the lateral resolution is currently limited to 150-200nm (sufficient for numerous microscale applications), it is reasonable to expect this resolution to improve in the future, perhaps eventually reaching the level of single molecules.

Biomolecular Self-Assembly

Assembling nanoscale machinery and electronics poses a problem for which we have yet to find an adequate solution, hence nanotechnology has yet to arrive. However, that a solution to the problem exists can be seen in all living systems. Composed of molecular building blocks such as nucleic acids, proteins and phospholipids, biological organisms enter the world equipped with all the necessary means for assembling these components into extremely well organized structures. The purpose of this treatise is to summarize the basic tools that nature has provided for the nanoscientist.

This vast supply of molecular components has been precisely designed not only to function as part of a complex molecular machine, but also with the ability to get themselves to the appropriate locations! Exactly how they find the right place is still under debate. While most agree that they bounce around until they stick somewhere, what exactly allows the speed and precision of most biomolecular self-assembly is still a mystery.

Nucleic Acids, Proteins and Bilayer Lipid Membranes

Perhaps the most specific biomolecular self-assembly occurs when two single strands of Deoxyribonucleic Acid (DNA) assemble into the infamous double helix. Based on Watson and Crick base pairing, two complimentary strands of DNA will find each other within an enormously complex mixture. Double helical DNA strands can be separated from one another through simple heating, resulting in single stranded DNA (ssDNA). If the solution is allowed to cool, the two strands will re-assemble into a stable double helix. While the Watson and Crick base pairing then allows the specific assembly of the double helix, which is stabilized by relatively strong *hydrophobic* interactions between the aromatic bases within the center of the helix also known as *pi-stacking*.

Nanoscientists can take advantage of the double helix's specific stability in numerous ways. One promising approach is to attach single-stranded DNA to small particles that one wishes to assemble. These components could range from diamondoid manifolds to metallic clusters that function as quantum dots. When DNA is used to create functional surfaces, the technique is called *DNA Directed Immobilization (DDI)*. Such patterned surfaces have applications for both nanoelectronic devices and biosensors. *Nanogen*, a biosensor manufacturer has a versatile method for that takes advantage of DNA's negative charge. By applying a voltage to various parts of a DNA containing solution, the concentrations of DNA over various parts surfaces can be regulated.

As the primary components of the molecular machinery that makes life possible, proteins have a vast array of resources available to them for biomolecular self-assembly. Take for instance the eukaryotic nucleus of a multicellular organism (such as a human). The organization of the information contained within this tiny compartment is so vast that scientists are only now beginning to attempt to understand it. Containing the compartment is the nuclear membrane, which is filled with proteins that pump nutrients in and messages out. Entwined within what is called the "nuclear matrix" is the entire human genome, containing about 3 meters worth of DNA, packed into an area of only a few cubic micrometers. Keeping it all under control are then *histones* and thousands of other (non-histone) proteins that carry out processes such as *transcription* and *translation*. Each of these proteins has the appropriate information encoded within its peptide sequence for getting it to the location where it can do its job. In order for each of our cells to produce those proteins required for its specific task, only certain genes can be transcribed in certain cells. Thus, a very important amount of information is contained not only in the sequence of the DNA, but in its spatial organization within the nucleus. For instance, when the nucleus divides, all the contents must be unraveled, copied, and then reassembled back into two daughters of the original organized state. All this is accomplished by means of biomolecular self-assembly.

Structural Proteins *are generally called "fibrous" because they often have a fiber like nature. With names such as collagen, keratin, elastin, tubulin and fibroin, fibrous proteins self-assemble (often covalently) into large polymers that provide various nano- meso- and even macro-scale structures. Even our most modern synthetic polymers can not compete with fibrous proteins such as fibroin (silk) when it comes to strength, flexibility and overall molecular scale order. Thus, serious efforts are underway to better understand and take advantage of fibrous self-assembly.*

Enzymes *(biological catalysts) are generally "globular" as opposed to "fibrous" and they often work together in the form of multi-enzyme complexes. This means that both intramolecular self-assembly and intermolecular self-assembly must take place in order for the enzyme to reach the conformation that allows it to do its job. Intramolecular self-assembly means that the newly synthesized protein chain must fold into a functional state. Most often this is accomplished by the aggregation of hydrophobic residues in the center of the protein, thus forcing hydrophilic residues to the surface of the protein. Once a single protein chain is adequately folded, it will often begin intermolecular self-assembly. In order to assemble a molecular machine known as a "multienzyme complex," the numerous subunits (or proteins) contact one another in ways that cause new surfaces to appear, thus allowing further subunit components to bind. Since all the information for this self-assembly is often contained within the protein sequence, it is possible to reproduce this assembly in a test tube. For instance, the molecular machine that builds proteins in E. Coli (the ribosome) can be taken apart into 58 isolated protein chains (as well as a few RNA molecules. If these components are mixed back together with the appropriate timing then functional ribosomes can be reassembled.*

Membrane bound proteins *are incorporated into the cellular membranes that surround cells and their compartments. In order for these proteins to perform their numerous functions (signal transduction, ion balance maintenance, molecular transport, electron transport, structure, etc.), they must be appropriately oriented and organized within the membrane. Determining their structure within the membrane is not an easy task for biochemists, since the membranes are often destroyed during purification. However, membrane bound proteins often have a common structural characteristic that suggests a mechanism for their assembly. That is, several hydrophobic residues exposed on a certain area of the protein. Since the center of the membrane is hydrophobic, it can be expected that these hydrophobic residues stick to the inside of the membrane. Again, the hydrophobic effect is most effective for this purpose. Finally, once the molecule is incorporated into the membrane, changes in the membranes constitution as well as interactions with neighboring membrane bound proteins govern the organization of these proteins.*

Cellular membranes define the boundries between various cellular compartments (such as the nucleus, mitochondria, endoplasmic reticulum, etc.) as well as enclosing the cell from the external environment. Similar to soap bubbles, cellular membranes are made of specific types of *surfactants* that tend to form bilayers. The primary surfactants are phospholipids, steroids (primarily cholesterol), and glycolipids. Two nanoscale bodies of water can be effectively separated from one by the formation of a bilayer of these molecules. The main barrier is the hydrophobic region in the center of a lipid bilayer membrane. Water and polar molecules dissolved in water can not pass through such a nonpolar barrier without the assistance of ion channels or molecular pumps.

Biomimetic Self-Assembly

Langmuir-Blodgett (LB) films are formed when synthetic or naturally occurring surfactants are placed at the air-water (or liquid gas) interface where they can organize into a layer that is one molecule thick (a monolayer), two molecules thick (bilayer) or several molecules thick (multi-layer). The polar part of the surfactant molecule tends to be attached to the surface of the water while the nonpolar hydrophobic regions of neighboring molecules will stick together and orient themselves away from the surface. As the lateral pressure (2 dimensional pressure created by compressing the interface area) is changed, the film undergoes phase transitions corresponding to various levels of self-organization.

Self-Assembled Monolayers

Thiol self-assembly can be achieved by reacting thiol containing compounds with clean gold surfaces. Sulfhydryl groups on molecules will covalently bind to gold, thus allowing molecules to be arranged two dimensionally over a gold surface. This is extra useful because gold conducts electricity and makes for excellent electrical contacts, thus Scanning Tunneling Microscopy (STM) measurements can be made on such samples. For example, redox proteins can be mutagenically modified with cysteine residues (an amino acid that contains a thiol group). Exposing a gold surface to such engineered molecules results in self-assembled monolayers of protein.

DNA Mediated Energy Transfer

The central component of all self-replicating molecular machinery alive today is the molecule known as Deoxyribonucleic acid (DNA). The molecular structure of DNA that allows it to translate, transcribe and copy information was first described by Watson and Crick in 1953 when they ended their *Nature* paper with the following understatement:

> "It has not escaped our notice that the specific pairing we have postulated immediately suggests a possible copying mechanism for the genetic material."

50 years of extrapolation on that idea have led to some of the greatest medical advances in history, leading eventually to the Human Genome Project. Considering that, imagine what the last 50 years may have brought had that sentence read:

> "It has not escaped our notice that the aromatic stacking that we have postulated immediately suggests a possible mechanism for nanoelectronic circuits."

The idea of charge transfers through DNA was indeed proposed in the years following 1953. However, since the electronic properties of DNA were not easily determined, it took about 40 years for the first experimental evidence of DNA mediated charge transfers to be published. Early skepicism claiming that the charge transfers occured though the backbone, or even through the surrounding solution have since been quenched by data showing that strong stacking within the double helix results in the fastest electron-transfer kinetics. Further research on electron transfers in DNA have indicated that the DNA sequence has predictable effects on the conductance of the ubiquitous biopolymer. Thus, depending on the sequence of the stretch of DNA propagating the electron transfer, DNA can either act as a *conductor*, a *semiconductor*, or an *insulator*. Most of the medically funded research on DNA mediated charge transfers focus on understanding how damaged DNA is repaired. For instance, when ionizing radiation or UV light strikes DNA, radicals can be formed within the DNA that may travel long distances before covalently binding atoms that are not normally connected. Studies involving both naturally occurring repair enzymes and mutagenically designed proteins are currently underway.

Some Resources on Computational Chemistry

Reviews in Computational Chemistry by Lipkowitz and Boyd is an excellent series of books which does just that. Unlike most review series, it includes introductory and tutorial as well as more advanced material.

The World-Wide Web Virtual Library: Chemistry is an excellent starting place to look for any chemistry related information.

The Computational Chemistry Archive is more specifically focused on issues surrounding computational chemistry.

Infochem has many links to Molecular Modelling Software.

Chemistry Software and Information Resources of the National HPCC.

QCPE (The Quantum Chemistry Program Exchange) provides a wide range of computational chemistry software, often free.

NetSci's list of molecular modeling software and their computational chemistry page

A list of chemistry resources on the internet is maintained at Rensselaer.

NIH has a page on molecular modeling and has a guide to software.

An online-text which discusses aspects of quantum mechanics, the Born-Oppenheimer approximation, and molecular mechanics.

An introduction to molecular modeling

I've had some direct experience with a few software packages:

MSI's Cerius 2 provides an excellent but relatively expensive package which runs on several commercial platforms (SGI, RS6000, and the like).

Hypercube's Hyperchem under Windows on the PC provides excellent capabilities at the price.

CambridgeSoft provides a package that runs on the Mac and under Windows. It is more limited in its modeling capabilities: it can effectively do energy minimizations on structures with perhaps a hundred atoms (their minimizer does only steepest descent). They provide truly excellent facilities for making black & white pictures of molecules, and equally excellent facilities for drawing journal-quality 2-dimensional chemical structures using standard notation.

Gaussian is an ab initio program that usually costs over $10,000 but which is now available on the PC for under $1,000. While it has an antediluvian user interface (designed for punch cards) it is the de facto standard against which competing companies (like the startup Schrodinger) compare themselves and is well known in the computational chemistry literature. They can be reached by e-mail at info@gaussian.com.

While I haven't had direct experience with it, Jay Ponder (who gave an excellent talk on molecular mechanics at the First Foresight conference on Molecular Nanotechnology) has made available a molecular mechanics and dynamics package written in Fortran 77 which is free (including source code). To quote their description:

> TINKER is a complete package for performing molecular mechanics calculations. It is intended to serve as a platform for algorithm development and parameterization, while still being efficient enough for most production work. The currently available potentials include MM2 and MM3, AMBER-95, AMBER/OPLS, CHARMM22 and our own TINKER potential. Many other literature potentials can be implemented via external parameter files; parameter sets for ENCAD and MMFF-94 are under development.
>
> Programs are provided to perform many functions including the following: (1) energy minimization over Cartesian or torsional coordinates via conjugate gradient, variable metric or truncated Newton methods, (2) molecular dynamics with periodic boundaries and options to control temperature and pressure, (3) normal mode vibrational analysis, (4) distance geometry including a new, efficient random atom-pair metrization, (5) simulated annealing with a choice of cooling protocols, (6) analysis and breakdown of single point potential energies, (7) verification of analytical derivatives of the standard potential and/or user defined functions, (8) location of a transition state between two minima, (9) free energy perturbation calculations, (10) fitting of intermolecular potential parameters to crystal structure and thermodynamic data, and (11) diffusion equation method and Gaussian density annealing approaches to optimization via smoothed potential energy surfaces.

This list is short and biased, but it should provide a starting point for anyone interested in looking into computational chemistry software.

Protein Based Molecular Machines

This section focuses on the study of the following main protein based molecular machines:

(i) ATP Synthase

(ii) The Kinesin, Myosin, Dynein and Flagella Molecular Motors

Synthesis of ATP is carried out by an enzyme, known as ATP Synthase. The inner mitochondrial membrane contains the ATP Synthase. The ATP Synthase is actually a combination of two motors functioning together as described. This enzyme consists of a proton-conducting F_0 unit and a catalytic F_1 unit. The figure also illustrates the subunits in side the two motor components. F_1 constitutes of a_3b_3gde subunits. F_0 has three different protein molecules, namely, subunit a, b and c. The g-subunit of F_1 is attached to the c subunit of F_0 and is hence rotated along with it. The a_3b_3 subunits are fixed to the b-subunit of F_0 and hence do not move. Further the b-subunit is held inside the membrane by a subunit of F_0 (shown in the figure by Walker).

1. Reversibility of the ATP Synthase: There are two directions in ATP Synthase system and these two directions correspond to two different functionalities and behavior. This two-way behavior is because of the reversible nature of the ATP-ADP cycle and the structure of the ATP Synthase. Let us term the forward direction as when the F_0 drives the g-subunit (because of proton motive force) of F_1 and hence ATP Synthesis takes place. And the backward direction is when hydrolysis of ATP counter-rotated the ã-subunit and hence the F_0 motor and leads to pumping back the protons. Therefore the forward direction is powered by the proton motive force and the backward direction is powered by the ATP hydrolysis. Which particular direction is being followed depends upon the situation and the environmental factors around the ATP Synthase.

2. Coupling of Proton Flow (F_0) and the ATP synthesis and hydrolysis (in F_1): Boyer proposed a model which predicted that the F_0 and F_1 motors are connected through the g-subunit. Further he proposed that this connection was mechanical in nature. The following figure illustrates the electrostatic surface potential on a_3b_3 subunits. The red colour represents a negatively charged surface and blue a positively charged surface.

3. Boyer's binding Change Mechanism: Boyer isolated the F_1 part of the ATP Synthase complex. It was found that the alpha and beta subunits alternate in this cylindrical part of the F_1 structure. As per this model each a and b pair forms a catalytic site. The rotation of the g subunit induces structural conformation in the a_3b_3 subunits. Although the three catalytic units are identical in their chemistry but they are functionally very different at any given point in time. These conformal changes induce a change in the binding affinities of all the three catalytic sites towards the ATPase reactants (ADP, P_i, ATP etc.).

The three catalytic sites could be in three distinct forms. O form stands for open which implies that it has got very low affinity for reactants; L form stands for loose, which implies that it would bind the reactants loosely but is still catalytically inactive; T form stands for tight, which binds the reactant tightly and is catalytically active.

One of the sites being bound to ATP at T. The site for ADP and P_i would probably be L as it is more binding than the O site. Now when the energy is input to the system the sites are changed. T becomes O; O becomes L and L becomes T. Due to these changes the ATP which was previously in T is now released. Further the ADP and P_i are now tightly bound due to the conversion of L site to the T site. This tight binding of ADP and P_i allows them to spontaneously get converted into ATP. Hence, this model proposed that the proton flow from the F_0 is coupled to the site inter conversions in the F_1 unit which triggers synthesis-hydrolysis of ATP. Boyer's theory was supported when Walker and his group solved the structure of the F_1-ATPase motor. The high resolution structure thus obtained gave hints towards many experiments which proved the fact that g subunit indeed rotated against the alpha and beta subunits.

4. *F_1-ATPase a true nano rotary motor*: Till today the exact mechanism of the molecular motor characterized by F_1-ATpase has not been fully determined. Research by Kinosita's lab is a step towards this goal and proposes some very conclusive models for the same. The results obtained show not only the various methods through which we can analyze these nano devices, but also predicts many characteristics for these.

What is known till now is that g subunit rotates inside the alpha-beta hexamer, but whether the rotation is continuous or is random was not known. Kinosita's lab solved this problem by imaging the F_1-ATPase molecule. The objective of their experiment was to determine the uniqueness of the rotary motion and its characteristics. The experiment that was performed by this group. They attached a micrometer long actin filament to the g subunit. This actin filament was fluorescently label, so that its fluorescence could be measured under a microscope. Hydrolysis of the ATP (when introduced in the experiment) led to the rotation of the g subunit and in effect the rotation of the actin filament. As reported by the authors, not all the actin filaments were observed to have rotation. But some percentage of them did rotate and that too in a unique direction and without having much reversibility in the direction. This direct imaging proved that the structure solved by Walker and group was indeed correct and there exists rotary motion between g subunit and the alpha and beta hexamer.

The imaging of the actin filament that was obtained. The rotation is not restricted to some particular rotational angle but is continuously progressive is a particular direction. As the rotational motion is a continuous one, possibility of it being a twisting motion is ruled out. Hence, there should be no direct linkage between the g subunit and the alpha-beta hexamer in F_1-ATPase. It was further observed by the authors that the rotation of the actin filament was in certain steps. These step sizes were approximately equal to 120 degrees.

Rotational rates: Rotational rates were dependent upon the length of the actin filament. Rotational rates were found to be inversely proportional to the filament length. This could be attributed to the fact that the hydrodynamic friction is proportional to the cube of the length of the filament, therefore, longer length implied that the frictional forces were higher and rotational speeds in effect slower. As reported in the principle article, if the filament were to rotate at the speed of 6 rev per sec and the length of the actin filament were 1 mm, then as per the relation.

Efficiency of F_1: Work done in one step, i.e., 120 degree rotation is equal to the torque generated (=40 pN nm) times the angular displacement for that step (=2pi/3). This comes to about 80pN nm. Free energy obtained by the hydrolysis of one ATP molecule.

Here, the standard free energy change per ATP molecule = -50 pN nm at pH 7. Thermal energy at room temperature = 4.1 pN nm. At intracellular conditions, [ATP] and $[P_i]$ is in the order of 10^{-3} M. Therefore, for [ADP] of 50 mM, DG = -90.604 pN nm. Comparing this value of the free energy obtained by the hydrolysis of one molecule of ATP with the mechanical work done by the motor (=80 pN nm), we can deduce that the F_1-ATPase motor works on efficiency close to 100 per cent!

- *Chemical synthesis of ATP powered by mechanical energy:* In another study conducted by Hiroyasu Itoh, Akira Takahashi, Kengo Adachi, Hiroyuki Noji, Ryohei Yasuda, Masasuke Yoshida & Kazuhiko Kinosita Jr, evidence has been provided to justify the claim that the chemical synthesis of ATP occurs when propelled by mechanical energy. This basic nature of the F_1-ATPase was known for quite some while, but it is first time that it is being experimentally verified. To prove this concept, the F_1 was bound to the glass surface through histidine residues attached at the end of the g subunits. A magnetic bead coated with streptavidin was attached to the g subunit.

Electric magnets were used to rotate this bead attached to the g subunit. The rotation resulted in appearance of ATP in the medium (which was initially immersed in ADP). Thus, the connection between the syntheses of ATP as a result of the mechanical energy input is established. The exact mechanism of the F_1-ATPase rotation is still an active area of research today and many groups are working towards finding it. The key to solving the mechanism is solving the transient conformation of the catalytic sites and the g subunit when rotation is taking place. What is not clear is the correspondence between the chemical reactions at the catalytic sites and their influence on the rotation of the g subunit. Which event triggers the rotation and which not has still to be exactly determined? Many models have been predicted, but they all still elude the reality of the rotational mechanism.

With modern microscopic tools, we view a cell as a set of many different moving components powered by molecular machines rather than a static environment. Molecular motors that move unidirectionally along protein polymers (actin or microtubules) drive the motions of muscles as well as much smaller intracellular cargoes. In addition to the F_0-F_1-ATPase motors inside the cell, there are linear transport motors present as tiny vehicles known as motor proteins that transport molecular cargoes that also require ATP for functioning. These minute cellular machines exist in three families—the kinesins, the myosins and the dyneins. The cargoes can be organelles, lipids or proteins etc. They play an important role in cell division and motility. There are over 250 kinesin-like proteins, and they are involved in processes as diverse as the movement of chromosomes and the dynamics of cell membranes. The only part they have in common is the catalytic portion known as the motor domain. They have significant differences in their location within cells, their structural organization, and the movement they generate. Muscle myosin, whose study dates back to 1864, has served as a model system for understanding motility for decades. Kinesin, however, was discovered rather recently using in vitro motility assays in 1985. Conventional Kinesin is a highly processive motor that can take several hundred steps on a microtubule without detaching whereas muscle myosin executes a single "stroke" and then dissociates. A detailed analysis and modeling of these motors has been done. Kinesin and myosin make up for an interesting comparison. Kinesin is microtubule-based; it binds to and carries cargoes along microtubules whereas myosin is actin-based. The motor domain of kinesin weighs one third the size of that of myosin and one tenth of that of dynein. Before the advent of modern microscopic and analytic techniques, it was believed that these two have little in common. However, the crystal structures available today indicate that they probably originated from a common ancestor.

Myosin is a diverse superfamily of motor proteins. Myosin-based molecular machines transport cargoes along actin filaments—the two stranded helical polymers of the protein actin, about 5-9 nm in diameter. They do this by hydrolyzing ATP and utilizing the energy released. In addition to transport, they are also involved in the process of force generation during muscle contraction, wherein thin actin filaments and thick myosin filaments slide past each other. Not all members of the myosin superfamily have been characterized as of now. However, much is known about the structure and function. Myosin molecules were first sighted through electron microscope protruding out from thick filaments and interacting with the thin actin filaments in late 1950s. Since then it was known that ATP plays a role in myosin related muscle movement along actin. However, the exact mechanism was unknown, which was explained later in 1971 by Lymn and Taylor.

- *Structure of Myosin Molecular Motor:* A myosin molecule binding to an actin polymer. Myosin molecule has a size of about 520 kilodaltons (kD) including two 220 kD heavy chains and light chains of sizes between 15-22 kD. They can be visualized as two identical globular 'motor heads', also known as motor domains, each comprising of a catalytic domain (actin, nucleotide as well as light chain binding sites) and about 8 nm long lever arms. The heads, also sometimes referred to as S1 regions (subfragment 1) are shown in blue, while the lever arms or the light chains, in yellow. Both these heads are connected via a coiled coil made of two a-helical coils (grey) to the thick base filament. The light chains have considerable sequence similarity with the protein 'calmodulin' and troponin C, and are sometimes referred to as calmodulin-like chains. They act as links to the motor domains and do not play any role in their ATP binding activity but for some exceptions. The motor domain in itself is sufficient for moving actin filaments. Three-dimensional structures of myosin head revealed that it is a pear-shaped domain, about 19 nm long and 5 nm in maximum diameter.

- *Function of Myosin Molecular Motor:* A crossbridge-cycle model for the action of myosin on actin has been widely accepted since 1957. Since the atomic structures of actin monomer and myosin were resolved this model has been refined into a 'lever-arm model' which is now acceptable. Only one motor head is able to connect to the actin filament at a time, the other head remains passive. Initially the catalytic domain in the head has ADP and P_i bound to it and as a result, its binding with actin is weak. With the active motor head docking properly to the actin-binding site, the P_i has to be released. As soon as this happens, the lever arm swings counterclockwise due to a conformational change. This pushes the actin filament down by about 10 nm along its longitudinal axis. The active motor head now releases its bound ADP and another ATP molecule by way of Brownian motion quickly replaces it, making the binding of the head to the actin filament weak again. The myosin motor then dissociates from the actin filament, and a new cycle starts. However, nanomanipulation of single S1 molecules (motor domains) show that myosin can take multiple steps per ATP molecule hydrolyzed, moving in 5.3 nm steps and resulting in displacements of 11 to 30 nm.

Kinesin and Dynein family of proteins are involved in cellular cargo transport along microtubules as opposed to actin in the case of myosin. Microtubules are 25 nm diameter tubes made of protein tubulin and are present in the cells in an organized manner. Microtubules have polarity; one end being the plus (fast growing) end while the other end is the minus (slow growing) end. Kinesins move from minus end to plus end, while dyneins move from plus end to the minus end of the microtubules. Microtubule arrangement varies in different cell systems. In nerve axons, they are arranged longitudinally in such a manner that their plus ends point away from the cell body and into the axon. In epithelial cells, their plus end points towards the basement membrane. They deviate radially out of the cell center in fibroblasts and macrophages with the plus end protruding outwards. Like myosin, kinesin is also an ATP-driven motor. One unique characteristic of kinesin family of proteins is their processivity—they bind to microtubules and literally 'walk' on it for many enzymatic cycles before detaching. Also, each of the globular heads/motor domains of kinesin is made of one single polypeptide unlike myosin (heavy and light chains and dynein heavy, intermediate and light chains).

- *Structure of Kinesin Molecular Motor:* A lot of structural information about kinesin is now available through the crystal structures. The motor domain contains a folding motif similar to that of myosin and G proteins. The two heads or the motor domains of kinesin are linked via 'neck linkers' to a long coiled coil, which extends up to the cargo . They interact with the a and b-subunits of the tubulin hetrodimer along the microtubule protofilament. The heads have the nucleotide and the microtubule binding domains in them.
- *Function of Kinesin Molecular Motor:* While kinesin is also a two-headed linear motor, its modus operandi is different from myosin in the sense that both its head work together in a coordinated manner rather than one was being left out. The kinesin walk. Each of the motor heads is near the microtubule in the initial state with each motor head carrying an ADP molecule. When one of the heads loosely binds to the microtubule, it looses its ADP molecule to facilitate a stronger binding. Another ATP molecule replaces the ADP which facilitates a conformational change such that the neck region of the bound head snaps forward and zips on to the head. In the process it pulls the other ADP carrying motor head forward by about 16 nm so that it can bind to the next microtubule-binding site. This results in the net movement of the cargo by about 8 nm. The second head now binds to the microtubule by losing its ADP, which is promptly replaced by another ATP molecule due to Brownian motion. The first head meanwhile hydrolyses the ATP and loses the resulting P_i. It is then snapped forward by the second head while it carries its ADP forward. Hence coordinated hydrolysis of ATP in the two motor heads is the key to the kinesin processivity. Kinesin is able to take about 100 steps before detaching from the microtubule while moving at 1000 nm/sec and exerting forces of the order of 5-6 pN at stall.

Dynein superfamily of proteins was introduced in 1965. Dyneins exist in two isoforms, the cytoplasmic and the axonemal. Cytoplasmic dyneins are involved in cargo movement, while axonemal dyneins are involved in producing bending motions of cilia and flagella.

- *Structure of Dynein Molecular Motor:* The structure consists of two heavy chains in the form of globular heads, three intermediate chains and four light intermediate chains. Recent studies have exposed a linker domain connecting the 'stem' region below the heads to the head itself. Also from the top of the heads the microtubule binding domains (the stalk region, not visible in the figure) protrude out. The ends of these stalks have smaller ATP sensitive globular domains which bind to the microtubules.

Cytoplasmic dynein is associated with a protein complex known as dynactin, which contains ten subunits. Some of them are shown in the figure as p150, p135, actin related protein 1 (Arp1), actin, dynamitin, capping protein and p62 subunit. These play an important regulatory role in the binding ability of dynein to the microtubules. The heavy chains forming the two globular heads contain the ATPase and microtubule motor domains.

One striking difference that dynein exhibits compared to kinesins and myosins is that dynein has AAA (ATPases Associated with a variety of cellular Activities) modules, which indicate that its mode of working will be entirely different from kinesins and myosins. This puts dyneins into the AAA superfamily of mechanoenzymes. The dynein heavy chains contain six tandemly linked AAA modules with the head having a ring-like domain organization, typical of AAA superfamily. Four of these are nucleotide binding motifs, named P1-P4, but only P1 (AAA1) is able to hydrolyse ATP.

- *Function of Dynein Molecular Motor:* Since dynein is larger and more complex structure as compared to other motor proteins, its mode of operation is not as well known. However, very recently, Burgess et al., have used electron microscopy and image processing to show the structure of a flagellar dynein at the start and end of its power stroke; giving some insight into its possible mode of force generation. When the dynein contains bound ADP and V_i (vandate), it is in the pre-power stroke conformation. The state when it has lost the two, known as the apo-state is the more compact post power stroke state. There is a distinct conformational change involving the stem, linker, head and the stalk that produces about 15 nm of translation onto the microtubule bound to the stalk.

Unicellular organisms, such as, E. coli have an interesting mode of motility (for reviews see. They have a number of molecular motors, about 45 nm in diameter, that drive their 'feet' or the flagella that help the cell to swim. Motility is critical for cells, as they often have to travel from a less favorable to a more favorable environment. The flagella are helical filaments that extend out of the cell into the medium and perform a function analogous to what the oars perform to a boat. The flagella and the motor assembly are called a flagellum. The flagella motors impart a rotary motion into the flagella. In addition to a rotary mechanism, the flagella machines consist of components such as rate meters, particle counters, and gearboxes. These are necessary to help the cell decide which way to go, depending on the change of concentration of nutrients in the surroundings. The rotary motion imparted to the flagella needs to be modulated to ensure the cell is moving in the proper direction as well as all flagella of the given cell are providing a concerted effort towards it. When the motors rotate the flagella in a counterclockwise direction as viewed along the flagella filament from outside, the helical flagella create a wave away from the cell body. Adjacent flagella subsequently intertwine in a propulsive corkscrew manner and propel the bacteria. When the motors rotate clockwise, the flagella fly apart, causing the bacteria to tumble, or change its direction. These reversals occur randomly, giving the bacterium a 'random walk', unless of course, there is a preferential direction of motility due to reasons mentioned earlier. The flagella motors allow the bacteria to move at speeds of as much as 25 mm/s with directional reversals occurring approximately 1 per second. A number of bacterial species in addition to E. coli., depend on flagella motors for motility. Some of these are Salmonella enterica serovar Typhimurium (Salmonella), Streptococcus, Vibrio spp., Caulobacter, Leptospira, Aquaspirrilum serpens and Bacillus. The rotation of flagella motors is stimulated by a flow of ions through them which is a result of a build-up of a transmembrane ion gradient. There is no direct ATP-involvement; however the proton gradient needed for the functioning of flagella motors can be produced by ATPase.

- *Structure of the Flagella Motors:* A complete part list of the flagella motors may not be available as of now. Continued efforts dating back to early 1970s have however revealed much of their structure, composition, genetics and function. Newer models of the motor function are still being proposed with an aim to explain observed experimental phenomena. That means that we do not fully understand the functioning of this motor. A typical flagella motor from E. coli. consists of about 20 different proteins, while there are yet more that are involved in the assembly and functioning. There are 14 Flg-type proteins named FlgA to FlgN; 5 Flh-type proteins called FlhA to FlhE; 19 Fli-type proteins named FliA to FliT; MotA and MotB making a total of 40 related proteins. The name groups Flg, Flh, Fli and Flg originate from the names of the corresponding genes. Out of these the main structural proteins are FliC or the filament; FliD (filament cap); FliF or the MS-ring; FliG; FliM and FliN (C-ring); FlgB, FlgC and FlgF (proximal rod); FlgG (distal rod); FlgH (L-ring); FlgI (P-ring); FlgK and FlgL (hook-filament junction); and MotA-MotB (torque generating units). Earlier it was believed that the M and S are two separate rings and M was named after membrane and S after supramembranous. Now they are jointly called the MS-ring as it has been found that they are two domains of the same protein FliF. The C-ring is named after cytoplasmic, while the names of the P and L-rings come from 'peptidoglycan' and 'lipopolysaccharide' respectively. The FlhA,B, FliH,I,O,P,Q,R constitute the 'transport apparatus'.

The hook and filament part of the flagellum is located outside the cell body, while the motor portion is embedded in the cell membrane with parts (the C-ring and the transport apparatus) that are inside the inner membrane in the cytoplasmic region. MotA and MotB are arranged in a circular array embedded in the inner membrane, with the MS-ring at the center. Connected to the MS-ring is the proximal end of a shaft, to which the P-ring, which is embedded in the peptidoglycan layer, is attached.

Moving further outwards, there is the L-ring embedded in the outer cell membrane followed by the distal shaft end that protrudes out of the cell. To this end there is an attachment of the hook and the filament, both of which are polymers of hook-protein and flagellin respectively.

- *Function of the Flagella Motors:* The flagellar motors in most cases are powered by protons flowing through the cell membrane (protonmotive force, defined earlier) barring exceptions such as certain marine bacteria, for example, the Vibrio spp., which are driven by Na+ ions. There are about 1200 protons required to rotate the motor by one rotation. A complete explanation of how this proton flow is able to generate torque is not available as of today. From what is known, the stator units of MotA and MotB play an important role in torque generation. They form a MotA/MotB complex which when oriented properly binds to the peptidoglycan and opens proton channels through which protons can flow. It is believed that there are eight such channels per motor. The protonmotive force is a result of the difference of pH in the outside and the inside of the cell. The E. coli cells like to maintain a pH of 7.6-7.8 on their inside, so depending on the pH of the surroundings, the protonmotive force will vary, and hence the speed of rotation of their motors. To test how the speed of rotation depends on the protonmotive force, the motors were powered by external voltage with markers acting as heavy loads attached to them. It was found that the rotation indeed depends directly on the protonmotive force. According to the most widely accepted model, MotA/MotB complex interacts with the rotor via binding sites. The passage of protons through a MotA/MotB complex (stator or torque generator) moves it so that they bind to the next available binding site on the rotor, thereby stretching their linkage. When the linkage recoils, the rotor assembly has to rotate by one step. Hence, whichever complex receives protons from the flux will rotate the rotor, generating torque. The torque-speed dependence of the motor has been studied in detail and indicates the torque range of about 2700 pN-nm to 4600 pN-nm.

As mentioned earlier, nature chose DNA mainly as an information carrier. There was no mechanical work assigned to it. Energy conversion, trafficking, sensing etc. were the tasks assigned mainly to proteins. Probably for this reason, DNA turns out to be a simpler structure—with only four kinds of nucleotide bases Adenosine, Thiamine, Guanine, and Cytosine (A, T, G & C) attached in a linear fashion that take a double helical conformation when paired with a complementary strand. Such structural simplicity vis-à-vis proteins—made of twenty odd amino acids with complex folding patterns—results in a simpler structure and predictable behavior. There are certain qualities that make DNA an attractive choice for the construction of artificial nano machines. In recent years, DNA has found use in not only mechanochemical, but nano electronic systems as well. A DNA double-helical molecule is about 2 nm in diameter and has 3.4-3.6 nm helical pitch no matter what its base composition is; a structural uniformity not achievable with protein structures if one changes their sequence. Furthermore, double stranded DNA (ds-DNA) has a respectable persistence length of about 50 nm which provides it enough rigidity to be a candidate component of molecular machinery. Single stranded DNA (ss-DNA) is very flexible and cannot be used where rigidity is required; however this flexibility allows its application in machine components like hinges or nanoactuators. Its persistence length is about 1 nm covering up to 3 base pairs at 1M salt concentration.

Other than the above structural features there are two important and exclusive properties that make DNA suitable for molecular level constructions. These are—molecular recognition and self-assembly. The nucleotide bases A and T on two different ss-DNA have affinity towards each other, so do G and C. Effective and stable ds-DNA structures are only formed if the base orders of the individual strands are complementary. Hence, if two complementary single strands of DNA are in a solution, they will eventually recognize each other and hybridize or zip-up forming a ds-DNA. This property of molecular recognition and self-assembly has been exploited in a number of ways to build complex molecular structures. In the mechanical perspective, if the free energy released by hybridization of two complementary DNA strands is used to lift a hypothetical load, a force capacity of 15 pN can be achieved, comparable to that of other molecular machines like kinesin (5 pN).

Dr. Nadrian Seeman and colleagues built the first artificial DNA based structure in the form of a cube in 1991. They then went on to create more complex structures such as knots and Borromean rings. In addition to these individual constructs, two dimensional arrays with the help of the double-crossover (DX) DNA molecule. This DX molecule gave the structural rigidity required to create a dynamic molecular device, the B-Z switch. DNA double helices can be of three types—the A, B or the Z-DNA. The B-DNA is the natural, right-handed helical form of DNA, while the A-DNA is a shrunken, low-humidity form of the B-DNA. Z-DNA can be obtained from certain C-G base repeat sequences occurring in B-DNA that can take a left-handed double helical form. The CG repeated base pair regions can be switched between the left and the right-handed conformations by changing ionic concentration. The switch was designed in such a way that it had three cyclic strands of DNA, two of them wrapped around a central strand that had the CG-repeat region in the middle. On the two free ends of the side strands fluorescent dyes were attached in order to monitor the conformational change. With the change in ionic concentration the central CG-repeat sequence could alternate between the B and the Z modes bidirectionally which was observed through fluorescence resonance energy transfer (FRET) spectroscopy.

In the year 2000, Dr. Bernard Yurke and colleagues made an artificial DNA based molecular machine that also accepted DNA as a fuel. The machine, called DNA tweezers, consisted of three strands of DNA labeled A, B and C. Strands B and C are partially hybridized on to the central strand A with overhangs on both ends. This conformation of the machine is the open conformation. When an auxiliary fuel strand F is introduced, that is designed to hybridize with both overhang regions, the machine attains a closed conformation. The fuel strand is then removed from the system by the introduction of its exact complement, leaving the system to go back to its original open conformation. This way, a reversible motion is produced, which can be observed by attaching fluorescent tags to the two ends of the strand A. In this case the 5' end was labeled with the dye TET (tetrachloro-fluorescein phosphoramidite) and the 3' end was labeled with TAMRA (carboxy-tertamethylrhodamine). Aside from the creation of a completely new molecular machine, this showed a way of selective fueling of such machines. The fuel strands are sequence-specific, so they will work on only those machines towards which they are directed, and not trigger other machines surrounding them. This machine was later improved to form a Three-State Device which had two robust states and one flexible intermediate state. A variation of the tweezers came about as the DNA-scissors.

In the past two decades, chemists have been able to create, modify and control many different types of molecular machines. Many of these machines carry a striking resemblance with our everyday macro-scale machines such as gears, propellers, shuttles etc. Not only this, all of these machines are easy to synthesize artificially and are generally more robust than the natural molecular machines. Most of these machines are organic compounds of Carbon, Nitrogen and Hydrogen, with the presence of a metal ion being required occasionally. Electrostatic interactions, covalent and hydrogen bonding play essential role in the performance of these machines. Such artificial chemical machines are controllable in various ways—chemically, electrochemically and photochemically (through irradiation by light). Some of them are even controllable by more than one ways, rendering more flexibility and enhancing their utility. A scientist can have more freedom with respect to the design of chemical molecular machines depending on the performance requirements and conditions. Rotaxanes and Catenanes make the basis of many of the molecular machines described in this section. These are families of interlocked organic molecular compounds with a distinctive shape and properties that guide their performance and control.

Rotaxane family of molecular machines is characterized by two parts —a dumb-bell shaped compound with two heavy chemical groups at the ends and a light, cyclic component, called macrocycle, interlocked between the heads. It has been shown that a reversible switch can be made with a rotaxane setup. For this, one needs to have two chemically active recognition sites in the neck region of the dumb-bell. In this particular example, the thread was made of polyether, marked by recognition sites hydroquinol units and terminated at the ends by large triisoproplylsilyl groups. A tetracationic bead was designed and self-assembled into the system that interacts with the recognition sites.

The macrocycle has a natural low energy state on the first recognition site, but can be switched among the two sites reversibly upon application of suitable stimuli. Depending on the type of rotaxane setup the stimuli can be chemical, electrochemical or photochemical. The stereo-electronic properties of the recognition sites can be altered protonation or deprotonation, or by oxidation or reduction, thereby changing their affinity towards the macrocycle. In a recent example, light induced acceleration of rotaxane motion was achieved by photoisomerization, while similar controls through alternating current (oscillating electric fields) was shown before.

The catenanes are also special type of interlocked structures that represent a growing family of molecular machines. They are synthesized by supramolecular assistance to molecular synthesis. The general structure of a catenane is that of two interlocked ring-like components that are non-covalently linked via a mechanical bond, that is, they are held together without any valence forces. Both the macrocyclic components have recognition sites that are atoms of groups that are redox-active or photochemically reactive. It is possible to have both rings with similar recognition sites. In such a scenario, one of the rings may rotate inside the other with the conformations stabilized by noncovalent interactions, but the two states of the inner ring differing by 180° will be undistinguishable (degenerate). For better control and distinguishable molecular conformations, it is desirable to have different recognition sites within the macrocycles. Then they can be controlled independently through their own specific stimuli. The stereo-electronic property of one recognition site within a macrocycle can be varied such that at one point it has more affinity to the sites on the other ring. At this instant, the force balance will guide the rotating macrocycle for a stable conformation that requires that particular site to be inside the other macrocycle. Similarly, with other stimulus, this affinity can be turned off, or even reversed along with the affinity of the second recognition site on the rotating macrocycle increased towards those on the static one. There is a need for computational modeling, simulation and analysis of such molecular machine motion. Like rotaxanes, catenanes also can be designed for chemical, photochemical or electrochemical control.

For both rotaxane and catenane based molecular machines, it is desirable to have recognition sites such that they can be easily controlled externally. Hence it is preferable to build sites that are either redox-active or photo-active. Catenanes can also be self assembled. An example of catenane assembled molecular motors is the electronically controllable bistable switch. An intuitive way of looking at catenanes is to think of them as molecular equivalents of ball and socket and universal joints.

Pseudorotaxanes are structures that contain a ring-like element and a thread-like element that can be 'threaded' or 'dethreaded' onto the ring upon application of various stimuli. Again, the stimuli can be chemical, photochemical or electrochemical. These contain a promise of forming molecular machine components analogous to switches and nuts and bolts from the macroscopic world.

Many other molecular devices have been reported in the past four decades that bear a striking resemblance to macroscopic machinery. Chemical compounds behaving as bevel gears and propellers that were reported in the late 1960s and early 1970s are still being studied today. A molecular propeller can be formed when two bulky rings such as the aryl rings are connected to one central atom, often called the focal atom. Clockwise rotation of one such ring induces a counterclockwise rotation of the opposite ring about the bond connecting it to the central atom. It is possible to have a three-propeller system as well. Triptycyl and amide ring systems have been shown to observe a coordinated gear-like rotation. 'Molecular Turnstiles' which are rotating plates inside a macrocycle, have been created. Such rotations however were not controllable. A rotation of a molecular ring about a bond could be controlled by chemical stimuli, and this was shown by Kelly et al., when they demonstrated a molecular brake. A propeller-like rotation of a 9-triptycyi ring system, which was used in gears, this time connected to a 2, 2'-bipyridine unit could be controlled by the addition and subsequent removal of a metal. Thus, free rotations along single bonds can be stopped and released at will.

Another type of molecular switch is the 'chiroptical molecular switch'. Another large cyclic compound was found to be switchable between its two stable isomeric forms P and M' (right and left handed) stimulated by light. Depending on the frequency of light bombarded on it the cis and the trans conformations of the compound 4-(2'-meth-oxythioxanthylidene)]-7-methyl-1,2,3,4-tetrahydrophenanthrene can be interconverted. Allowing a slight variation to this switch, a striking molecular motor driven by light and/or heat was introduced by Koumura and colleagues. As opposed to the rotation around a single bond in the ratchet described above, the rotation was achieved around a carbon-carbon double bond in a helical alkene. Ultraviolet light or the change in temperature could trigger a rotation involving four isomerization steps in the compound (3R,3'R)-(P,P)-trans-1,1',2,2',3,3',4,4'-octahydro-3,3'-dimethyl-4,4',-biphenanthrylidene. A second generation motor along with 8 other motors from the same material is now operational. This redesigned motor has distinct upper and lower portions and it operates at a higher speed. This motor also provides a good example of how controlled motion at the molecular level can be used to produce a macroscopic change in a system that is visible to the naked eye. The light-driven motors when inside liquid crystal (LC) films can produce a color change by inducing a reorganization of mesogenic molecules.

In this section we present two protein based motors that are at initial developmental stages and yet possess some very original and interesting characteristics.

The idea of Viral Protein Linear motors stems from the fact that families of retroviruses like the influenza virus and the HIV-1 has a typical mechanism of infecting a human cell. When such a virus comes near the cell it is believed that due to the environment surrounding the cell it experiences a drop in pH of its surroundings.

This is a kind of a signal to the virus that its future host is near. The drop of pH changes the energetics of the outer (envelope glycoprotein) protein of the viral membrane in such a way that there is a distinct conformational change in a part of it. A triple stranded coiled coil domain of the membrane protein changes conformation from a loose random structure to a distinctive a-helical conformation. It is proposed to isolate this domain from the virus and trigger the conformational change by variation of pH in vitro. Once this is realized, attachments can be added to the N or C (or both) terminals of the peptide and a reversible linear motion can be achieved.

In a recent development, large plant proteins that can change conformation when stimulated by positively charged ions were separated from their natural environment and shown to exert forces in orthogonal directions. Proteins from sieve elements of higher plants that are a part of the microfluidics system of the plant were chosen to build a new protein molecular machine element. These elements change conformations in the presence of Ca2+ ions and organize themselves inside the tubes so as to stop the fluid flow in case there is a rupture downstream. This is a natural defense mechanism seen in such plants. The change in conformation is akin to a balloon inflating and extending in its lateral as well as longitudinal directions. These elements, designated as 'forisomes' adhered on to glass tubes were shown to reversibly swell in the presence of Ca^{2+} ions and shrink in their absence, hence performing a pulling or pushing action in both directions. Artificially prepared protein bodies like the forisomes could be a useful molecular machine component in a future molecular assembly, producing forces of the order of micronewtons. Unlike the ATP-dependant motors discussed previously, these machine elements are more robust because they can perform in the absence of their natural environment as well.

On Physics, Fundamentals, and Nanorobots: A Rebuttal to Smalley's Assertion that Self-Replicating Mechanical Nanorobots Are Simply Not Possible

The September 2001 issue of *Scientific American* was devoted to nanotechnology. It had an article by Nobelist Richard E. Smalley in which he said: "Self-replicating, mechanical nanobots are simply not possible in our world." In support of this remarkable assertion, two principal objections are advanced: the "fat fingers" problem and the "sticky fingers" problem. "Both these problems are fundamental," Smalley writes, "and neither can be avoided." Here we show that neither problem represents a fundamental barrier that will prevent the construction of self-replicating, mechanical nanorobots. Further, Smalley's objections refer to designs that he himself proposed, not the designs of Drexler, whom he attempted to refute.

The "Fat Fingers" Problem

Smalley has previously stated that: "Chemistry is the concerted motion of at least 10 atoms." As this excluded a large part of well-known chemistry, it is not surprising that he now more circumspectly says: "In an ordinary chemical reaction five to 15 atoms near the reaction site engage in an intricate three-dimensional waltz that is carried out in a cramped region of space measuring no more than a nanometer on each side."

The importance of the number of reactants lies in Smalley's argument that "There just isn't enough room in the nanometer-size reaction region to accommodate all the fingers of all the manipulators necessary to have complete control of the chemistry." Having trimmed the required number of fat fingers from ten to five, Smalley's argument is still on shaky ground. The argument collapses when we observe that chemical reactions often involve two reactants, such as in the controlled vacuum conditions used by the scanning tunneling microscope (STM). Two reactants can be brought together with controlled trajectories if one reactant is bound to a substrate and the second reactant is positioned and moved by a single "finger" as has already been done experimentally. For example, Ho and Lee physically bound a CO molecule to an iron atom on a silver substrate using an STM. Other approaches are also possible, Brenner *et al.* provided a molecular dynamics simulation of the hydrogen abstraction reaction from a diamond substrate.

As noted elsewhere, if steric constraints near the tool tip make it unexpectedly difficult to manipulate particular individual atoms or small molecules with sufficient reliability, a simple alternative is to rely upon conventional solution or gas phase chemistry for the bulk synthesis of nanoparts consisting of 10-100 atoms. These much larger nanoparts can then be bound to a positional device and assembled into larger (molecularly precise) structures without further significant steric constraints. This is the approach taken by the ribosome in the synthesis of proteins. Individual amino acids are sequentially assembled into an atomically precise polypeptide without the need to manipulate individual atoms. "Atomically precise" is a description of the precision of the final product, not a description of the manufacturing method. Complete control of every aspect of a chemical reaction is not actually required to build a nanorobot. Effective control that delivers a precise product is what is necessary. The "fat fingers" problem is *not* a fundamental barrier to the development of molecular assemblers or the nanorobots they enable.

The "Sticky Fingers" Problem

Smalley also advances the "sticky fingers" problem, which is the claim that: "...the atoms of the manipulator hands will adhere to the atom that is being moved. So it will often be impossible to release this minuscule building block in precisely the right spot....these problems are fundamental...." The existence of some reactions that don't work still leaves plenty of room for reactions that do. To argue that the "sticky fingers" problem is a fundamental barrier to building mechanical assemblers and nanorobots, Smalley must show that no set of reactions exists which allows the synthesis of a useful range of precise molecular structures. To consider but one approach, application of a voltage between a manipulator tool and the workpiece can cause the target atom or moiety to move to the desired position. Again, Ho and Lee have provided an experimental existence proof.

For a biological example, consider the ribosome. This ubiquitous biological molecular assembler suffers from neither the "fat finger" nor the "sticky finger" problem. If, as Smalley argues, both problems are "fundamental," then why would they prevent the development of mechanical assemblers and not biological assemblers? If the class of molecular structures known as proteins can be synthesized using positional techniques, then why would we expect there to be no other classes of molecular structures that can be synthesized using positional techniques? Upon observing experimentally that polymers such as proteins can be synthesized under programmatic control, what convincing evidence do we have that the programmatic synthesis of stiffer polycyclic structures such as diamond is "fundamentally" impossible, and that mechanical assemblers will never be built?

More directly, we can examine both those reactions that have been proposed specifically for use in a mechanical assembler and those reactions that take place using an SPM (Scanning Probe Microscope). *Nanosystems* discusses the mechanosynthetic reactions that might be used to synthesize some diamondoid structures of interest. Generally, these involve a single "finger," i.e., a probe tip with a functionalized end that would cause a site-specific reaction on a growing molecular work-piece. Merkle discusses several reactions which involve two, three, and even four reactants bound to the tips of molecular tools. A large and growing literature of research work with SPMs — both theoretical and experimental — supports the feasibility of site specific reactions involving a reactive tip structure interacting with a surface or with a molecule on a surface. In addition, SPMs are just the beginning of a growing list of tools for manipulating atoms and molecules.

Reductio ad Absurdum

One curious aspect of Smalley's comments is his stated replication time for a billion-atom "nanobot": one second. To obtain this blindingly fast replication speed, he adopts a 1 GHz atomic placement frequency — approximately one atom positioned every nanosecond. But Drexler's *Nanosystems*, the canonical textbook in this field, proposed a 1 MHz atomic placement frequency for mature molecular assembler systems with a corresponding replication time of ~10^3 seconds — figures that are commonly adopted in the supporting literature. Other proposals assume replication times up to ~10^5 seconds (~28 hours) for earlier and more primitive molecular assembler systems. Smalley's proposal of a one second replication time thus appears inconsistent with the extant literature, permitting him erroneously to extrapolate the replicative manufacture of a population of ~10^{18} nanorobots in just 60 seconds. This in turn leads to a rather obvious fallacy: Allowing ~500 zJ/bond and 1 bond/atom in Smalley's billion-atom molecular assembler, the power draw of a device population that occupies ~0.01 m^3 (~10^{27} atoms) during the last second of operation is ~1 GW, giving a power density of

~10^{11} W/m^3 which for diamond materials implies a mean autocatalytic graphitization decomposition time in vacuum of ~30 millisec, much briefer than the posited 1-second replication time — clearly a problem.

In rhetoric, this approach might be called *"reductio ad absurdum"* or perhaps a "strawman." In a serious scientific discussion, a discrepancy of three orders of magnitude between what has been proposed in the literature and what is criticized suggests at best an inadequate grasp of the proposal.

Smalley's article is accompanied by this categorical statement: "How soon will we see the nanometer-scale robots envisaged by K. Eric Drexler and other molecular nanotechnologists? The simple answer is never."

As Arthur C. Clarke once observed: "When a distinguished but elderly scientist states that something is possible, he is almost certainly right. When he states that something is impossible, he is very probably wrong."

In spite of Clarke's observation, we recommend that rather than focus on issues of credentials, age, cabals, or orthodoxy, we simply stick to the scientific facts and their implications for molecular nanotechnology. The most direct way of evaluating the feasibility of engineering molecular assemblers and nanorobots is to conduct serious design and engineering efforts, along with detailed modeling and analysis. This would be the fastest way to reveal any hypothetical "fundamental problem" that would make non-biological molecular assemblers infeasible to build.

There are two possibilities: either molecular assemblers and the nanorobots they enable are theoretically feasible within the framework of well-understood existing physical laws, or they are not. A substantial and growing body of work strongly supports their feasibility. There are many worthy molecular systems engineering challenges to overcome, but thus far, there has been no credible argument that these devices are infeasible. Molecular nanotechnology assemblers would change society's relationship to molecules and matter as fundamentally as the computer changed our relationship to bits and information. If they are feasible, we need to focus research on development pathways and safeguards.

Design and Control

Designing nanorobotic systems deal with vast variety of sciences, from quantum molecular dynamics, to kinematic analysis. The rules applicable to nanorobotics depend upon the nano material we intend to use. Nanomechanical robotic systems deal with science significantly different from the biological or inorganic nanorobotic systems. Eric Drexler, author of the famous book, *Nanosystems: Molecular machinery, manufacturing and Computation*, detailed on many areas of science which influences design of nanomechanical systems. In this review chapter on various laws which govern the designing of nanorobotics, we will concentrate on biological systems. We will consider that the components that details a nanorobot is made of biological components, such as, proteins and DNAs. There doesn't exist any particular guideline or a prescribed manner which details the methodology of designing a bio-nanorobot (bio-nanorobot implies nanorobots made up of biocomponents) up to the date. There are many complexities which are associated with using biocomponents (such as protein folding, presence of aqueous medium), but the advantages of using these are also quite considerable. These biocomponents offer immense variety and functionality at a scale where creating a manmade material with such capabilities would be extremely difficult. These bio components have been perfected by nature through millions of years of evolution and hence these are very accurate and efficient. As noted in the review section on Molecular Machines, F_1-ATPase is known to work at efficiencies which are close to 100 per cent. Such efficiencies, variety and form are not existent in any other form of material found today.

Also, the other significant advantages in using protein-based bio-nano composites is the development and refinement over the last 30 years of tools and techniques that enable researchers to mutate proteins in almost any way imaginable. These mutations can consist of anything from simple amino acid side-chain swapping, to amino acid insertions or deletions, incorporation of non-natural amino acids, and even the combination of unrelated peptide domains into whole new structures. An excellent example of this approach is the engineering of the F_1-ATPase, which is able rotate a nanopropeller in the presence of ATP. A computational algorithm was used to determine the mutations necessary to engineer an allosteric zinc-binding site into the F_1-ATPase using site-directed mutagenesis. The mutant F_1-ATPase was then shown to rotate an actin filament in the presence of ATP with average torque of 34 pN nm. This rotation could be stopped with the addition of zinc, and restored with the addition of a chelator to remove the zinc from the allosteric binding site. This type of approach can be used for the improvement of other protein-based nanocomponents. Hence, these bio components seem to be a very logical choice for designing nanorobots. Some of the core applications of nanorobots are in the medical field and using biocomponents for these applications seems to be a good choice as they both offer efficiency and variety of functionality. This idea is clearly inspired by nature's construction of nanorobots, bacteria and viruses which could be termed as intelligent organisms capable of movement, sensing and organized control. Hence, our scope would be limited to the usage of these bio components in the construction of bio-nanorobotics. A roadmap is proposed which details the main steps towards the design and development of bio-nanorobots.

Step 1: Bio-nanocomponents

Development of bio-nanocomponents from biological systems is the first step towards the design and development of an advanced bio-nanorobot, which could be used for future applications. Since the planned systems and devices will be composed of these components, we must have a sound understanding of how these behave and how could they be controlled. From the simple elements such as structural links to more advanced concepts such as motors, each component must be carefully studied and possibly manipulated to understand the functional limits of each one of them. DNA and carbon nanotubes are being fabricated into various shapes, enabling possibilities of constructing newer and complex devices. These nanostructures are potential candidates for integrating and housing the bio-nanocomponents within them. Proteins such as rhodopsin and bacteriorhodopsin are a few examples of such bio-nanocomponents. Both these proteins are naturally found in biological systems as light sensors. They can essentially be used as solar collectors to gather abundant energy from the sun. This energy could either be harvested (in terms of proton motive force) for later use or could be consumed immediately by other components, such as the ATP Synthase nano rotary motor. The initial work is intended to be on the bio-sensors, such as, heat shock factor. These sensors will form an integral part of the proposed bio-nano assemblies, where these will be integrated within a nano structure and will get activated, as programed, for gathering the required information at the nano scale. Tools and techniques from molecular modeling and protein engineering will be used to design these modular components.

Step 2: Assembled Bio-nanorobots

The next step involves the assembly of functionally stable bio-nanocomposites into complex assemblies. Some examples of such complex assemblies or bio-nanorobots. A bio-nanorobot with its 'feet' made of helical peptides and its body of carbon nano tubes, while the power unit is a biomolecular motor. A conceptual representation of modular organization of a bio-nanorobot. The modular organization defines the hierarchy rules and spatial arrangements of various modules of the bio-nanorobots such as: the inner core (the brain/ energy source for the robot); the actuation unit; the sensory unit; and the signaling and information processing unit. By the beginning of this phase a "library of bio-nanocomponents" will be developed, which will include various categories such as, actuation, energy source, sensory, signaling etc. Thereon, one will be able to design and develop such bio-nanosystems that will have enhanced mobile characteristics, and will be able to transport themselves as well as other objects to desired locations at nano scale. Furthermore, some bio-nanorobots will have the capability of assembling various bio-components and nanostructures from in situ resources to house fabrication sites and storage areas, while others will just manipulate existing structures by repairing damaged walls or making other renovations. There will also be robots that not only perform physical labor, but also sense the environment and react accordingly. There will be systems that will sense an oxygen deprivation and stimulate other components to generate oxygen, creating an environment with stable homoeostasis.

Step 3: Distributive Intelligence, Programing and Control

With the individual bio-nanorobots in full function, they will now need to collaborate with one another to further develop systems and "colonies" of similar and diverse nanorobots. This design step will lay the foundation to the concept of bio-nano swarms (distributive bio-nanorobots). Here work has to be performed towards control and programing of bio-nanoswarms. This will evolve concepts like distributive intelligence in the context of bio-nanorobots. Designing swarms of bio-nanorobots capable of carrying out complex tasks and capable of computing and collaborating amongst the group will be the focus. Therefore, the basic computational architectures needs to be developed and rules need to be evolved for the bio-nanorobots to make intelligent decisions at the nano scale.

To establish an interface with the macro world, the computers and electronic hardware have to be designed as well. The overall electronic communication architecture. From a location, humans should be able to control, monitor the behavior and action of these swarms. Also, the basic computational capabilities required for functioning of the swarms will be developed. A representative computational bio-nanocell, which will be deployed within a bio-nanorobot. This basic computational cell will initially be designed for data retrieval and storage at the nanoscale. This capability will enable us to program (within certain degrees of freedom) the swarm behavior in the bio-nanorobots. We will further be able to get their sensory data (from nano world) back to the macro world through these storage devices. This programming capability would form the core essence of a bio-nanorobotics system and hence enables them with immense power.

Step 4: Automatic Fabrication and Information Processing Machines

For carrying out complex missions, such as sensing, signaling and storing, colonies of these bio-nanorobotic swarms needs to be created. The next step in nanorobotic designing would see the emergence of automatic fabrication methodologies of such bio-nanorobots in vivo and in vitro. Capability of information processing, which will involve learning and decision making abilities, will be a key consideration of this step. This would enable bio-swarms to have capability of self-evolving based on the environment they will be subjected to. These swarms could be programed to search for alternate energy sources and would have an ability to adapt as per that resource. Energy management, self-repairing, and evolving will be some of the characteristics of these swarms.

Modular Organization: Modular organization defines the fundamental rule and hierarchy for constructing a bio-nanorobotic system. Such construction is performed through stable integration of the individual 'bio-modules or components', which constitute the bio-nanorobot. For example, if the entity ABCD, defines a bio-nanorobot having some functional specificity then, A, B, C, and D are said to be the basic bio-modules defining it. The basic construction will be based on the techniques of molecular modeling with emphasis on principles such as energy minimization on the hyper surfaces of the bio-modules; hybrid quantum-mechanical and molecular mechanical methods; empirical force field methods; and maximum entropy production in least time.

Modular organization also enables the bio-nanorobots with capabilities such as, organizing into swarms, a feature, which is extremely desirable for various applications. The conceptual representation of modular organization. A more realistic scenario in which all the modules are defined in some particular spatial arrangements based on their functionality and structure. A particular module could consist of other group of modules, just like a fractal structure (defined as fractal modularity). The concept of bio-nanocode has been devised, which basically describes the unique functionality of a bio-nanocomponent in terms of alphabetic codes. Each bio-nanocode represents a particular module defining the structure of the bio-nanorobot. For instance, a code like E-M-S will describe a bio-nanorobot having capabilities of energy storage, mechanical actuation and signaling at the nanoscale. Such representations will help in general classifications and representative mathematics of bio nanorobots and their swarms. The proposed capabilities of the bio-modules along with their targeted general applications. The bio-nanocode EIWR | | M | | S | | FG representing the bio-nanosystem.

The Universal Template—Bio-nano Stem System: The modular construction concept involves designing a universal template for bio-nano systems, which could be 'programed and grown' into any possible Bio-nano coded system. This concept mimics the embryonic stem cells found in the human beings, that are a kind of primitive human cells which give rise to all other specialized tissues found in a human foetus, and ultimately to all the three trillion cells in an adult human body. Our Bio-nano Stem system will act in a similar way. This universal growth template will be constituted of some basic Bio-nanocodes, which will define the Bio-Nano-stem system. This stem system will be designed in a manner that could enable it to be programed at run-time to any other required bio-module. One such variant of the Bio-nano stem system, having the Bio-nanocode: EIWR | | M | | S | | FG and having enhanced sensory abilities.

Molecular modeling techniques in sync with extensive experimentations would form the basis for designing these bio-nano systems. As per the roadmap, various bio-nano components would have to be designed and then these functional elements would have to be bound into assemblies which could be controlled and programed. Some of the molecular modeling techniques we are emphasizing are described below.

Energy Minimization Methods: The potential energy depends upon the co-ordinates of the molecular configuration considered. In this method, points on the hypersurface (potential energy surface) are calculated for which the function has the minimum value. Such geometries which correspond to the minimum energy are the stable states of the molecular configuration considered. By this method we can also analyze the change is configuration of the system from one minimum state to another. Methods such as the Newton-Raphson and Quasi-Newton are employed to calculate these minima. This method of finding minimum energy points in the molecule is used to prepare for other advanced calculations such as molecular dynamics or Monte-Carlo simulations. It is further used to predict various properties of the system under study.

- *Molecular Dynamics Simulation Methods:* To begin predicting the dynamic performance (i.e. energy and force calculation) of a biocomponent (say a peptide) molecular dynamics (MD) is performed. This method utilizes Newton's Law of motion through the successive configuration of the system to determine its dynamics. Another variant of molecular dynamics employed in the industry is the Monte Carlo simulation, which utilizes stochastic approach to generate the required configuration of a system. Simulations are performed based on the calculation of the free energy that is released during the transition from one configuration state to the other (for example using the MD software CHARMM, Chemistry at Harvard Molecular Mechanics). In MD, the feasibility of a particular conformation of a biomolecule is dictated by the energy constraints. Hence, a transition from one given state to another must be energetically favorable, unless there is an external impetus that helps the molecule overcome the energy barrier. When a macromolecule changes conformation, the interactions of its individual atoms with each other—as well as with the solvent—constitute a very complex force system. With one of the aspects of CHARMM, it is possible to model say, a peptide based on its amino acid sequence and allow a transition between two known states of the protein using Targeted Molecular Dynamics (TMD). TMD is used for approximate modeling of processes spanning long time-scales and relatively large displacements. Say to achieve a large conformational change the biomolecule is 'forced' towards a final configuration 'F' from an initial configuration 'I' by applying constraints. The constraint is in the form of a bias in the force field. Molecular kinematic simulations are also being used to study the geometric properties and conformational space of the bio molecules (peptides). The kinematic analysis is based on the development of direct and inverse kinematic models and their use towards the workspace analysis of the bio molecules.

This computational study calculates all geometrically feasible conformations of the bio molecule i.e. all conformations that can be achieved without any atom interference. This analysis suggests the geometric paths that could be followed by a bio molecule during the transition from the initial to a final state, while molecular dynamics narrow downs the possibilities by identifying the only energetically feasible paths. The workspace analysis also characterizes the geometrically feasible workspace in terms of dexterity.

• *Modular Pattern Recognition and Clustering Function*: The instantaneous value of the property A of a molecular system can be written as, $A\ (p^n\ (t),\ r^n\ (t))$, where $(p^n\ (t)$ and $(r^n\ (t)$ represent the n momenta and positions of a molecular system at time t. This instantaneous value would vary with respect to the interactions between the particles. There are two parts to this function (termed as M function). One part (function A), evaluates and hence recognizes the equivalent modules (in term of properties) in the molecular system. The second part of the function (D) forms the cluster (like a bioisosteres, which are atom, molecules or functional groups with similar physical and chemical properties) of various modular patterns according to the characteristic behaviors as identified by function A and also tracks their variations with rest to time.

$M^o\{A(p^n(t), r^n(t), C^n[n]), D(A(p,r,C,t), f(x, y, \ldots), Bn(t))\}$

Where, A is the first part of the function; (p^n (t) and (r^n (t) represents the n momenta and positions of a molecular system at time t; and $C^n[n]$ is an n dimensional matrix element for the individual components of the molecular system, which would store the categorized values of the modular patterns. D is the second part of the function, and it takes on function A recursively. It also maps these clusters based on a fitness function and stores the time-variant value in sub-function B.

This function is based on the hypothesis that in a complex molecular system, there are certain parts which have similar properties and behavior as the system goes from one state to the other. By identifying these property patterns we can considerably reduce the simulation and computational time frames and can have better predictability of the system. Having established the design for the bio-nano components, computational studies would focus on determining the overall design and architecture of the bio-nano assemblies. One of the methods used in the industry is of Molecular Docking.

- *Molecular Docking and Scoring Functions:* Molecular docking method is very important for the design of nanorobotic systems. This method is utilized two fit two molecules together in 3D space. The method of molecular docking could be used to computationally design the bio-nanoassemblies. The algorithms like DOCK, genetic algorithms and distance geometry are being specifically explored. Many drug design companies are using depth-first systematic conformational search algorithms for docking, and therefore, is a primary candidate for current research. Docking algorithm generates a large number of solutions. Therefore, we require scoring functions to refine the results so produced. Scoring functions in use today approximate the "binding free energy" of the molecules. The free energy of binding can be written as an additive equation of various components reflecting the various contributions to binding. A typical equation would be:

$$DG_{bind} = DG_{solvent} + DG_{cont} + DG_{int} + DG_{rot} + DG_{t/r} + DG_{vib}$$

Where, $DG_{solvent}$ is the solvent effects contribution, cDG_{conf} is due to conformational changes in the biomolecule, DG_{int} due to specific molecular interactions, DG_{rot} due to restrictive internal rotation of the binding bio-molecules, $DG_{t/r}$ is due to loss in the translational and rotational free energies due to binding, and DG_{vib} is due to vibrational mode variations. Once the overall architecture to bind the bio-nanocomponents is carried out through molecular docking, we would need to combine the components into real assemblies. The molecular component binding is the next step in the design scheme.

Connecting bio-nano components in a binding site: There are typically two approaches to design connectors for binding the bio-nanocomponents. One method searches the molecular database and selects a connector based on geometrical parameters. CAVEAT is a traditional routine which employs this method. The other approach is to design a connector from the scratch atom by atom. The geometrical parameters being known (as the configuration and the architecture was defined by molecular docking) these connectors are designed. A typical method employs designing molecular templates (which do not interfere with the basic geometry and design of the binding bio-nanocomponents) for defining a basic connector for binding the bio-nanocomponents. Feasibility of designing these connectors in real life is one challenge faced by molecular modelers. Minimum energy criterion is employed via Metropolis Monte-Carlo simulation annealing method to select the basic structure of the final connector. Genetic algorithm approaches are also employed to generate such designs.

The step 1 in the roadmap requires that we have a library of biocomponents which are characterized based on their potentials for using them for designing and developing nanorobots. Various molecular techniques could be employed for assessing the usefulness of the biocomponents. For developing a nanorobot, we require the knowledge of the structure of the biocomponents used and the degree of conformational change that each element is capable of undergoing, and the environmental cues (e.g., T, pH) that induce these structural changes. Based on the computational methods detailed in the previous section, the following experimental techniques could be used to obtain this information.

Circular dichroism spectroscopy (CD): CD spectroscopy can be utilized to gain insights into the secondary structure of each bio molecules such as peptides. In this technique, polarized far-UV light is used to probe the extent of secondary structure formation in solutions. This technique represents a proven method for estimating the structural composition of the nanorobotic elements during a change in environmental conditions. The effect of temperature on the structure of the nanorobotic elements can be evaluated by cooling the samples to +1°C and then heating to +85°C, with a five minute equilibration at each temperature interval. Intervals could be determined by first conducting CD scans with broad intervals, with subsequent narrowing as the location of the transition is identified. The effect of pH and ionic strength on the transitions could be monitored using separate, equilibrated preparations for each condition.

Förster Resonance Energy Transfer (FRET): Another method that can be used to rapidly assess the extent of conformational changes by the nanorobotic elements is FRET. In order to employ this method, the nanorobotic components have to be engineered to have cysteine residues at both ends. One end of the biocomponent then has to be labeled with a donor dye molecule (Alexa Fluor 488, Molecular Probes), and the other end labeled with an acceptor dye molecule (Alexa Fluor 594). The resulting FRET signal can be monitored with a laser (488 nm), whereby the signal changes would depend on the spatial separation of the ends of the biocomponent.

• *Nuclear Magnetic Resonance (NMR):* NMR is utilized to determine the exact 3-D structure of the nanorobotic components. Although CD spectroscopy can be used to estimate the changes in the structural composition of the biocomponents, it does not provide the exact 3-D configuration. This information can be obtained through the use NMR spectroscopy and the labeling of the biocomponent with NMR-active atoms. In addition to determining the endpoint structures of the components, NMR can also be used to follow the kinetics of the biomolecule conformational changes, in real time.

• *Laser-Based Optical Tweezers (LBOT) and Single Molecule Fluorescence:* LBOT and single molecule fluorescence could be used to estimate the forces exerted by the nanorobotic elements, and the structural changes that occur during the genesis of these forces, after exposure to different stimuli. LBOT can provide force and displacement results with a resolution of sub-picoNewton and approximately one nanometer, respectively. Single molecule fluorescence measurements, based on the previously described FRET method, can simultaneously provide information about the structural state of the molecule at each point of the force-displacement curve.

To correlate force measurements with structural changes, combination of the LBOT method with FRET can be used. The recent development of this technique has enabled the simultaneous assessment of nanoscale structural changes and their associated biomechanical forces.

For scanning and manipulating matter at the nanoscale, Scanning Tunneling Microscopy (STM), Scanning Electron Microscopy (SEM), Atomic Force Microscopy (AFM) and Scanned-Probe Microscope (SPM) seem to be the common tools. Current work is mainly focused on using atomic force microscopes (AFM) nanoprobes for teleoperated physical interactions and manipulation at the nanoscale. Precise manipulation could help scientists better understand the principles of nanorobots. To achieve this it is essential to visualize the atom-to-atom interaction in real-time and see the results in a fully immersive 3D environment. Also, to facilitate user input in nanorobotic systems it is essential to develop voice, gesture and touch recognition features in addition to the conventional visualization and manipulation techniques. Virtual Reality (VR) technology is applied here, which not only provides immersive visualization but also gives an added functionality of navigation and interactive manipulation of molecular graphical objects. VR technology comes to our aid by providing the experience of perception and interaction with the nano world through the use of sensors, effectors and interfaces in a simulated environment. These interfaces transform the signals occurring at nanoscale processes into signals at macro-level and vice-versa. The requirement is that the communication with the nano-world must be at high-level and in real-time, preferably in a natural, possibly intuitive 'language'. Considering the nano-specific problems related to task application, tools and the interconnection technologies it leads to many flexible nano manipulation concepts. They can range from pure master/slave tele-operation (through 3-D visual/VR and/or haptic force feedback), over shared autonomy control (where for example some degrees of freedom are tele-operated and other are operating autonomously) to fully autonomous operation.

In order to precisely control and manipulate biomolecules, we need tools that can interact with these objects at the nanoscale in their native environments. Existing bio-nanomanipulation techniques can be classified as non-contact manipulations including laser trapping and electro-rotation, and contact manipulation referred to as mechanical stylus-, AFM- or STM-based nanomanipulation. The rapid expansion of AFM studies in biology/biotechnology results from the fact that AFM techniques offer several unique advantages: first, they require little sample preparation, with native biomolecules usually being imaged directly; second, they can provide a 3-D reconstruction of the sample surface in real space at ultra-high resolution; third, they are less destructive than other techniques (e.g. electron microscopy) commonly employed in biology; and fourth, they can operate in several environments, including air, liquid, and vacuum. Rather than drying the sample, one can image quite successfully with AFM in fluid. The operation of AFM in aqueous solution offers an unprecedented opportunity for imaging biological molecules and cells in their physiological environments and for studying biologically important dynamic processes in real time.

Currently, these bio-nanomanipulations are conducted manually; however, long training, disappointingly low success rates from poor reproducibility in manual operations, and contamination call for the elimination of direct human involvement. Furthermore, there are many sources of spatial uncertainty in AFM manipulation, e.g., tip effects, thermal drift, slow creeping motion, and hysteresis. To improve the bio-nanomanipulation techniques, automatic manipulation must be addressed. Visual tracking of patterns from multiple views is a promising approach, which is currently investigated in autonomous embryo pronuclei DNA injection. Interactive nanomanipulation can be improved by imaging 3D viewpoint in a virtual environment. Construction of a VR space in an off-line operation mode for trajectory planning combined with a real time operation mode for vision tracking of environmental change ensures a complete 'immersed' visual display. An example of 3D bio-micro/nano manipulation system with a 3D VR model of the environment including the biocell, and carrying out the user viewpoint change in the virtual space.

Molecular Dynamics Simulations in Virtual Environment: Molecular Dynamics (MD) simulations of complex molecular systems require enormous computational power and produces large amount of data in each step. The resulting data includes number of atoms per unit volume, atomic positions; velocity of each atom, force applied on each atom and the energy contents. These results of MD simulations need to be visualized to give the user a more intuitive feel of what is happening. Haptic interaction used in conjunction with VR visualization helps the scientist to control/monitor the simulation progress and to get feedback from the simulation process as well.

The control of nanorobotic systems could be classified in two categories:

(i) Internal control mechanisms.

(ii) External control mechanisms.

The other category could be the hybrid of internal and external control mechanisms.

This type of control depends upon the mechanism of biochemical sensing and selective binding of various biomolecules with various other elements. This is a traditional method, which has been in use since quite sometime for designing biomolecules. Using the properties of the various biomolecules and combining with the knowledge of the target molecule that is to be influenced, these mechanisms could be effective. But again, this is a passive control mechanism where at run time these biomolecules cannot change their behavior. Once programed for a particular kind of molecular interaction, these molecules stick to that. Here lies the basic issue in controlling the nanorobots which are supposed to be intelligent and hence programed and controled so that they could be effective in the ever dynamic environment. The question of actively controling the nanorobots using internal control mechanism is a difficult one. We require an 'active' control mechanism for the designed nanorobots such that they can vary their behavior based on situations they are subjected to, similar to the way macro robots perform. For achieving this internal control, the concept of molecular computers could be utilized. Leonard Adleman (from the University of Southern California) introduced DNA computers a decade ago to solve a mathematical problem by utilizing DNA molecules.

Professor Ehud Shapiro's lab (at Israel's Weizmann Institute) has devised a biomolecular computer which could be an excellent method for an internal 'active' control mechanism for nanorobots. They have recently been successful in programing the biomolecular computer to analyze the biological information, which could detect and treat cancer (prostate and a form of lung cancer) in their laboratory. The molecular computer has an input and output module which acting together can diagnose a particular disease and in response produce a drug to cure that disease. It uses novel concept of software (made up of DNAs) and hardware (made up of enzymes) molecular elements. This molecular computer is in generalized form and can be used for any disease which produces a particular pattern of gene expression related to it.

This type of control mechanism employs affecting the dynamics of the nanorobot in its work environment through the application of external potential fields. Researchers are actively looking at using MRI as an external control mechanism for guiding the nano particles. Professor Sylvain Martel's Nanorobotics Laboratory (at École Polytechnique de Montréal, Canada) is actively looking at using MRI system as a mean of propulsion for nanorobots. An MRI system is capable of generating variable magnetic field gradients which can exert force on the nanorobot in the three dimensions and hence control its movement and orientation. Professor Martel's laboratory is exploring effect of such variable magnetic field on a ferromagnetic core that could probably be embedded into the nanorobots.

Other possibilities being explored are in the category of 'hybrid' control mechanisms where the target is located and fixed by an external navigational system but the behavior of the nanorobot is determined locally through an active internal control mechanism. The use of nanosensors and evolutionary agents to determine the nanorobots behavior is suggested by the mentioned reference.

Manipulating matter at molecular scale and influencing their behavior (dynamics and properties) is the biggest challenges for the nanorobotic systems. This field is still in very early stages of development and still a lot has to be figured out before any substantial outcome is produced. The recent explosion of research in nanotechnology, combined with important discoveries in molecular biology have created a new interest in bio nanorobotic systems. The preliminary goal in this field is to use various biological elements—whose function at the cellular level creates a motion, force or a signal—as nanorobotic components that perform the same function in response to the same biological stimuli but in an artificial setting. In this way proteins and DNA could act as motors, mechanical joints, transmission elements, or sensors. If all these different components were assembled together they can form nanorobots and nanodevices with multiple degrees of freedom, with ability to apply forces and manipulate objects in the nanoscale world, transfer information from the nano to the macroscale world and even travel in a nanoscale environment. Ability to determine the structure, behavior and properties of the nanocomponents is the first step which requires focused research thrust. Only when the preliminary results on these nanocomponents are achieved, steps towards actually building complex assemblies could be thought of. Still problems like protein (basic bio nano component folding, precise mechanism behind working of the molecular motors like ATP Synthase have still has to be solved. Active control of nanorobots has to be further refined. Hybrid control mechanisms, where in, a molecular computer and external (navigational) control system work in sync to produce the precise results seems very promising. Further, concepts like swarm behavior in context of nanorobotics is still have to be worked out. As it would require colonies of such nanorobots for accomplishing a particular task, concept of co-operative behavior, distributed intelligence has to be evolved. The future of bio-nanorobots (molecular robots) is bright. We are at the dawn of a new era in which many disciplines will merge including robotics, mechanical, chemical and biomedical engineering, chemistry, biology, physics and mathematics so that fully functional systems will be developed. However, challenges towards such a goal abound. Developing a complete database of different biomolecular machine components and the ability to interface or assemble different machine components are some of the challenges to be faced in the near future.

REFERENCES

A. Carbone & N.C. Seeman, Molecular Tiling and DNA Self-Assembly, Aspects of Molecular Computing, Lecture Notes in Computer Science 2340, Springer-Verlag, Berlin, 61-83 (2004).

Akis, R., D. Ferry, C. Musgrave, "Kinetic Lattice Monte Carlo Simulations of Processes on the Silicon (100) Surface," *Physica E-Low Dimensional Systems and Nanostructures* 19, 183-187 (2003).

Albrecht, T.R., Akamine, S., Carver, T.E., and Quate, C.F. (1990) Microfabrication of cantilever styli for the atomic force microscope. J. Vac. Sci. Technol. A 8(4), 3386-3396

Albrecht, T.R., Grütter, P., Horne, D., and Rugar, D. (1991) Frequency modulation detection using high-Q cantilevers for enhanced force microscope sensitivity. J. Appl. Phys. 69(2), 668-673

Alexander, S., Hellemans, L., Marti, O., Schneir, J., Elings, V., Hansma, P.K., Longmiro, M., and Gurley, J. (1989) An atomic-resolution atomic-force microscope implemented using an optical lever. J. Appl. Phys. 65(1), 164-167

B. Ding, R. Sha & N.C. Seeman, Pseudohexagonal 2D DNA Crystals from Double Crossover Cohesion, J. Am. Chem. Soc. 126, 10230-10231 (2004).

B. Liu, N.B. Leontis and N.C. Seeman, Bulged 3-arm DNA Branched Junctions as Components for Nanoconstruction, Nanobiology 3, 177-188 (1994). N.C. Seeman, Molecular Craftwork with DNA, The Chemical Intelligencer 1(3), 38-47 (1995).

B.H. Robinson and N.C. Seeman, The Design of a Biochip: A Self-Assembling Molecular-Scale Memory Device. Protein Engineering 1, 295-300 (1987).

Binnig, G., Quate, C.F., and Gerber, Ch. (1986) Atomic force microscope. Phys. Rev. Lett. 56(9), 930-933

C. Mao, T. LaBean, J.H. Reif and N.C. Seeman, Logical Computation Using Algorithmic Self-Assembly of DNA Triple Crossover Molecules, Nature 407, 493-496 (2000).

C. Mao, W. Sun and N.C. Seeman, Construction of Borromean Rings from DNA, Nature 386, 137-138 (1997).

C. Mao, W. Sun and N.C. Seeman, Designed Two-Dimensional DNA Holliday Junction Arrays Visualized by Atomic Force Microscopy, Journal of the American Chemical Society 121, 5437-5443 (1999).

C. Mao, W. Sun, Z. Shen and N.C. Seeman, A DNA Nanomechanical Device Based on the B-Z Transition, Nature 397, 144-146 (1999).

E. Winfree, F. Liu, L. A. Wenzler, and N.C. Seeman, Design and Self-Assembly of Two-Dimensional DNA Crystals, Nature 394, 539-544 (1998).

E. Winfree, X. Yang and N.C. Seeman, Universal Computation via Self-assembly of DNA: Some Theory and Experiments, In: DNA Based Computers II, ed.by L.F. Landweber and E.B. Baum, Am. Math. Soc., Providence, pp. 191-213 (1998).

F. Liu, H. Wang and N.C. Seeman, Short Extensions to Sticky Ends for DNA Nanotechnology and DNA-Based Computation, Nanobiology 4, 257-262 (1999).

F. Liu, R. Sha and N.C. Seeman, Modifying the Surface Features of Two-Dimensional DNA Crystals, Journal of the American Chemical Society. 121, 917-922 (1999).

F. Mathieu, S. Liao, C. Mao, J. Kopatsch, T. Wang, and N.C. Seeman, Six-Helix Bundles Designed from DNA, NanoLetters 5, 661-665 (2005).

Filler, M., C. Mui, G. Wang, C. Musgrave, and S. Bent, "Competition and Selectivity in the Reaction of Nitriles on Ge (100)-2x1," *Journal of the American Chemical Society* 125, 4928-4936 (2003).

G. Serrano & N.C. Seeman, Nanotecnologia basada en ADN, Revista de Química 19, 11-20 (2004), in Spanish.

Gallego-Juárez, J.A. (1989) Piezoelectric ceramics and ultrasonic transducers. J. Phys. E: Sci. Instrum. 22, 804-816

H. Qiu, J.C. Dewan and N.C. Seeman, A DNA Decamer with a Sticky End: The Crystal Structure of d-CGACGATCGT. J. Mol. Biol. 267, 881-898 (1997).

H. Yan and N.C. Seeman, Edge-Sharing Motifs in DNA Nanotechnology. Journal of Supramolecular Chemistry 1, 229-237 (2003).

H. Yan, X. Zhang, Z. Shen and N.C. Seeman, A Robust DNA Mechanical Device Controlled by Hybridization Topology, Nature 415, 62-65 (2002).

Hoh, J.H. and Hansma, P.K. (1992) Atomic force microscopy for high-resolution imaging in cell biology. Trends Cell Bio. 2, 208-213

J. Chen and N.C. Seeman, The Synthesis from DNA of a Molecule with the Connectivity of a Cube, Nature 350, 631-633 (1991).

J. D. Le, Y. Pinto, N.C. Seeman, K. Musier-Forsyth, T. A. Taton and R.A. Kiehl, Self-Assembly of Nanoelectronic Component Arrays by In Situ Hybridization to 2D DNA Scaffolding, NanoLetters 4, 2343-2347 (2004).

J. Qi, X. Li, X.P. Yang and N.C. Seeman, The Ligation of Triangles Built from Bulged Three-Arm DNA Branched Junctions, Journal of the American Chemical Society, 118, 6121-6130 (1996).

J.H. Reif, T.H. LaBean & N.C. Seeman, Challenges and Applications for Self-Assembled DNA Nanostructures, Sixth International Workshop on DNA-Based Computers, DNA 2000, Leiden, The Netherlands, (June, 2000) ed. A. Condon, G. Rozenberg. Springer-Verlag, Berlin Heidelberg, Lecture Notes in Computer Science 2054, 173-198, (2001).

Kang, J. and C. Musgrave, "A Quantum Chemical Study of the Self-Directed Growth Mechanism of Styrene and Propylene Molecular Nanowires on the Silicon (100) 2x1 Surface," *Journal of Chemical Physics* 116, 9907-9913 (2002).

Kang, J. and C. Musgrave, "The Mechanism of Atomic Layer Deposition of SiO 2 on the Silicon (100)-2x1 Surface Using SiC1 4 and H 2 O as Precursors," *Journal of Applied Physics* 91, 3408-3414 (2002).

Keller, D.J. and Chih-Chung, C. (1992) Imaging steep, high structures by scanning force microscopy with electron beam deposited tips. Surf. Sci. 268, 333-339

L. Zhu, P.S. Lukeman, J. Canary & N.C. Seeman, Nylon/DNA: Single-Stranded DNA with Covalently Stitched Nylon Lining, J. Am. Chem. Soc. 125, 10178-10179 (2003).

M.L. Petrillo, C.J. Newton, R.P. Cunningham, R.-I. Ma, N.R. Kallenbach and N.C. Seeman. The Ligation and Flexibility of 4-Arm DNA Junctions. Biopolymers, 27, 1337-1352 (1988).

Meyer, G. and Amer, N.M. (1988) Novel optical approach to atomic force microscopy. Appl. Phys. Lett. 53(12), 1045-1047

Meyer, G. and Amer, N.M. (1990) Simultaneous measurement of lateral and normal forces with an optical-beam-deflection atomic force microscope. Appl. Phys. Lett. 57(20), 2089-2091

Mui, C. and C. Musgrave, "The Hydroxylation and Oxidation of the Ge(100)-2x1 Surface by Hydrogen Peroxide and Water," *Langmuir*, 20, 7604-7609 (2004).

Mui, C., S. Bent and C. Musgrave, "A Quantum Chemistry Based Statistical Mechanical Model of Hydrogen Deposition from Si(100)-2x1, Ge(100)-2x1, and SiGe Alloy Surfaces," *Journal of Physical Chemistry B*, 108, 18243-18253, (2004).

Mui, C., Y. Widjaja, J. Kang, and C. Musgrave, "Surface Reaction Mechanisms for Atomic Layer Deposition of Silicon Nitride," *Surface Science,* 557, 159-170, (2004).

N. Jonoska, P. Sa-Ardyen and N.C. Seeman, Computation by Self-Assembly of DNA Graphs, Journal of Genetic Programming and Evolvable Machines 4, 123-137 (2003).

N. Jonoska, S. Liao, & N.C. Seeman, Transducers with programmable input by DNA Self-assembly, Aspects of Molecular Computing, Lecture Notes in Computer Science 2340, Springer-Verlag, Berlin, 219-240 (2004).

N.C. Seeman & A.M. Belcher, Emulating Biology: Nanotechnology from the Bottom Up, Proceedings of the National Academy of Sciences (USA) 99 (supp. 2), 6451-6455 (2002).

N.C. Seeman and N.R. Kallenbach, Design of Immobile Nucleic Acid Junctions. Biophysical Journal 44, 201-209 (1983).

N.C. Seeman and N.R. Kallenbach, Nucleic Acid Junctions, The Tensors of Life? In: Nucleic Acids: The Vectors of Life, ed. by B. Pullman and J. Jortner, D. Reidel, Dordrecht, 183-200 (1983).

N.C. Seeman and N.R. Kallenbach, Nucleic Acid Junctions: A Successful Experiment in Macromolecular Design. In: Molecular Structure: Chemical Reactivity and Biological Activity, ed. by J.J. Stezowski J.-L. Huang and M.-C. Shao, Oxford University Press, Oxford, (1988) pp. 189-194.

N.C. Seeman and P.S. Lukeman, Nucleic Acid Nanostructures, Reports on Progress in Physics 68, 237-270 (2005).

N.C. Seeman, At the Crossroads of Chemistry, Biology and Materials: Structural DNA Nanotechnology, Chemistry & Biology 10, 1151-1159 (2003).

N.C. Seeman, B. Ding, S. Liao, T. Wang, W.B. Sherman, P.E. Constantinou, J. Kopatsch, C. Mao, R. Sha, F. Liu, H. Yan & P.S. Lukeman, Experiments in Structural DNA Nanotechnology: Arrays and Devices, Proc. SPIE; Nanofabrication: Technologies, Devices and Applications 5592, 71-81 (2005).

N.C. Seeman, Biochemistry and Structural DNA Nanotechnology: An Evolving Symbiotic Relationship, Biochemistry 42, 7259-7269 (2003).

N.C. Seeman, C. Mao, F. Liu, R. Sha, X. Yang, L. Wenzler, X. Li, Z. Shen, H. Yan, P. Sa-Ardyen, X. Zhang, W. Shen, J. Birac, P. Lukeman, Y. Pinto, J. Qi, B. Liu, H. Qiu, S.M. Du, H. Wang, W. Sun, Y. Wang, T.-J. Fu, Y. Zhang, J.E. Mueller and J. Chen, Nicks, Nodes, and New Motifs for DNA Nanotechnology, Frontiers of Nano-Optoelectronic Systems, ed. by L. Pavesi & E. Buzanova, Kluwer, Dordrecht, 177-198 (2000).

N.C. Seeman, De Novo Design of Sequences for Nucleic Acid Structure Engineering, Journal of Biomolecular Structure and Dynamics 8, 573-581 (1990).

N.C. Seeman, Design and Engineering of Nucleic Acid Nanoscale Assemblies, Current Opinion in Structural Biology, 6, 519-526 (1996).

N.C. Seeman, DNA in a Material World, Nature 421, 33-37 (2003).

P. Sa-Ardyen, A.V. Vologodskii and N.C. Seeman, The Flexibility of DNA Double Crossover Molecules. Biophysical Journal 84, 3829-3837 (2003).

P.S. Lukeman, A. Mittal, & N.C. Seeman, Two Dimensional PNA/DNA Arrays: Estimating the Helicity of Unusual Nucleic Acid Polymers, Chemical Communications 2004, 1694-1695 (2004).

Putman, C.A.J., De Grooth, B.G., Van Hulst, N.F., and Greve, J. (1992) A detailed analysis of the optical beam deflection technique for use in atomic force microscopy. J. App. Phys. 72(1), 6-12

R. Sha, F. Liu and N.C. Seeman, Atomic Force Measurement of the Interdomain Angle in Symmetric Holliday Junctions, Biochemistry 41, 5950-5955 (2002).

R. Sha, F. Liu, D.P. Millar N.C. Seeman, Atomic Force Microscopy of Parallel DNA Branched Junction Arrays, Chemistry & Biology 7, 743-751 (2000).

R.-I. Ma, N.R. Kallenbach, R.D. Sheardy, M.L. Petrillo and N.C. Seeman, 3-Arm Nucleic Acid Junctions Are Flexible. Nucleic Acids Research 14, 9745-9753 (1986).

Reviews in Computational Chemistry, Vol 2, Kenny B. Lipkowitz and Donald B. Boyd, VCH Publishers, 1991.

Rod Logic and Thermal Noise in the Mechanical Nanocomputer, by K. Eric Drexler, in *Proceedings of the 3rd International Symposium on Molecular Electronic Devices*, F.L. Carter, R.E. Siatkowski, H. Wohltjen, eds., North-Holland 1988.

S. Liao and N.C. Seeman, Translation of DNA Signals into Polymer Assembly Instructions, Science 306, 2072-2074 (2004).

S. Xiao, F. Liu, A. Rosen, J.F. Hainfeld, N.C. Seeman, K.M. Musier-Forsyth & R.A. Kiehl, Self-Assembly of Nanoparticle Arrays by DNA Scaffolding, J. Nanoparticle Research 4, 313-317 (2002).

Senosiain, J., C. Musgrave and D. Golden, "Temperature and Pressure Dependence of the Reaction of OH and CO: Master Equation Modeling on a High Level Potential Energy Surface," *International Journal of Chemical Kinetics* 35, 464-474 (2003).

Sha, R., Zhang, X., Liao, S., Constantinou, P.E., Ding, B., Wang, T., Garibotti, A.V., Zhong, H., Israel, L.B., Wang, X., Wu, G., Chakraborty, B., Chen, J., Zhang, Y., Mao, C., Yan, H., Kopatsch, J., Zheng, J., Lukeman, P.S., Sherman, W.B. and Seeman, N.C., Motifs and Methods in Structural DNA Nanotechnology, Proc. Intl. Conf. Nanomaterials, NANO 2005, July 13-15, 2005, Mepco Schlenk Engineering College, Srivakasi, India, V. Rajendran, ed., pp. 3-10 (2005).

T. LaBean, H. Yan, J. Kopatsch, F. Liu, E. Winfree, J.H. Reif and N.C. Seeman, The Construction of DNA Triple Crossover Molecules, Journal of the American Chemical Society 122, 1848-1860 (2000).

W. Shen, M. Bruist, S. Goodman & N.C. Seeman, A Nanomechanical Device for Measuring the Excess Binding Energy of Proteins that Distort DNA, Angew. Chem. Int. Ed. 43, 4750-4752 (2004); Angew.Chem. 116, 4854-4856 (2004).

W.B. Sherman and N.C. Seeman, A Precisely Controlled DNA Bipedal Walking Device, NanoLetters 4, 1203-1207 (2004); Erratum, 4, 1801-1801.

Weisenhorn, A.L., Hansma, P.K., Albrecht, T.R., and Quate, C.F. (1989) Forces in atomic force microscopy in air and water. Appl. Phys. Lett. 54(26), 2651-2653

Widjaja, Y. and C. Musgrave, "Atomic Layer Deposition of Hafnium Oxide: A Detailed Reaction Mechanism from First Principles," *Journal of Chemical Physics* 117, 1931-1934 (2002).

Widjaja, Y. and C. Musgrave, "Indirect Adsorbate-Adsorbate Interactions Mediated Through the Surface Electronic Structure of the Si(100)-(2x1) Surface," *Journal of Chemical Physics* 120, 1555-1559 (2004).

Widjaja, Y. and C. Musgrave, "Quantum Chemical Study of the Elementary Reactions in Zirconium Oxide Atomic Layer Deposition," *Applied Physics Letters* 81, 304-306 (2002).

X. Yang, B. Liu, A. Vologodskii, B. Kemper and N.C. Seeman, Torsional Control of Double Stranded DNA Branch Migration. Biopolymers 45, 69-83 (1998).

X. Yang, L.A. Wenzler, J. Qi, X. Li and N.C. Seeman, Ligation of DNA Triangles Containing Double Crossover Molecules, Journal of the American Chemical Society 120, 9779-9786 (1998).

X. Zhang, H. Yan, Z. Shen and N.C. Seeman, Paranemic Cohesion of Topologically-Closed DNA Molecules, J Am. Chem. Soc.124, 12940-12941 (2002).

X.J. Li, X.P. Yang, J. Qi, and N.C. Seeman, Antiparallel DNA Double Crossover Molecules as Components for Nanoconstruction, Journal of the American Chemical Society, 118, 6131-6140 (1996).

Xu, Y. and C. Musgrave, "A DFT Study of the Atomic Layer Deposition of Al 2 O 3 on SAMs: The Effect of SAM Termination," *Chemistry of Materials,* 16, 646 (2004).

Y. Wang, J.E. Mueller, B. Kemper, and N.C. Seeman, The Assembly and Characterization of 5-Arm and 6-Arm DNA Junctions, Biochemistry 30, 5667-5674 (1991).

Y. Zhang and N.C. Seeman, A Solid-Support Methodology for the Construction of Geometrical Objects from DNA, Journal of the American Chemical Society 114, 2656-2663 (1992).

Y. Zhang and N.C. Seeman, The Construction of a DNA Truncated Octahedron, Journal of the American Chemical Society 116, 1661-1669 (1994).

3

Nanotech Products & Applications: An Overview

Bottom-up Methods for Making Nanotechnology Products

There are two general ways available to produce nanomaterials, as shown in the following figure. The first way is to start with a bulk material and then break it into smaller pieces using mechanical, chemical or other form of energy (top-down). An opposite approach is to synthesise the material from atomic or molecular species via chemical reactions, allowing for the precursor particles to grow in size (bottom-up). Both approaches can be done in either gas, liquid, supercritical fluids, solid states, or in vacuum. Most of the manufacturers are interested in the ability to control: a) particle size b) particle shape c) size distribution d) particle composition e) degree of particle agglomeration.

What Processes are used for Bottom-up Manufacturing?

Methods to produce nanoparticles from atoms are chemical processes based on transformations in solution e.g. sol-gel processing, chemical vapour deposition (CVD), plasma or flame spraying synthesis, laser pyrolysis, atomic or molecular condensation. These chemical processes rely on the availability of appropriate "metal-organic" molecules as precursors. Sol-gel processing differs from other chemical processes due to its relatively low processing temperature. This makes the sol-gel process cost-effective and versatile. In spraying processes the flow of reactants (gas, liquid in form of aerosols or mixtures of both) is introduced to high-energy flame produced for example by plasma spraying equipment or carbon dioxide laser. The reactants decompose and particles are formed in a flame by homogeneous nucleation and growth. Rapid cooling results in formation of nanoscale particles.

These are chemical processes to materials based on transformations in solution such as sol-gel processing, hydro or solvo thermal syntheses, Metal Organic Decomposition (MOD), or in the vapour phase chemical vapour deposition (CVD). Most chemical routes rely on the availability of appropriate "metal-organic" molecules as precursors. Among the various precursors of metal oxides, namely metal b-diketonates and metal carboxylates, metal alkoxides are the most versatile. They are available for nearly all elements and cost-effective synthesis from cheap feedstock have been developed for some.

How to Control the Construction and Growth of the Nanoparticles

Two general ways are available to control the formation and growth of the nanoparticles. One is called arrested precipitation and depends either on exhaustion of one of the reactants or on the introduction of the chemical that would block the reaction. Another method relies on a physical restriction of the volume available for the growth of the individual nanoparticles by using templates.

Sol-gel Process

The sol gel technique is a long-established industrial process for the generation of colloidal nanoparticles from liquid phase, that has been further developed in last years for the production of advanced nanomaterials and coatings. Sol-gel-processes are well adapted for oxide nanoparticles and composites nanopowders synthesis. The main advantages of sol-gel techniques for the preparation of materials are low temperature of processing, versatility, and flexible rheology allowing easy shaping and embedding. They offer unique opportunities for access to organic-inorganic materials. The most commonly used precursors of oxides are alkoxides, due to their commercial availability and to the high liability of the M-OR bond allowing *facile tailoring in situ* during processing.

Aerosol-Based Processes

Aerosol-based processes are a common method for the industrial production of nanoparticles. Aerosols can be defined as solid or liquid particles in a gas phase, where the particles can range from molecules up to 100 μm in size. Aerosols were used in industrial manufacturing long before the basic science and engineering of the aerosols were understood. For example, carbon black particles used in pigments and reinforced car tires are produced by hydrocarbon combustion; titania pigment for use in paints and plastics is made by oxidation of titanium tetrachloride; fumed silica and titania formed from respective tetrachlorides by flame pyrolysis; optical fibres are manufactured by similar process.

Traditionally, spraying is used either to dry wet materials or to deposit coatings. Spraying of the precursor chemicals onto a heated surface or into the hot atmosphere results in precursor pyrolysis and formation of the particles. For example, a room temperature electro-spraying process was developed at Oxford University to produce nanoparticles of compound semiconductors and some metals. In particular, CdS nanoparticles were produced by generating aerosol micro-droplets containing Cd salt in the atmosphere containing hydrogen sulphide.

Chemical Vapour Deposition (CVD)

CVD consists in activating a chemical reaction between the substrate surface and a gaseous precursor. Activation can be achieved either with temperature (Thermal CVD) or with a plasma (PECVD: Plasma Enhanced Chemical Vapour Deposition). The main advantage is the nondirective aspect of this technology. Plasma allows to decrease significantly the process temperature compared to the thermal CVD process. CVD is widely used to produce carbon nanotubes.

Atomic or Molecular Condensation

This method is used mainly for metal containing nanoparticles. A bulk material is heated in vacuum to produce a stream of vaporised and atomised matter, which is directed to a chamber containing either inert or reactive gas atmosphere. Rapid cooling of the metal atoms due to their collision with the gas molecules results in the condensation and formation of nanoparticles. If a reactive gas like oxygen is used then metal oxide nanoparticles are produced.

Using Gas-Phase Condensation to Produce Metal Nanopowders

The theory of gas-phase condensation for the production of metal nanopowders is well known, having been first reported in 1930. Gas-phase condensation uses a vacuum chamber that consists of a heating element, the metal to be made into nano-powder, powder collection equipment and vacuum hardware.

How the Gas-Phase Condensation Process Works

The process utilises a gas, which is typically inert, at pressures high enough to promote particle formation, but low enough to allow the production of spherical particles. Metal is introduced onto a heated element and is rapidly melted. The metal is quickly taken to temperatures far above the melting point, but less than the boiling point, so that an adequate vapour pressure is achieved. Gas is continuously introduced into the chamber and removed by the pumps, so the gas flow moves the evaporated metal away from the hot element. As the gas cools the metal vapour, nanometer-sized particles form. These particles are liquid since they are still too hot to be solid. The liquid particles collide and coalesce in a controlled environment so that the particles grow to specification, remaining spherical and with smooth surfaces. As the liquid particles are further cooled under control, they become solid and grow no longer. At this point the nanoparticles are very reactive, so they are coated with a material that prevents further interaction with other particles (agglomeration) or with other materials.

Supercritical Fluid Synthesis

Methods using supercritical fluids are also powerful for the synthesis of nanoparticles. For these methods, the properties of a supercritical fluid (fluid forced into supercritical state by regulating its temperature and its pressure) are used to form nanoparticles by a rapid expansion of a supercritical solution. Supercritical fluid method is currently developed at the pilot scale in a continuous process.

Spinning to Make Thin Polymer Fibers

An emerging technology for the manufacture of thin polymer fibers is based on the principle of spinning dilute polymer solutions in a high voltage electric field. Electro spinning is a process by which a suspended drop of polymer is charged with thousands of volts. At a characteristic voltage the droplet forms a Taylor cone, and a fine jet of polymer releases from the surface in response to the tensile forces generated by interaction of an applied electric field, with the electrical charge carried by the jet. This produces a bundle of polymer fibers. The jet can be directed to a grounded surface and collected as a continuous web of fibers ranging in size from a few μm's to less than 100 nm.

Using Templates to Form Nanoparticles

Any material containing regular nano-sized pores or voids can be used as a template to form nanoparticles. Examples of such templates include porous alumina, zeolites, di-block co-polymers, dendrimers, proteins and other molecules. The template does not have to be a 3D object. Artificial templates can be created on a plane surface or a gas-liquid interface by forming self-assembled monolayers.

Self-Assembly of Nanoparticles

Nanoparticles of a wide range of materials—including a variety of organic and biological compounds, but also inorganic oxides, metals, and semiconductors—can be processed using chemical self-assembly techniques. These techniques exploit selective attachment of molecules to specific surfaces, biomolecular recognition and self-ordering principles (e.g. the preferential docking of DNA strands with complementary base pairs) as well as well-developed chemistry for attaching molecules onto clusters and substrates (e.g. thiol (-SH) end groups) and other techniques like reverse micelle, sonochemical, and photochemical synthesis to realise 1-D, 2-D and 3-D self-assembled nanostructures. The molecular building blocks act as parts of a jigsaw puzzle that join together in a perfect order without an obvious driving force present.

Molecular Nanotechnology Offers Visions for the Future

Long-term and visionary nanotechnological conceptions, however, go far beyond these first approaches. This applies in particular to the development of biomimetic materials with the ability of self-organisation, self-healing and self-replication by means of molecular nanotechnology. One objective here is the combination of synthetic and biological materials, architectures and systems, respectively, the imitation of biological processes for technological applications. This field of nanobiotechnology is at present still in the state of basic research, but is regarded as one of the most promising research fields for the future.

Nanotech Applications: An Overview

Laboratories throughout the world are rapidly gaining atomically precise control over matter. As this control extends to an ever wider variety of materials, processes and devices, opportunities for applications relevant to NASA's missions will be created. This document surveys a number of future molecular nanotechnology capabilities of aerospace interest. Computer applications, launch vehicle improvements, and active materials appear to be of particular interest. We also list a number of applications for each of NASA's enterprises. If advanced molecular nanotechnology can be developed, almost all of NASA's endeavors will be radically improved. In particular, a sufficiently advanced molecular nanotechnology can arguably bring large scale space colonization within our grasp. This study describes potential aerospace applications of molecular nanotechnology, defined as the thorough three-dimensional structural control of materials, processes and devices at the atomic scale.

The inspiration for molecular nanotechnology comes from Richard P. Feynman's 1959 visionary talk at Caltech in which he said, "The problems of chemistry and biology can be greatly helped if our ability to see what we are doing, and to do things on an atomic level, is ultimately developed—a development which I think cannot be avoided." Indeed, scanning probe microscopes (SPMs) have already given us this ability in limited domains. Synthetic chemistry, biotechnology, "laser tweezers" and other developments are also bringing atomic precision to our endeavors. An expanded version of Drexler's MIT PhD thesis, examines one vision of molecular nanotechnology in considerable technical detail. It proposes the development of programmable molecular assembler/replicators. These are atomically precise machines that can make and break chemical bonds using mechanosynthesis to produce a wide variety of products under software control, including copies of themselves. Interestingly, living cells exhibit many properties of assembler/replicators. Cells make a wide variety of products, including copies of themselves, and can be programmed with DNA. Replication is one approach to building large systems, such as human rated launch vehicles, from molecular machines manipulating matter one or a few atoms at a time. Note that biological replication is responsible for systems as large as redwood trees and whales. Another approach to nanotechnology is supramolecular self-assembly, where molecular systems are designed to attract each other in a particular orientation to form larger systems. Hollow spheres large enough to be visible in a standard light microscope have been created this way using self-assembling lipids.

There are many other examples and this field is rapidly advancing. Biological systems can do most of what molecular nanotechnology strives to accomplish — atomically precise products, active materials, reproduction, etc. However, biological systems are extremely complex and molecular nanotechnology seeks simpler systems to understand, control and manufacture. Also, biological systems usually work at fairly mild temperature and pressure conditions in solution — conditions that are not found in most aerospace environments. Today, extremely precise atomic and molecular manipulation is common in many laboratories around the world and our abilities are rapidly approaching Feynman's dream. The implications for aerospace development are profound and ubiquitous. A number of applications are mentioned here and a few are described in some detail with references. From this sample of applications it should be clear that although molecular nanotechnology is a long term, high risk project, the payoff is potentially enormous — vastly superior computers, aerospace transportation, sensors and other technologies; technologies that may enable large scale space exploration and colonization. This chapter is organized into two sections. In the first, we examine three technologies—computers, aerospace transportation, and active materials—useful to nearly all NASA missions.

In the second, we investigate some potential molecular nanotechnology payoffs for each area identified in NASA's strategic plan. Some of these applications are under investigation by nanotechnology researchers at NASA Ames. Some of the applications described below have relatively near-term potential and working prototypes may be realized within three to five years. This is certainly not true in other cases. Indeed, many of the possible applications of nanotechnology that we describe here are, at the present time, rather speculative and futuristic. However, each of these ideas have been examined at least cursorily by competent scientists, and as far as we know all of them are within the bounds of known physical laws. We are not suggesting that their achievement will be easy, cheap or near-term. Some may take decades to realize; some other ideas may be scrapped in the coming years as insuperable barriers are identified. But we feel that they are worth mentioning here as illustrations of the potential future impact of nanotechnology. The applicability of manufacturing at an ever smaller scale is nowhere more self-evident than in computer technology. Indeed, Moore's law (an observation not a physical law) says that computer chip feature size decreases exponentially with time, a trend that predicts atomically precise computers by about 2010-2015. This capability is being approached from many directions. Here we will concentrate on those under development by NASA Ames and her partners.

Carbon nanotubes can be viewed as rolled up sheets of graphite from 0.7 to many nanometers in diameter. The smaller tubes are single molecules. Dai placed carbon nanotubes on an SPM tip thus extending our ability to manipulate a single molecule with sub-angstrom accuracy. Not only are the tips atomically precise, but they should have approximately the same chemistry as C_{60}, and thus be functionalizable with a wide variety of molecular fragments. Functionalizing carbon nanotube tips will allow mechanical manipulation of many molecular systems on various surfaces with sub-angstrom accuracy. One particularly intriguing possibility along this line is to utilize a carbon nanotube SPM tip to engrave patterns on a silicon surface. It should be possible to create features a few nanometers across. These would be perhaps 100 times finer than the current state of the art in commercial semiconductor photolithography. Further, in contrast to approaches such as electron microscope lithography for which the speed of operation now appears to be an insuperable obstacle for industrial production, nanotube SPM-based lithography can be accelerated by utilizing an array with thousands of SPM tips simultaneously engraving different parts of a silicon surface. Also, nanotube SPM lithography could provide a practical means to explore various futuristic electronic device technology ideas, such as quantum cellular automata, which require exceedingly small feature sizes. Needless to say, if these ideas pan out, they could literally revolutionize computer device technology, paving the way for systems that are many times more powerful and more compact than any available today. For the near term, it should be noted that the semiconductor industry is a major market for SPM products. These are used to examine production equipment. High performance carbon nanotube tips should be of substantial value. NASA Ames is collaborating with Dr. Dai, now at Stanford, to develop these tips. It is possible to store data on long chain molecules (for example, DNA) and it may be possible to read these data with carbon nanotube tipped SPMs. Existing DNA synthesis techniques can be used to write data. If the different DNA base pairs can be distinguished with a carbon nanotube tipped SPM, then the data can be read non-destructively (current techniques allow a destructive read).

However, the difference between base pairs is not great. If the base pairs cannot be distinguished, techniques for attaching modified enzymes to specific base pair sequences could be used. Certain enzymes (DNA (cytosine-5) methyltransferases) attach themselves onto a specific sequence of base pairs with a covalent bond. The enzyme then performs its operation and breaks the bond. Smith modified the enzyme such that the initial covalent bond was formed but the subsequent operation was disrupted. The result is that DNA synthesized with the target base pair sequences at the desired location can force precise placement of the enzymes. The presence of an enzyme could represent 1 and its absence 0. Enzymes are sufficiently large that distinguishing their presence should be straightforward. If the DNA/enzyme approach proves impossible, a wide variety of other polymer systems could be examined. Bauschlicher computationally studied storing data in a pattern of fluorine and hydrogen atoms on the diamond surface. If write-once data could be stored this way, 10^{15} bytes/cm^2 is theoretically possible.

By comparison, the new DVD write-once disks now coming on the market hold about 10^8 bytes/cm^2. Bauschlicher compared the interaction of different probe molecules with a one dimensional model of the diamond surface. This study found some molecules whose interaction energies with H and F are sufficiently different that the force differential should be detectable by an SPM. These studies were extended to include a two dimensional model of the diamond surface and two other systems besides F/H. Other surfaces, such as Si, and other probes, such as those including transition metal atoms, have also been investigated. Among the better probes was C_5H_5N (pyridine). Quantum calculations suggest that pyridine is stable when attached to C_{60} in the orientation necessary for sensing the difference between hydrogen and fluorine. Half of C_{60} can form the end cap of a (9,0) or (5,5) carbon nanotube, and carbon nanotubes have been attached to an SPM tip. Thus, it might be possible using today's technology to build a system to read the diamond memory surface. Avouris has shown that individual hydrogen atoms can be removed from a silicon surface. If this could be accomplished in a gas that donates fluorine to vacancies on a diamond surface, the data storage system could be built. Thummel computationally investigated methods for adding a fluorine at the radical sites where a hydrogen atom had been removed from a diamond surface. As mentioned before, carbon nanotubes can be described as rolled up sheets of graphite. Different tubes can have different helical windings depending on how the graphite sheet is connected to itself. Theory suggests that single-walled carbon nanotubes can have metallic or semiconductor properties depending on the helical winding of the tube. Chico, Han, Menon, and others have computationally examined the properties of some of hypothetical devices that might be made by connecting tubes with different electrical properties. Such devices are only few nanometers across — 100 times smaller than current computer chip features. Several authors, including Tour have described methods to produce conjugated macromolecules of precise length and composition. This technique was used to produce molecular electronic devices in mole quantities. The resultant single molecular wires were tested experimentally and found to be conducting. The three and four terminal devices have been examined computationally and look promising. The features of these components are approximately 3 angstroms wide, about 750 times smaller than current silicon technology can produce. Helical logic is a theoretical proposal for a future computing technology using the presence or absence of individual electrons (or holes) to encode 1s and 0s.

The electrons are constrained to move along helical paths, driven by a rotating electric field in which the entire circuit is immersed. The electric field remains roughly orthogonal to the major axis of the helix and confines each charge carrier to a fraction of a turn of a single helical loop, moving it like water in an Archimedean screw. Each loop could in principle hold an independent carrier, permitting high information density. One computationally universal logic operation involves two helices, one of which splits into two "descendant" helices. At the point of divergence, differences in the electrostatic potential resulting from the presence or absence of a carrier in the adjacent helix controls the direction taken by a carrier in the splitting helix. The reverse of this sequence can be used to merge two initially distinct helical paths into a single outgoing helical path without forcing a dissipative transition. Because these operations are both logically and thermodynamically reversible, energy dissipation can be reduced to extremely low levels. ... It is important to note that this proposal permits a single electron to switch another single electron, and does not require that many electrons be used to switch one electron. The energy dissipated per logic operation can likely be reduced to less than 10^{-27} joules at a temperature of 1 Kelvin and a speed of 10 gigahertz, though further analysis is required to confirm this. Irreversible operations, when required, can be easily implemented and should have a dissipation approaching the fundamental limit of $\ln 2 \times kT$.

One study not conducted by Ames or partners is particularly worth mentioning since it places a loose lower bound on the computational capabilities of molecular nanotechnology. Drexler designed a number of computer components using small diamondoid rods with knobs that allow or prevent movement to accomplish computation. While this tiny mechanical Babbage Machine is probably not an optimal computational engine, its calculated performance for a desktop computer is 10^{18} MIPS — about a million times more powerful than the largest supercomputer that exists today. Note that with very fast computation energy use and heat dissipation become a severe problem. One approach to addressing this issue is reversible logic. Drexler proposed a nanotechnology based on diamond and investigated its potential properties. In particular, he examined applications for materials with a strength similar to that of diamond (69 times strength/mass of titanium). This would require a very mature nanotechnology constructing systems by placing atoms on diamond surfaces one or a few at a time in parallel. Assuming diamondoid materials, McKendree predicted the performance of several existing single-stage-to-orbit (SSTO) vehicle designs. The predicted payload to dry mass ratio for these vehicles using titanium as a structural material varied from < 0 (the vehicle won't work) to 36 per cent, i.e., the vehicle weighs substantially more than the payload.

With hypothetical diamondoid materials the ratios varied from 243 per cent to 653 per cent, i.e., the payload weighs far more than the vehicle. Using a very simple cost model ($1000 per vehicle kilogram) sometimes used in the aerospace industry, he estimated the cost per kilogram launched to low-Earth-orbit for diamondoid structured vehicles should be $153-412. This would meet NASA's 2020 launch to orbit cost goals. Estimated costs for titanium structured vehicles varied from $16,000-59,000/kg. Although this cost model is probably adequate for comparison, the absolute costs are suspect. Drexler used a more speculative methodology to estimate that a four passenger SSTO weighing three tons including fuel could be built using a mature nanotechnology. Using McKendree's cost model, such a vehicle would cost about $60,000 to purchase — the cost of today's high-end luxury automobiles. These studies assumed a fairly advanced nanotechnology capable of building diamondoid materials. In the nearer term, it may be possible to develop excellent structural materials using carbon nanotubes. Carbon nanotubes have a Young's modulus of approximately one terapascal— comparable to diamond. Issacs and Pearson proposed a space elevator—a cable extending from the Earth's surface into space with a center of mass at geosynchronous altitude. If such a system could be built, it should be mechanically stable and vehicles could ascend and descend along the cable at almost any reasonable speed using electric power (actually generating power on the way down). The first incredibly difficult problem with building a space elevator is strength of materials. Maximum stress is at geosynchronous altitude so the cable must be thickest there and taper exponentially as it approaches Earth. Any potential material may be characterized by the taper factor—the ratio between the cable's radius at geosynchronous altitude and at the Earth's surface. For steel the taper factor is tens of thousands—clearly impossible. For diamond, the taper factor is 21.9 including a safety factor. Diamond is, however, brittle. Carbon nanotubes have a strength in tension similar to diamond, but bundles of these nanometer-scale radius tubes shouldn't propagate cracks nearly as well as the diamond tetrahedral lattice. Thus, if the considerable problems of developing a molecular nanotechnology capable of making nearly perfect carbon nanotube systems approximately 70,000 kilometers long can be overcome, the first serious problem of a transportation system capable of truly large scale transfers of mass to orbit can be solved. The next immense problem with space elevators is safety—how to avoid dropping thousands of kilometers of cable on Earth if the cable breaks. Active materials may help by monitoring and repairing small flaws in the cable and/or detecting a major failure and disassembling the cable into small elements. Drexler calculates that lightsails made of 20 nm aluminum in tension should achieve an outward acceleration of ~14 km/s per day at Earth orbit with no payload and minimal structural overhead. For comparison, the delta V from low Earth to geosynchronous orbit is 3.8 km/s. Lightsails generate thrust by reflecting sunlight. Tension is achieved by rotating the sail. The direction of thrust is normal to the sail and away from the Sun. By directing thrust along or against the velocity vector, orbits can be lowered or raised. This form of transportation requires no reaction mass and generates thrust continuously, although the instantaneous acceleration is small so sails cannot operate in an atmosphere and must be large for even moderate payloads. Today, the smallest feature size in production systems is about 250 nanometers — the smallest feature size in computer chips. Since atoms are an angstrom or so across and carbon nanotubes have a diameter as small as 0.7 nanometers, atomically precise molecular machines can be smaller than current MEMS devices by two to three orders of magnitude in each dimension, or six to nine orders of magnitude smaller in volume (and mass).

For example, the size of the kinesin motor, which transports material in cells, is 12 nm. Han computationally demonstrated that molecular gears fashioned from single-walled carbon nanotubes with benzyne teeth should operate well at 50-100 gigahertz. These gears are about two nanometers across. Han computationally demonstrated cooling the gears with an inert atmosphere. Srivastava simulated powering the gears using alternating electric fields generated by a single simulated laser. In this case, charges were added to opposite sides of the tube to form a dipole. To make active materials, a material might be filled with nano-scale sensors, computers, and actuators so the material can probe its environment, compute a response, and act. Although this document is concerned with relatively simple artificial systems, living tissue may be thought of as an active material. Living tissue is filled with protein machines which gives living tissue properties (adaptability, growth, self-repair, etc.) unimaginable in conventional materials. Active materials can theoretically be made entirely of machines. These are sometimes called swarms since they consist of large numbers of identical simple machines that grasp and release each other and exchange power and information to achieve complex goals. Swarms change shape and exert force on their environment under software control. Although some physical prototypes have been built, at least one patent issued, and many simulations run, swarm potential capabilities are not well analyzed or understood. We briefly discuss some concepts here. Michael proposes brick-shaped machines of various sizes that slide past each other to assume a variety of shapes. He has generated a large number of videos showing computer simulations of simple motions. Although his web site contains rather extravagant claims, this work has received a U.K. patent. Yim built a small swarm with macroscopic (size in inches) components called polypod, built a simulator of polypod, and programmed it to move in various ways to study locomotion. There are two brick shaped components in polypod, one of which has two prismatic joints linked by a revolute joint. The second component is a cubic connector with no mechanical motion.

Polypod is programmed by tables for each member of the swarm. Each member is programmed to move at various speeds in each degree of freedom for certain amounts of time. The swarm components are implicitly synchronized so there is no clock signal. Hall proposes a swarm with 10 micron dodecahedral components each with 12 arms that can move in and out, rotate a little, and grab and release each other. This concept is called the "utility fog." Hall estimates that the utility fog would have a density of 0.2, tensile strength of 1000 psi in action and 100,000 psi in a passive mode, and have a maximum shear rate of 100 km/second/meter. Bishop proposes a swarm consisting of 100 nanometer brick-shaped components that slide past each other to change shape. Globus proposes a swarm with two kinds of components — edges and nodes. The terms "node" and "edge" are chosen to correspond to those in graph theory. The roughly spherical nodes are capable of attaching to five edges (for a tetrahedral geometry with one free edge per node) and rotating each edge in pitch and yaw. The rod-like edges are capable of changing length, rotating around their long axis, and attaching/detaching to/from nodes. Component design, power distribution and control software are significant challenges for swarm development. Consider that with 10 micron components a cubic meter of swarm would contain about 10^{15} devices, each with an internal computer communicating with its neighbors to accomplish a global task. NASA's mission is divided into five enterprises: Mission to Planet Earth, Aeronautics, Human Exploration and Development of Space, Space Science, and Space Technology. We will examine some potential nanotechnology applications in each area. The Earth Observing System (EOS) will use satellites and other systems to gather data on the Earth's environment. The EOS data system will need to process and archive >terabyte per day for the indefinite future. Simply storing this quantity of data is a significant challenge—each day's data would fill about 1,000 DVD disks. With projected nanocomputer processing speeds of 10^{18} MIPS, a million calculations on each byte of one day's data would take one second on the desktop. Given a mature nanotechnology, it should be possible to build sensors in balloon-borne systems approximately the size of bacteria. With replication based manufacturing, these should be quite inexpensive. If the serious communication and control problems can be solved, one can imagine spreading billions of tiny lighter-than-air vehicles into the atmosphere to measure wind currents and atmospheric composition. A similar approach might be taken in the oceans—note that the oceans are full of floating microscopic living organisms that can sense and react to their environment. Smart dust might sense the environment, note the location via a GPS-like system, and store that information until close enough to a data-collection point to transfer the data to the outside world. The strength of materials and computational capabilities previously discussed for space transportation should also allow much more advanced aircraft. Stronger, lighter materials can obviously make aircraft with greater lift and range. More powerful computers are invaluable in the design stage and of great utility in advanced avionics. MEMS technology has been used to replace traditional large control structures on aircraft with large numbers of small MEMS controlled surfaces. This control system was used to operate a model airplane in a windtunnel. Nanotechnology should allow even finer control—finer control than exhibited by birds, some of which can hover in a light breeze with very little wing motion. Nanotechnology should also enable extremely small aircraft. A reasonably advanced nanotechnology should be able to make simple atomically precise materials under software control. If the control is at the atomic level, then the full range of shapes possible with a given material should be achievable. Aircraft construction requires complex shapes to accommodate aerodynamic requirements. With molecular nanotechnology, strong complex-shaped components might be manufactured by general purpose machines under software control. The aeronautics mission is responsible for launch vehicle development. Payload handling is an important function. Very efficient payload handling might be accomplished by a very advanced swarm. The sequence begins by placing each payload on a single large swarm located next to the shuttle orbiter. The swarm forms itself around the payloads and then moves them into the payload bay, arranging the payloads to optimize the center of gravity and other considerations.

The swarm holds the payload in place during launch and may even damp out some launch vibrations. On orbit, satellites can be launched from the payload bay by having the swarm give them a gentle push. The swarm can then be left in orbit, perhaps at a space station, and used for orbital operations. This scenario requires a very advanced swarm that can operate in an atmosphere and on orbit in a vacuum. Besides the many and obvious difficulties of developing a swarm for a single environment, this provides additional challenges. Note that a simpler swarm might be used for aircraft payload handling. Aerospace vehicles often require complex checkout procedures to insure safety and reliability. This is particularly true of reusable launch vehicles. A very advanced swarm with some special purpose appendages might be placed on a vehicle. It might then spread out over the vehicle and into all crevices to examine the state of the vehicle in great detail. Nanotechnology-enabled Earth-to-orbit transportation has the greatest potential to revolutionize human access to space by dropping the current $10,000 per pound cost of launch, but this was discussed above. Other less dramatic technologies include: Space structures with a long design life (such as space station modules) need high-reliability materials that do not degrade. Active materials might help. The machines monitor structural integrity at the sub-micrometer scale. When a portion of the material becomes defective, it could be disassembled and then correctly reassembled. It should be noted that bone works somewhat along these lines. It is constantly being removed and added by specialized cells.

To effect timely repairs, space stations require a large store of spare parts and tools that are rarely used. A mature nanotechnology might create a "matter compiler," a machine that converts raw materials into a wide variety of products under software control. Contemporary examples of very limited matter compilers are numerically controlled machines and polypeptide sequencers. With a substantially more capable nanotechnology-based matter compiler, a space station crew could simply make spare parts and tools as needed. The programs could be stored on-board or on the ground. New tools invented on Earth could be transferred as software to the station for manufacture. Once used, unneeded tools and broken parts could be ionized in a solar furnace, transferred using controlled magnetic fields, and the constituent atoms stored for later manufacture into new products.

An advanced nanotechnology might be able to build filters that dynamically modify themselves to attract the contaminant molecules detected by the air and water quality sensors. Once attached to the filter, the filter could in principle move the offending molecules to a molecular laboratory for modifications to useful or at least inert products. A swarm might implement such an active filter if it was able to dynamically manufacture proteins that could bind contaminant molecules. The protein and bound contaminant might then be manipulated by the swarm for transportation. With a sufficiently advanced nanotechnology it might even be possible to directly generate food by non-biological means. Then agriculture waste in a self-sufficient space colony could be converted directly to useful nutrition. Making this food attractive will be a major challenge.

Sleeping crew members in the shuttle experience considerable pain and sleep disruption when the reaction control system fires and they collide with the cabin walls. If crew members were connected to the walls by a swarm, the swarm could absorb most or all of the force before the crew member struck the wall. The swarm could then gradually return the crew member to center (without the oscillations associated with bungee cords) in preparation for the next firing.

For resupply, spacecraft docking is a frequent necessity in space station operations. When two spacecraft are within a few meters of each other, a swarm could extend from each, meet in the middle, and form a stable connection before gradually drawing the spacecraft together. A swarm could support space-suited astronauts in simulated partial-g environments by holding them up appropriately. The swarm moves in response to the astronaut's motion providing the appropriate simulation of partial or 0 gravities. Tools and other objects are also manipulated by the swarm to simulate non-standard gravity.

Active nanotechnology materials (see active materials) might enable construction of a skin-tight space suit covering the entire body except the head (which is in a more conventional helmet). The material senses the astronaut's motions and changes shape to accommodate it. This should eliminate or substantially reduce the limitations current systems place on astronaut range of motion.

In situ resource utilization is undoubtedly necessary for large scale colonization of the solar system. Asteroids are particularly promising for orbital use since many are in near Earth orbits. Moving asteroids into low Earth orbit for utilization poses a safety problem should the asteroid get out of control and enter the atmosphere. Very small asteroids can cause significant destruction. The 1908 Tunguska explosion, which was calculated to be a 60 meter diameter stony asteroid, leveled 2,200 km^2 of forest. Hills calculated that 4 meter diameter iron asteroids are near the threshold for ground damage. Both these calculations assumed high collision speeds. At a density of 7.7 g/cm^3 Babadzhanov, a 3 meter diameter asteroid should have a mass of about 110 tons. Rabinowitz estimates that there are about one billion ten meter diameter near Earth asteroids and there should be far more smaller objects. For colonization applications one would ideally provide the same radiation protection available on Earth. Each square meter on Earth is protected by about 10 tons of atmosphere. Therefore, structures orbiting below the van Allen belts would like 10 $tons/meter^2$ surface area shielding mass. This would dominate the mass requirements of any system and require one small asteroid for each 11 $meter^2$ of colony exterior surface area. A 10,000 person cylindrical space colony such as Lewis One with a diameter of almost 500 meters and a length of nearly 2000 meters would require a minimum of about 90,000 retrieval missions to provide the shielding mass. The large number of missions required suggests that a fully automated, replicating nanotechnology may be essential to build large low Earth orbit colonies from small asteroids. A nanotechnology swarm along with an atomically precise lightsail is a promising small asteroid retrieval system. Lightsail propulsion insures that no mass will be lost as reaction mass. The swarm can control the lightsail by shifting mass. When a target asteroid is found, the swarm spreads out over the surface to form a bag. The interface to the sail must be active to account for the rotation of the asteroid—which is unlikely to have an axis-of-rotation in the proper direction to apply thrust for the return to Earth orbit. The active interface is simply swarm elements that transfer between each other to allow the sail to stay in the proper orientation. Of course, there are many other possibilities for nanotechnology based retrieval vehicles.

Extraterrestrial materials brought into orbit could be fed into a high-temperature solar furnace and partially ionized. Magnetic fields might then be used to separate the nuclei. These are fed in appropriate quantities to a matter-compiler to build the products desired.

Several authors, including Freitas have speculated that a sufficiently advanced nanotechnology could examine and repair cells at the molecular level. Should this capability become available—presumably driven by terrestrial applications—the small size and advanced capabilities of such systems could be of great utility on long duration space flights and on self-sufficient colonies.Self-replicating systems permit efforts of great scope to be pursued economically. Adjusting the environment on another planet to suit the tastes of humans is one such undertaking. Heating and cooling can be achieved by (among many other methods) using space-based mirrors. Chemical modifications of the planetary surface and atmosphere can be achieved in relatively short periods by the use of self-replicating systems that absorb sunlight and raw materials, and convert them into the desired products. Much as plants changed the environment of the earth to what we see today, so self-replicating molecular manufacturing systems might more rapidly convert the environments of other planets. As interstellar trips might last many years, the ability to conserve supplies by maintaining some crew members in a suspended state would be useful. An extremely advanced nanotechnology might use molecular manipulations of each cell to provide (a) better methods of slowing or suspending the metabolic activity of crew members and (b) better methods of restoring metabolic activity to a normal state when the destination is reached. Molecular manufacturing should enable the creation of very precise mirrors. Unlike lightsail applications, telescope mirrors require a very precise and somewhat complicated shape. A swarm with special purpose appendages capable of bonding to the mirror might be able to achieve and maintain the desired shape. A very advanced nanotechnology would be capable of imaging and then removing the surface atoms of an extra-terrestrial sample. By removing successive surface layers the location of each atom in the sample might be recorded, destroying the sample in the process. This data could then be sent to Earth. Besides requiring a very advanced nanotechnology, there is a more fundamental—but not necessarily fatal—problem: as the outside layer of atoms is removed the next layer may rearrange itself so the sample is not necessarily perfectly recorded. Solar Power—Low Earth orbit spacecraft generally depend on solar cells and batteries for power.

For energy collection, molecular manufacturing can be used to make solar photovoltaic cells at least as efficient as those made in the laboratory today. Efficiencies can therefore be > 30 per cent. In space applications, a reflective optical concentrator need consist of little more than a curved aluminum shell < 100 nanometers thick (photovoltaic cells operate with higher efficiency at high optical power densities). A metal fin with a thickness of 100 nanometers and a conduction path length of 100 microns can radiate thermal energy at a power density as high as 1000 W/m^2 with a temperature differential from base to tip of < 1 K. Accordingly, solar collectors can consist of arrays of photovoltaic cells several microns in thickness and diameter, each at the focus of a mirror of ~100 micron diameter, the back surface of which serves as a ~100 micron diameter radiator. If the mean thickness of this system is ~1 micron, the mass is $\sim 10^{-3}$ kg/m^2 and the power per unit mass, at Earth's distance from the Sun, where the solar constant is ~1.4 kW/m^2, is > 10^5 W/kg." By comparison, the U.S. built Photovoltaic

Panel Module solar cells currently used on the Mir Space Station and planned for use on the International Space Station generate about 118 W/kg. Fuel Cells—A critical component in hydrogen/oxygen fuel cells is the PEM (Proton Exchange Membrane). This membrane must (a) permit the passage of protons while (b) blocking everything else. Present membranes do a rather poor job. One group at Ames is designing and computationally testing PEMs to study possible energy mechanisms in early life. While these studies are not meant to design optimal membranes for fuel cell use, the basic knowledge and approach may be of value. Another proposal is to design a diamond membrane a few nanometers thick with "proton pores." The pores might be lined with fluorine, oxygen and nitrogen to create a region with a high proton affinity. In addition, a positionally controlled platinum might be held at the mouth of the pore to verify that H_2 can be catalytically split into H+ and e–, and that the barrier for migration of the H+ into the pore is modest in size. Nanotechnology must provide precise control over the manufacturing process of the diamondoid PEM since the pores must be made very precisely. Studies of H_2 absorption and packing in carbon nanotubes and nanoropes are in progress at NASA Ames and elsewhere. Nanotubes provide large pore sizes and nanoropes have different pore sizes depending on interstitial and other locations. Dillon estimated that the single walled nanotubes in their sample contained 5 to 10 per cent by weight of H_2.

The nanotubes were about 0.1 to 0.2 per cent by weight of the total sample. Computational studies at Ames suggest that to store 7-10 per cent H_2 in single walled nanotubes at room temperature the H_2s must be stored inside the tubes, not merely adsorbed on the walls. This work suggests that carbon nanotubes might be developed into an excellent H_2 storage medium within 3-5 years. Calculations with oxygen suggest that a diamondoid sphere ~0.1 microns in diameter should easily hold oxygen at ~1,000 atmospheres. While higher pressures are feasible, they offer declining returns. At higher pressures, the pressure-volume relationship becomes severely non-linear and the density approaches a limiting value. Other gases might also be stored if diamondoid spheres can be built, but the analysis has not been done. High strength light-weight materials will allow greater efficiency of energy storage as angular momentum. Many kinds of ultraminiature electromechanical devices have utility on a miniaturized space craft. It has been shown that manipulating carbon nanotubes changes their electrical properties. This might be exploited to build nanometer scale strain devices. This may be achievable within 3-5 years, and simulations along these lines are in progress. Similar results have been achieved experimentally with C_{60}. The electrical properties of a C_{60} molecule were changed by applying pressure to the molecule with an SPM tip. Smaller, lighter spacecraft are cheaper to launch (current costs are about $10,000/lb) and generally cheaper to build. Diamondoid structural materials can radically reduce structural mass, miniaturized electronics can shrink the avionics and reduce power consumption, and atomically precise materials and components should shrink most other subsystems. Thermal protection is crucial for atmospheric reentry and other tasks. The carbon nanotubes under investigation at

NASA Ames and elsewhere may play a significant role. Most production processes for carbon nanotubes create a tangled mat of nanotubes that has a very low mass-to-volume ratio. Like graphite, the tubes should withstand high temperatures but the tangled mat should prevent them from ablating. This may lead to high temperature applications. Many of the applications discussed here are speculative to say the least. However, they do not appear to violate the laws of physics.

Something similar to these applications at these performance levels should be feasible if we can gain complete control of the three-dimensional structure of materials, processes and devices at the atomic scale. How to gain such control is a major, unresolved issue. However, it is clear that computation will play a major role regardless of which approach —positional control with replication, self-assembly, or some other means —is ultimately successful. Computation has already played a major role in many advances in chemistry, SPM manipulation, and biochemistry. As we design and fabricate more complex atomically precise structures, modeling and computer aided design will inevitably play a critical role. Not only is computation critical to all paths to nanotechnology, but for the most part the same or similar computational chemistry software and expertise supports all roads to molecular nanotechnology. Thus, even if NASA's computational molecular nanotechnology efforts should pursue an unproductive path, the expertise and capabilities can be quickly refocused on more promising avenues as they become apparent. As nanotechnology progresses we may expect applications to become feasible at a slowly increasing rate. However, if and when a general purpose programmable assembler/replicator can be built and operated, we may expect an explosion of applications. From this point, building new devices will become a matter of developing the software to instruct the assembler/replicators. Development of a practical swarm is another potential turning point. Once an operational swarm that can grow and divide has been built, a large number of applications become software projects. It is also important to note that the software for swarms and assembler/replicators can be developed using simulators — even before operational devices are available.

Nanotechnology advocates and detractors are often preoccupied with the question "When?" There are three interrelated answers to this question:

(a) Nobody knows. There are far too many variables and unknowns. Beware of those who have excessive confidence in any date.

(b) The time-to-nanotechnology will be measured in decades, not years. While a few applications will become feasible in the next few years, programmable assembler/replicators and swarms will be extremely difficult to develop.

(c) The time-to-nanotechnology is very sensitive to the level of effort expended. Resources allocated to developing nanotechnology are likely to be richly rewarded, particularly in the long term.

Application of Superconductors

An element, inter-metallic alloy, or compound that will conduct electricity without resistance below a certain temperature. Resistance is undesirable because it produces losses in the energy flowing through the material. Once set in motion, electric current will flow forever in a closed loop of superconducting material – making it the closest thing to perpetual motion in nature. Scientists refer to superconductivity as a "microscopic quantum phenomenon". Superconductivity is a phenomenon which occurs in certain materials and is characterized by the absence of electrical resistivity. Until recently, this phenomenon had been restricted to metals and alloys with transition temperatures of less than 23K. In 1986, superconductivity was discovered in a ceramic material. This precipitated an onrush of ceramic based superconductors with transition temperatures as high as 120K. The ceramic-based materials are commonly known as high temperature (HTc) superconductors while the metallic and alloy materials are called low temperature (LTc) superconductors.

Currently, only low temperature superconductors are of interest to the magnet designer and manufacturer. Superconductors are divided into two types depending on their characteristic behavior in the presence of a magnetic field. Type I superconductors are comprised of pure metals, whereas Type II superconductors are comprised primarily of alloys or intermetallic compounds. Both, however, have one common feature: below a critical temperature, T_c, their resistance vanishes. The critical temperature at which the resistance vanishes in a superconductor is reduced when a magnetic field is applied. The maximum field that can be applied to a superconductor at a particular temperature and still maintain superconductivity is call the critical field, or H_c. This field varies enormously between Type I and Type II superconductors. The maximum critical field (H_c) in any Type I superconductor is about 2000 Gauss (0.2 Tesla), but in Type II materials superconductivity can persist to several hundred thousand Gauss (H_{c2}). At fields greater than H_c in a Type I superconductor and greater than H_{c2} in a Type II superconductor, the conductor reverts to the normal state and regains its normal state resistance. A Type I superconductor excludes the applied magnetic field from the center of the sample by establishing circulating currents on its surface that counteract the applied field. Type II superconductors, however, permit the field to penetrate through the sample in quantized amounts of flux. These quanta are comprised of circulating vortices of current and the flux contained in the vortices. The total flux in a vortex is 2×10^{-7} Gauss-cm^2. Great numbers of these vortices, or fluxoids as they are frequently called, can exist in a superconductor. For example, at a field intensity of 80 kilogauss (8 Tesla) there are 4×10^{11} fluxoids/cm^2. These fluxoids and their interactions with defects in the superconductor give rise to the high current carrying capabilities of superconducting magnets. Properties of superconducting materials are altered locally by the presence of defects in the materials.

A fluxoid encompassing or adjacent to such a defect in the material has its energy altered and its free motion through the superconductor is inhibited. This phenomenon, known as flux pinning, causes a field gradient in the superconductor and gives rise to a net current in the material. In the absence of defects in a Type II superconductor, no bulk current can be conducted without a transition into the normally conducting resistive state. Since the pinning force is small, fluxoids can be broken loose from their pinning centers, resulting in a net creep of the flux through a conductor as a function of time. This results in an effective voltage in a Type II superconductor. If the current density is low and the magnetic field is not intense, flux creep is insignificant and the induced voltage and effective resistance of the conductor will be essentially zero. At very high fields and high current densities, fluxoids will migrate rapidly, giving rise to a phenomenon called flux flow.

The maximum pinning force and the maximum current density sustainable by a conductor increases with reduced temperature. Similarly, as the magnetic induction, B, increases, the sustainable current density decreases. Consequently, the current density, magnetic field, and critical temperature are all interdependent. By increasing any of these parameters to a sufficiently high value, superconductivity can be destroyed and the conductor will revert to a normal, non-superconducting state. One of the most crucial factors in determining the performance of the final magnet is the design of the conductor. This design affects the ultimate field achieved by the magnet, the rate at which the magnet can be energized and the drift rate in the persistent mode of operation. Several phenomena observed in magnets are caused by the conductor itself. One of the earliest phenomena observed in superconducting magnets wound with single filament conductors was flux jumping. This phenomena arises from current induced in the conductor by the presence of the transverse field generated by the magnet. If a superconductor is placed transverse to the magnetic field, currents are induced in the conductor which shield the bulk of the conductor from the external magnetic field. These circulating currents extend for a finite length along the conductor, flowing in one direction on one side of the conductor and returning on the other side to complete the circuit. If the heat produced by the penetration of the magnet field into the superconductor cannot escape to the surface rapidly enough, then the temperature increase inside the superconductor triggers a runaway condition known as a flux jump. Consequently, the conductor is driven into the resistive state at fields and currents low in comparison with the critical values. Resistive heat is then dissipated in the small normal zone which increases in temperature causing the normal zone to expand and propagate both along the length of the conductor and transverse to it. This results in the magnet being discharged as the energy in the magnet is dissipated in the resistive portion of the conductor. To decrease the problem of transition to the normal state, it is common practice to shunt the conductor with a low resistivity normal metal by embedding the superconductor in a copper matrix to form a composite conductor. The copper provides additional heat capacity as well as providing a path for the magnet current while the superconductor is driven normal during a flux jump. If the resistance of the copper is low enough, the temperature of the conductor can remain below the critical temperature at the ambient field, and superconductivity will resume after the currents in the superconductor have decayed. Embedding the superconductor in a low resistivity metal matrix is effective in reducing the chance of a flux jump which can cause a magnet quench. Magnets constructed with this type of material are dissipative and the heat generated during a flux jump must be conducted to the helium bath. Thus, the magnetic field must be changed slowly to allow time for the heat to be conducted to and dissipated in the liquid helium. Also, the diamagnetic currents in the superconductor contribute to the field generated by the magnet and can reduce its homogeneity.

If, instead of one superconducting filament, many fine filaments of superconductor are used, the heat generated in individual filaments can easily be conducted a short distance to the filament surface thus avoiding flux jumps. Consequently, conductors are made in which many fine filaments of superconductor are coextruded and drawn in a matrix of either copper or aluminum stabilizers. Although this has the desired effect of avoiding flux jumping, circulating currents can again be formed if the conductors are parallel in the highly conductive normal matrix. In this case, the circulating current is between two or more filaments in parallel and the current crosses over through the normally conductive matrix. This gives rise to diamagnetism and unequal distribution of currents in the filaments that limits the rate at which the magnet can be charged. Problems arising from constructing the superconductor from filaments have been largely circumvented in modern conductors by twisting the filaments in the conductor. This causes the flux from the external magnetic field to be alternated through successive superconducting filaments, thereby reducing the unequal distribution of currents between the superconducting filaments and reducing the diamagnetism of the conductor. This reduction in diamagnetism or hysteresis has two desirable effects. First, it reduces the amount of energy dissipated in the magnet and permits it to be charged more rapidly. Secondly, the reduced diamagnetism causes the current in the magnet and the magnetic field generated by the magnet to be more linearly related. Such conductors are known as intrinsically stabilized conductors.

Superconductors, materials that have no resistance to the flow of electricity, are one of the last great frontiers of scientific discovery. Not only have the limits of superconductivity not yet been reached, but the theories that explain superconductor behavior seem to be constantly under review. In 1911 superconductivity was first observed in mercury by Dutch physicist Heike Kamerlingh Onnes of Leiden University (shown above). When he cooled it to the temperature of liquid helium, 4 degrees Kelvin (-452F, -269C), its resistance suddenly disappeared. The Kelvin scale represents an "absolute" scale of temperature. Thus, it was necessary for Onnes to come within 4 degrees of the coldest temperature that is theoretically attainable to witness the phenomenon of superconductivity. Later, in 1913, he won a Nobel Prize in physics for his research in this area. The next great milestone in understanding how matter behaves at extreme cold temperatures occurred in 1933. German researchers Walter Meissner (above) and Robert Ochsenfeld discovered that a superconducting material will repel a magnetic field (below graphic). A magnet moving by a conductor induces currents in the conductor. This is the principle upon which the electric generator operates. But, in a superconductor the induced currents exactly mirror the field that would have otherwise penetrated the superconducting material—causing the magnet to be repulsed. This phenomenon is known as strong diamagnetism and is today often referred to as the "Meissner effect" (an eponym). The Meissner effect is so strong that a magnet can actually be levitated over a superconductive material. In subsequent decades other superconducting metals, alloys and compounds were discovered. In 1941 niobium-nitride was found to superconduct at 16 K. In 1953 vanadium-silicon displayed superconductive properties at 17.5 K. And, in 1962 scientists at Westinghouse developed the first commercial superconducting wire, an alloy of niobium and titanium (NbTi). High-energy, particle-accelerator electromagnets made of copper-clad niobium-titanium were then developed in the 1960s at the Rutherford-Appleton Laboratory in the UK, and were first employed in a superconducting accelerator at the Fermilab Tevatron in the US in 1987. The first widely-accepted theoretical understanding of superconductivity was advanced in 1957 by American physicists John Bardeen, Leon Cooper, and John Schrieffer (above). Their *Theories of Superconductivity* became know as the BCS theory - derived from the first letter of each man's last name—and won them a Nobel prize in 1972. The mathematically-complex BCS theory explained superconductivity at temperatures close to absolute zero for elements and simple alloys. However, at higher temperatures and with different superconductor systems, the BCS theory has subsequently become inadequate to fully explain how superconductivity is occurring. Another significant theoretical advancement came in 1962 when Brian D. Josephson (above), a graduate student at Cambridge University, predicted that electrical current would flow between 2 superconducting materials - even when they are separated by a non-superconductor or insulator. His prediction was later confirmed and won him a share of the 1973 Nobel Prize in Physics. This tunneling phenomenon is today known as the "Josephson effect" and has been applied to electronic devices such as the SQUID, an instrument capabable of detecting even the weakest magnetic fields. (Below SQUID graphic courtesy Quantum Design.) The 1980's were a decade of unrivaled discovery in the field of superconductivity. In 1964 Bill Little of Stanford University had suggested the possibility of organic (carbon-based) superconductors. The first of these theoretical superconductors was successfully synthesized in 1980 by Danish researcher Klaus Bechgaard of the University of Copenhagen and 3 French team members. $(TMTSF)_2PF_6$ had to be cooled to an incredibly cold 1.2K transition temperature (known as Tc) and subjected to high pressure to superconduct. But, its mere existence proved the possibility of "designer" molecules - molecules fashioned to perform in a predictable way.

Then, in 1986, a truly breakthrough discovery was made in the field of superconductivity. Alex Müller and Georg Bednorz (above), researchers at the IBM Research Laboratory in Rüschlikon, Switzerland, created a brittle ceramic compound that superconducted at the highest temperature then known: 30 K. What made this discovery so remarkable was that ceramics are normally insulators. They don't conduct electricity well at all. So, researchers had not considered them as possible high-temperature superconductor candidates. The Lanthanum, Barium, Copper and Oxygen compound that Müller and Bednorz synthesized, behaved in a not-as-yet-understood way. The discovery of this first of the superconducting copper-oxides (cuprates) won the 2 men a Nobel Prize the following year. It was later found that tiny amounts of this material were actually superconducting at 58 K, due to a small amount of lead having been added as a calibration standard - making the discovery even more noteworthy. Müller and Bednorz' discovery triggered a flurry of activity in the field of superconductivity. Researchers around the world began "cooking" up ceramics of every imaginable combination in a quest for higher and higher Tc's. In January of 1987 a research team at the University of Alabama-Huntsville substituted Yttrium for Lanthanum in the Müller and Bednorz molecule and achieved an incredible 92 K Tc. For the first time a material (today referred to as YBCO) had been found that would superconduct at temperatures warmer than liquid nitrogen - a commonly available coolant. Additional milestones have since been achieved using exotic—and often toxic—elements in the base perovskite ceramic. The current class (or "system") of ceramic superconductors with the highest transition temperatures are the mercuric-cuprates. The first synthesis of one of these compounds was achieved in 1993 by Prof. Dr. Ulker Onbasli at the University of Colorado and by the team of A. Schilling, M. Cantoni, J.D. Guo, and H.R. Ott of Zurich, Switzerland. The world record Tc of 138 K is now held by a thallium-doped, mercuric-cuprate comprised of the elements Mercury, Thallium, Barium, Calcium, Copper and Oxygen. The Tc of this ceramic superconductor was confirmed by Dr. Ron Goldfarb at the National Institute of Standards and Technology-Colorado in February of 1994. Under extreme pressure its Tc can be coaxed up even higher - approximately 25 to 30 degrees more at 300,000 atmospheres. The first company to capitalize on high-temperature superconductors was Illinois Superconductor (today known as ISCO International), formed in 1989. This amalgam of government, private-industry and academic interests introduced a depth sensor for medical equipment that was able to operate at liquid nitrogen temperatures (~ 77K). While no significant advancements in superconductor Tc's have been achieved in recent years, other discoveries of equal importance have been made. In 1997 researchers discovered that at a temperature very near absolute zero an alloy of gold and indium was both a superconductor and a natural magnet. Conventional wisdom held that a material with such properties could not exist! Since then, over a half-dozen such compounds have been found. Recent years have also seen the discovery of the first high-temperature superconductor that does NOT contain any copper (2000), and the first all-metal perovskite superconductor (2001). Also in 2001 a material that had been sitting on laboratory shelves for decades was found to be an extraordinary new superconductor. Japanese researchers measured the transition temperature of magnesium diboride at 39 Kelvin—far above the highest Tc of any of the elemental or binary alloy superconductors. While 39 K is still well below the Tc's of the "warm" ceramic superconductors, subsequent refinements in the way MgB_2 is fabricated have opened the door to its use in industrial applications.

Laboratory testing has found MgB_2 will outperform NbTi and Nb_3Sn wires in high magnetic field applications like MRI. Startling—and serendipitous—discoveries like these are forcing scientists to continually re-examine longstanding theories on superconductivity and to consider heretofore-unimagined combinations of elements.

Magnetic-levitation is an application where superconductors perform extremely well. Transport vehicles such as trains can be made to "float" on strong superconducting magnets, virtually eliminating friction between the train and its tracks. Not only would conventional electromagnets waste much of the electrical energy as heat, they would have to be physically much larger than superconducting magnets. A landmark for the commercial use of MAGLEV technology occurred in 1990 when it gained the status of a nationally-funded project in Japan. The Minister of Transport authorized construction of the Yamanashi Maglev Test Line which opened on April 3, 1997. In December 2003, the MLX01 test vehicle (shown above) attained an incredible speed of 361 mph (581 kph).

Although the technology has now been proven, the wider use of MAGLEV vehicles has been constrained by political and environmental concerns (strong magnetic fields can create a bio-hazard). The world's first MAGLEV train to be adopted into commercial service, a shuttle in Birmingham, England, shut down in 1997 after operating for 11 years. A Sino-German maglev is currently operating over a 30 km course at Pudong International Airport in Shanghai, China. The U.S. plans to have its first (non-superconducting) Maglev train in operation this year on a Virginia college campus. Click this link for a website that lists other uses for MAGLEV. An area where superconductors can perform a life-saving function is in the field of biomagnetism. Doctors need a non-invasive means of determining what's going on inside the human body. By impinging a strong superconductor-derived magnetic field into the body, hydrogen atoms that exist in the body's water and fat molecules are forced to accept energy from the magnetic field. They then release this energy at a frequency that can be detected and displayed graphically by a computer. Magnetic Resonance Imaging (MRI) was actually discovered in the mid 1940's. But, the first MRI exam on a human being was not performed until July 3, 1977. And, it took almost five hours to produce one image! Today's faster computers process the data in much less time.

The Korean Superconductivity Group within KRISS has carried biomagnetic technology a step further with the development of a double-relaxation oscillation SQUID (Superconducting quantum Interference Device) for use in Magnetoencephalography. SQUID's are capable of sensing a change in a magnetic field over a billion times weaker than the force that moves the needle on a compass (compass: 5e-5T, SQUID: e-14T.). With this technology, the body can be probed to certain depths without the need for the strong magnetic fields associated with MRI's. Probably the one event, more than any other, that has been responsible for putting "superconductors" into the American lexicon was the Superconducting Super-Collider project planned for construction in Ellis county, Texas. Though Congress cancelled the multi-billion dollar effort in 1993, the concept of such a large, high-energy collider would never have been viable without superconductors. High-energy particle research hinges on being able to accelerate sub-atomic particles to nearly the speed of light. Superconductor magnets make this possible. CERN, a consortium of several European nations, is doing something similar with its Large Hadron Collider (LHC) now under construction along the Franco-Swiss border.

Other related web sites worth visiting include the proton-antiproton collider page at Fermilab. This was the first facility to use superconducting magnets. Get information on the electron-proton collider HERA at the German lab pages of DESY (with English text). Lastly, Brookhaven National Laboratory features a page dedicated to its RHIC heavy-ion collider.

Electric generators made with superconducting wire are far more efficient than conventional generators wound with copper wire. In fact, their efficiency is above 99 per cent and their size about half that of conventional generators. These facts make them very lucrative ventures for power utilities. General Electric has estimated the potential worldwide market for superconducting generators in the next decade at around $20-30 billion dollars. Late in 2002 GE Power Systems received $12.3 million in funding from the U.S. Department of Energy to move high-temperature superconducting generator technology toward full commercialization. Other commercial power projects in the works that employ superconductor technology include energy storage to enhance power stability. American Superconductor Corp. received an order from Alliant Energy in late March 2000 to install a Distributed Superconducting Magnetic Energy Storage System (D-SMES) in Wisconsin. Just one of these 6 D-SMES units has a power reserve of over 3 million watts, which can be retrieved whenever there is a need to stabilize line voltage during a disturbance in the power grid. AMSC has also installed more than 22 of its D-VAR systems to provide instantaneous reactive power support. Recently, power utilities have also begun to use superconductor-based transformers and "fault limiters". The Swiss-Swedish company ABB was the first to connect a superconducting transformer to a utility power network in March of 1997. ABB also recently announced the development of a 6.4MVA (mega-volt-ampere) fault current limiter - the most powerful in the world. This new generation of HTS superconducting fault limiters is being called upon due to their ability to respond in just thousandths of a second to limit tens of thousands of amperes of current. Advanced Ceramics Limited is another of several companies that makes BSCCO type fault limiters. Intermagnetics General recently completed tests on its largest (15kv class) power-utility-size fault limiter at a Southern California Edison (SCE) substation near Norwalk, California. And, both the US and Japan have plans to replace underground copper power cables with superconducting BSCCO cable-in-conduit cooled with liquid nitrogen. By doing this, more current can be routed through existing cable tunnels. In one instance 250 pounds of superconducting wire replaced 18,000 pounds of vintage copper wire, making it over 7000 per cent more space-efficient.

An idealized application for superconductors is to employ them in the transmission of commercial power to cities. However, due to the high cost and impracticality of cooling miles of superconducting wire to cryogenic temperatures, this has only happened with short "test runs". In May of 2001 some 150,000 residents of Copenhagen, Denmark, began receiving their electricity through HTS (high-temperature superconducting) material. That cable was only 30 meters long, but proved adequate for testing purposes. In the summer of 2001 Pirelli completed installation of three 400-foot HTS cables for Detroit Edison at the Frisbie Substation capable of delivering 100 million watts of power. This marked the first time commercial power has been delivered to customers of a US power utility through superconducting wire. Intermagnetics General has announced that its IGC-SuperPower subsidiary has joined with BOC and Sumitomo Electric in a $26 million project to install an underground, HTS power cable in Albany, New York, in Niagara Mohawk Power Corporation's power grid. The 350-meter cable, believed to be four times the length of any previously installed HTS cable, will be designed to provide more power and operate at significantly lower loss levels than other HTS installations.

The National Science Foundation, along with NASA and DARPA and various universities, are currently researching "petaflop" computers. A petaflop is a thousand-trillion floating point operations per second. Today's fastest computing operations have only reached "teraflop" speeds - trillions of operations per second. Currently the fastest is the IBM Blue Gene/L running at 70.7 teraflops per second (multiple CPU's). The fastest single processor is a Lenslet optical DSP running at 8 teraflops. It has been conjectured that devices on the order of 50 nanometers in size along with unconventional switching mechanisms, such as the Josephson junctions associated with superconductors, will be necessary to achieve such blistering speeds. TRW researchers (now Northrop Grumman) have quantified this further by predicting that 100 billion Josephson junctions on 4000 microprocessors will be necessary to reach 32 petabits per second. These Josephson junctions are incorporated into field-effect transistors which then become part of the logic circuits within the processors. Recently it was demonstrated at the Weizmann Institute in Israel that the tiny magnetic fields that penetrate Type 2 superconductors can be used for storing and retrieving digital information.

It is, however, not a foregone conclusion that computers of the future will be built around superconducting In the electronics industry, ultra-high-performance filters are now being built. Since superconducting wire has near zero resistance, even at high frequencies, many more filter stages can be employed to achive a desired frequency response. This translates into an ability to pass desired frequencies and block undesirable frequencies in high-congestion rf (radio frequency) applications such as cellular telephone systems. ISCO International and Superconductor Technologies are companies currently offering such filters. Superconductors have also found widespread applications in the military. HTSC SQUIDS are being used by the US NAVY to detect mines and submarines. And, significantly smaller motors are being built for NAVY ships using superconducting wire and "tape". In mid-July, 2001, American Superconductor unveiled a 5000-horsepower motor made with superconducting wire (below).

The military is also looking at using superconductive tape as a means of reducing the length of very low frequency antennas employed on submarines. Normally, the lower the frequency, the longer an antenna must be. However, inserting a coil of wire ahead of the antenna will make it function as if it were much longer. Unfortunately, this loading coil also increases system losses by adding the resistance in the coil's wire. Using superconductive materials can significantly reduce losses in this coil. The Electronic Materials and Devices Research Group at University of Birmingham (UK) is credited with creating the first superconducting microwave antenna. Applications engineers suggest that superconducting carbon nanotubes might be an ideal nano-antenna for high-gigahertz and terahertz frequencies, once a method of achieving zero "on tube" contact resistance is perfected. The most ignominious military use of superconductors may come with the deployment of "E-bombs". These are devices that make use of strong, superconductor-derived magnetic fields to create a fast, high-intensity electro-magnetic pulse (EMP) to disable an enemy's electronic equipment. Such a device saw its first use in wartime in March 2003 when US Forces attacked an Iraqi broadcast facility.

Among emerging technologies are a stabilizing momentum wheel (gyroscope) for earth-orbiting satellites that employs the "flux-pinning" properties of imperfect superconductors to reduce friction to near zero. Superconducting x-ray detectors and ultra-fast, superconducting light detectors are being developed due to their inherent ability to detect extremely weak amounts of energy. Already Scientists at the European Space Agency (ESA) have developed what's being called the S-Cam, an optical camera of phenomenal sensitivity. And, superconductors may even play a role in Internet communications soon. In late February, 2000, Irvine Sensors Corporation received a $1 million contract to research and develop a superconducting digital router for high-speed data communications up to 160 Ghz. Since Internet traffic is increasing exponentially, superconductor technology is being called upon to meet this super need.

According to June 2002 estimates by the Conectus consortium, the worldwide market for superconductor products is projected to grow to near US $5 billion by the year 2010 and to US $38 billion by 2020. Low-temperature superconductors are expected to continue to play a dominant role in well-established fields such as MRI and scientific research, with high-temperature superconductors enabling the newer industries. The above ISIS graph gives a rough breakdown of the various markets in which superconductors are expected to make a contribution. All of this is, of course, contingent upon a linear growth rate. Should new superconductors with higher transition temperatures be discovered, growth and development in this exciting field could explode virtually overnight. Another impetus to the wider use of superconductors is political in nature. The reduction of green-house gas (GHG) emissions has becoming a topical issue due to the Kyoto Protocol which requires the European Union (EU) to reduce its emissions by 8 per cent from 1990 levels by 2012. Physicists in Finland have calculated that the EU could reduce carbon dioxide emissions by up to 53 million tons if high-temperature superconductors were used in power plants. The future melding of superconductors into our daily lives will also depend to a great degree on advancements in the field of cryogenic cooling. New, high-efficiency magnetocaloric-effect compounds such as gadolinium-silicon-germanium are expected to enter the marketplace soon. Such materials should make possible compact, refrigeration units to facilitate additional HTS applications.

Laboratories throughout the world are rapidly gaining atomically precise control over matter. As this control extends to an ever wider variety of materials, processes and devices, opportunities for applications relevant to NASA's missions will be created. This document surveys a number of future molecular nanotechnology capabilities of aerospace interest. Computer applications, launch vehicle improvements, and active materials appear to be of particular interest. We also list a number of applications for each of NASA's enterprises. If advanced molecular nanotechnology can be developed, almost all of NASA's endeavors will be radically improved. In particular, a sufficiently advanced molecular nanotechnology can arguably bring large scale space colonization within our grasp.

This study describes potential aerospace applications of molecular nanotechnology, defined as the thorough three-dimensional structural control of materials, processes and devices at the atomic scale. The inspiration for molecular nanotechnology comes from Richard P. Feynman's 1959 visionary talk at Caltech in which he said, "The problems of chemistry and biology can be greatly helped if our ability to see what we are doing, and to do things on an atomic level, is ultimately developed—a development which I think cannot be avoided." Indeed, scanning probe microscopes (SPMs) have already given us this ability in limited domains. Synthetic chemistry, biotechnology, "laser tweezers" and other developments are also bringing atomic precision to our endeavors. An expanded version of Drexler's MIT PhD thesis, examines one vision of molecular nanotechnology in considerable technical detail. It proposes the development of programmable molecular assembler/replicators. These are atomically precise machines that can make and break chemical bonds using mechanosynthesis to produce a wide variety of products under software control, including copies of themselves. Interestingly, living cells exhibit many properties of assembler/replicators. Cells make a wide variety of products, including copies of themselves, and can be programmed with DNA. Replication is one approach to building large systems, such as human rated launch vehicles, from molecular machines manipulating matter one or a few atoms at a time. Note that biological replication is responsible for systems as large as redwood trees and whales.

Another approach to nanotechnology is supramolecular self-assembly, where molecular systems are designed to attract each other in a particular orientation to form larger systems. Hollow spheres large enough to be visible in a standard light microscope have been created this way using self-assembling lipids. There are many other examples and this field is rapidly advancing. Biological systems can do most of what molecular nanotechnology strives to accomplish — atomically precise products, active materials, reproduction, etc. However, biological systems are extremely complex and molecular nanotechnology seeks simpler systems to understand, control and manufacture. Also, biological systems usually work at fairly mild temperature and pressure conditions in solution — conditions that are not found in most aerospace environments. Today, extremely precise atomic and molecular manipulation is common in many laboratories around the world and our abilities are rapidly approaching Feynman's dream. The implications for aerospace development are profound and ubiquitous. A number of applications are mentioned here and a few are described in some detail with references. From this sample of applications it should be clear that although molecular nanotechnology is a long term, high risk project, the payoff is potentially enormous — vastly superior computers, aerospace transportation, sensors and other technologies; technologies that may enable large scale space exploration and colonization.

This chapter is organized into two sections. In the first, we examine three technologies—computers, aerospace transportation, and active materials—useful to nearly all NASA missions. In the second, we investigate some potential molecular nanotechnology payoffs for each area identified in NASA's strategic plan. Some of these applications are under investigation by nanotechnology researchers at NASA Ames. Some of the applications described below have relatively near-term potential and working prototypes may be realized within three to five years. This is certainly not true in other cases. Indeed, many of the possible applications of nanotechnology that we describe here are, at the present time, rather speculative and futuristic. However, each of these ideas have been examined at least cursorily by competent scientists, and as far as we know all of them are within the bounds of known physical laws. We are not suggesting that their achievement will be easy, cheap or near-term. Some may take decades to realize; some other ideas may be scrapped in the coming years as insuperable barriers are identified. But we feel that they are worth mentioning here as illustrations of the potential future impact of nanotechnology.

The applicability of manufacturing at an ever smaller scale is nowhere more self-evident than in computer technology. Indeed, Moore's law (an observation not a physical law) says that computer chip feature size decreases exponentially with time, a trend that predicts atomically precise computers by about 2010-2015. This capability is being approached from many directions. Here we will concentrate on those under development by NASA Ames and her partners.

Carbon nanotubes can be viewed as rolled up sheets of graphite from 0.7 to many nanometers in diameter. The smaller tubes are single molecules. Dai placed carbon nanotubes on an SPM tip thus extending our ability to manipulate a single molecule with sub-angstrom accuracy. Not only are the tips atomically precise, but they should have approximately the same chemistry as C_{60}, and thus be functionalizable with a wide variety of molecular fragments. Functionalizing carbon nanotube tips will allow mechanical manipulation of many molecular systems on various surfaces with sub-angstrom accuracy. One particularly intriguing possibility along this line is to utilize a carbon nanotube SPM tip to engrave patterns on a silicon surface. It should be possible to create features a few nanometers across. These would be perhaps 100 times finer than the current state of the art in commercial semiconductor photolithography. Further, in contrast to approaches such as electron microscope lithography for which the speed of operation now appears to be an insuperable obstacle for industrial production, nanotube SPM-based lithography can be accelerated by utilizing an array with thousands of SPM tips simultaneously engraving different parts of a silicon surface. Also, nanotube SPM lithography could provide a practical means to explore various futuristic electronic device technology ideas, such as quantum cellular automata, which require exceedingly small feature sizes. Needless to say, if these ideas pan out, they could literally revolutionize computer device technology, paving the way for systems that are many times more powerful and more compact than any available today. For the near term, it should be noted that the semiconductor industry is a major market for SPM products. These are used to examine production equipment. High performance carbon nanotube tips should be of substantial value. NASA Ames is collaborating with Dr. Dai, now at Stanford, to develop these tips.

It is possible to store data on long chain molecules (for example, DNA) and it may be possible to read these data with carbon nanotube tipped SPMs. Existing DNA synthesis techniques can be used to write data. If the different DNA base pairs can be distinguished with a carbon nanotube tipped SPM, then the data can be read non-destructively (current techniques allow a destructive read). However, the difference between base pairs is not great. If the base pairs cannot be distinguished, techniques for attaching modified enzymes to specific base pair sequences could be used. Certain enzymes (DNA (cytosine-5) methyltransferases) attach themselves onto a specific sequence of base pairs with a covalent bond. The enzyme then performs its operation and breaks the bond. Smith modified the enzyme such that the initial covalent bond was formed but the subsequent operation was disrupted. The result is that DNA synthesized with the target base pair sequences at the desired location can force precise placement of the enzymes. The presence of an enzyme could represent 1 and its absence 0. Enzymes are sufficiently large that distinguishing their presence should be straightforward. If the DNA/enzyme approach proves impossible, a wide variety of other polymer systems could be examined.

Bauschlicher computationally studied storing data in a pattern of fluorine and hydrogen atoms on the (111) diamond surface. If write-once data could be stored this way, 10^{15} bytes/cm^2 is theoretically possible. By comparison, the new DVD write-once disks now coming on the market hold about 10^8 bytes/cm^2. Bauschlicher compared the interaction of different probe molecules with a one dimensional model of the diamond surface. This study found some molecules whose interaction energies with H and F are sufficiently different that the force differential should be detectable by an SPM. These studies were extended to include a two dimensional model of the diamond surface and two other systems besides F/H. Other surfaces, such as Si, and other probes, such as those including transition metal atoms, have also been investigated. Among the better probes was C_5H_5N (pyridine). Quantum calculations suggest that pyridine is stable when attached to C_{60} in the orientation necessary for sensing the difference between hydrogen and fluorine. Half of C_{60} can form the end cap of a (9,0) or (5,5) carbon nanotube, and carbon nanotubes have been attached to an SPM tip. Thus, it might be possible using today's technology to build a system to read the diamond memory surface. Avouris has shown that individual hydrogen atoms can be removed from a silicon surface. If this could be accomplished in a gas that donates fluorine to vacancies on a diamond surface, the data storage system could be built. Thummel computationally investigated methods for adding a fluorine at the radical sites where a hydrogen atom had been removed from a diamond surface.

As mentioned before, carbon nanotubes can be described as rolled up sheets of graphite. Different tubes can have different helical windings depending on how the graphite sheet is connected to itself. Theory suggests that single-walled carbon nanotubes can have metallic or semiconductor properties depending on the helical winding of the tube. Chico, Han, Menon, and others have computationally examined the properties of some of hypothetical devices that might be made by connecting tubes with different electrical properties. Such devices are only few nanometers across — 100 times smaller than current computer chip features.

Several authors, including Tour have described methods to produce conjugated macromolecules of precise length and composition. This technique was used to produce molecular electronic devices in mole quantities. The resultant single molecular wires were tested experimentally and found to be conducting. The three and four terminal devices have been examined computationally and look promising. The features of these components are approximately 3 angstroms wide, about 750 times smaller than current silicon technology can produce.

Helical logic is a theoretical proposal for a future computing technology using the presence or absence of individual electrons (or holes) to encode 1s and 0s. The electrons are constrained to move along helical paths, driven by a rotating electric field in which the entire circuit is immersed. The electric field remains roughly orthogonal to the major axis of the helix and confines each charge carrier to a fraction of a turn of a single helical loop, moving it like water in an Archimedean screw. Each loop could in principle hold an independent carrier, permitting high information density. One computationally universal logic operation involves two helices, one of which splits into two "descendant" helices. At the point of divergence, differences in the electrostatic potential resulting from the presence or absence of a carrier in the adjacent helix controls the direction taken by a carrier in the splitting helix. The reverse of this sequence can be used to merge two initially distinct helical paths into a single outgoing helical path without forcing a dissipative transition. Because these operations are both logically and thermodynamically reversible, energy dissipation can be reduced to extremely low levels. ... It is important to note that this proposal permits a single electron to switch another single electron, and does not require that many electrons be used to switch one electron. The energy dissipated per logic operation can likely be reduced to less than 10^{-27} joules at a temperature of 1 Kelvin and a speed of 10 gigahertz, though further analysis is required to confirm this. Irreversible operations, when required, can be easily implemented and should have a dissipation approaching the fundamental limit of $\ln 2 \times kT$.

One study not conducted by Ames or partners is particularly worth mentioning since it places a loose lower bound on the computational capabilities of molecular nanotechnology. Drexler designed a number of computer components using small diamondoid rods with knobs that allow or prevent movement to accomplish computation. While this tiny mechanical Babbage Machine is probably not an optimal computational engine, its calculated performance for a desktop computer is 10^{18} MIPS — about a million times more powerful than the largest supercomputer that exists today. Note that with very fast computation energy use and heat dissipation become a severe problem. One approach to addressing this issue is reversible logic.

Drexler proposed a nanotechnology based on diamond and investigated its potential properties. In particular, he examined applications for materials with a strength similar to that of diamond (69 times strength/mass of titanium). This would require a very mature nanotechnology constructing systems by placing atoms on diamond surfaces one or a few at a time in parallel. Assuming diamondoid materials, McKendree predicted the performance of several existing single-stage-to-orbit (SSTO) vehicle designs. The predicted payload to dry mass ratio for these vehicles using titanium as a structural material varied from < 0 (the vehicle won't work) to 36 per cent, i.e., the vehicle weighs substantially more than the payload. With hypothetical diamondoid materials the ratios varied from 243 per cent to 653 per cent, i.e., the payload weighs far more than the vehicle. Using a very simple cost model ($1000 per vehicle kilogram) sometimes used in the aerospace industry, he estimated the cost per kilogram launched to low-Earth-orbit for diamondoid structured vehicles should be $153-412. This would meet NASA's 2020 launch to orbit cost goals. Estimated costs for titanium structured vehicles varied from $16,000-59,000/kg. Although this cost model is probably adequate for comparison, the absolute costs are suspect. Drexler used a more speculative methodology to estimate that a four passenger SSTO weighing three tons including fuel could be built using a mature nanotechnology. Using McKendree's cost model, such a vehicle would cost about $60,000 to purchase — the cost of today's high-end luxury automobiles. These studies assumed a fairly advanced nanotechnology capable of building diamondoid materials. In the nearer term, it may be possible to develop excellent structural materials using carbon nanotubes. Carbon nanotubes have a Young's modulus of approximately one terapascal— comparable to diamond.

Issacs and Pearson proposed a space elevator—a cable extending from the Earth's surface into space with a center of mass at geosynchronous altitude. If such a system could be built, it should be mechanically stable and vehicles could ascend and descend along the cable at almost any reasonable speed using electric power (actually generating power on the way down). The first incredibly difficult problem with building a space elevator is strength of materials. Maximum stress is at geosynchronous altitude so the cable must be thickest there and taper exponentially as it approaches Earth. Any potential material may be characterized by the taper factor—the ratio between the cable's radius at geosynchronous altitude and at the Earth's surface. For steel the taper factor is tens of thousands—clearly impossible. For diamond, the taper factor is 21.9 including a safety factor. Diamond is, however, brittle. Carbon nanotubes have a strength in tension similar to diamond, but bundles of these nanometer-scale radius tubes shouldn't propagate cracks nearly as well as the diamond tetrahedral lattice. Thus, if the considerable problems of developing a molecular nanotechnology capable of making nearly perfect carbon nanotube systems approximately 70,000 kilometers long can be overcome, the first serious problem of a transportation system capable of truly large scale transfers of mass to orbit can be solved. The next immense problem with space elevators is safety—how to avoid dropping thousands of kilometers of cable on Earth if the cable breaks. Active materials may help by monitoring and repairing small flaws in the cable and/or detecting a major failure and disassembling the cable into small elements.

Drexler calculates that lightsails made of 20 nm aluminum in tension should achieve an outward acceleration of ~14 km/s per day at Earth orbit with no payload and minimal structural overhead. For comparison, the delta V from low Earth to geosynchronous orbit is 3.8 km/s. Lightsails generate thrust by reflecting sunlight. Tension is achieved by rotating the sail. The direction of thrust is normal to the sail and away from the Sun. By directing thrust along or against the velocity vector, orbits can be lowered or raised. This form of transportation requires no reaction mass and generates thrust continuously, although the instantaneous acceleration is small so sails cannot operate in an atmosphere and must be large for even moderate payloads.

Today, the smallest feature size in production systems is about 250 nanometers — the smallest feature size in computer chips. Since atoms are an angstrom or so across and carbon nanotubes have a diameter as small as 0.7 nanometers, atomically precise molecular machines can be smaller than current MEMS devices by two to three orders of magnitude in each dimension, or six to nine orders of magnitude smaller in volume (and mass). For example, the size of the kinesin motor, which transports material in cells, is 12 nm. Han computationally demonstrated that molecular gears fashioned from single-walled carbon nanotubes with benzyne teeth should operate well at 50-100 gigahertz. These gears are about two nanometers across. Han computationally demonstrated cooling the gears with an inert atmosphere. Srivastava simulated powering the gears using alternating electric fields generated by a single simulated laser. In this case, charges were added to opposite sides of the tube to form a dipole. To make active materials, a material might be filled with nano-scale sensors, computers, and actuators so the material can probe its environment, compute a response, and act. Although this document is concerned with relatively simple artificial systems, living tissue may be thought of as an active material. Living tissue is filled with protein machines which gives living tissue properties (adaptability, growth, self-repair, etc.) unimaginable in conventional materials.

Active materials can theoretically be made entirely of machines. These are sometimes called swarms since they consist of large numbers of identical simple machines that grasp and release each other and exchange power and information to achieve complex goals. Swarms change shape and exert force on their environment under software control. Although some physical prototypes have been built, at least one patent issued, and many simulations run, swarm potential capabilities are not well analyzed or understood. We briefly discuss some concepts here. Michael proposes brick-shaped machines of various sizes that slide past each other to assume a variety of shapes. He has generated a large number of videos showing computer simulations of simple motions. Although his web site contains rather extravagant claims, this work has received a U.K. patent.

Yim built a small swarm with macroscopic (size in inches) components called polypod, built a simulator of polypod, and programmed it to move in various ways to study locomotion. There are two brick shaped components in polypod, one of which has two prismatic joints linked by a revolute joint. The second component is a cubic connector with no mechanical motion. Polypod is programmed by tables for each member of the swarm. Each member is programmed to move at various speeds in each degree of freedom for certain amounts of time. The swarm components are implicitly synchronized so there is no clock signal. Hall proposes a swarm with 10 micron dodecahedral components each with 12 arms that can move in and out, rotate a little, and grab and release each other. This concept is called the "utility fog." Hall estimates that the utility fog would have a density of 0.2, tensile strength of 1000 psi in action and 100,000 psi in a passive mode, and have a maximum shear rate of 100 km/second/meter. Bishop proposes a swarm consisting of 100 nanometer brick-shaped components that slide past each other to change shape. Globus proposes a swarm with two kinds of components — edges and nodes. The terms "node" and "edge" are chosen to correspond to those in graph theory. The roughly spherical nodes are capable of attaching to five edges (for a tetrahedral geometry with one free edge per node) and rotating each edge in pitch and yaw. The rod-like edges are capable of changing length, rotating around their long axis, and attaching/detaching to/from nodes. Component design, power distribution and control software are significant challenges for swarm development. Consider that with 10 micron components a cubic meter of swarm would contain about 10^{15} devices, each with an internal computer communicating with its neighbors to accomplish a global task.

NASA's mission is divided into five enterprises: Mission to Planet Earth, Aeronautics,

Human Exploration and Development of Space, Space Science, and Space Technology. We will examine some potential nanotechnology applications in each area.

The Earth Observing System (EOS) will use satellites and other systems to gather data on the Earth's environment. The EOS data system will need to process and archive >terabyte per day for the indefinite future. Simply storing this quantity of data is a significant challenge—each day's data would fill about 1,000 DVD disks. With projected nanocomputer processing speeds of 10^{18} MIPS, a million calculations on each byte of one day's data would take one second on the desktop.

Given a mature nanotechnology, it should be possible to build sensors in balloon-borne systems approximately the size of bacteria. With replication based manufacturing, these should be quite inexpensive. If the serious communication and control problems can be solved, one can imagine spreading billions of tiny lighter-than-air vehicles into the atmosphere to measure wind currents and atmospheric composition. A similar approach might be taken in the oceans—note that the oceans are full of floating microscopic living organisms that can sense and react to their environment. Smart dust might sense the environment, note the location via a GPS-like system, and store that information until close enough to a data-collection point to transfer the data to the outside world.

The strength of materials and computational capabilities previously discussed for space transportation should also allow much more advanced aircraft. Stronger, lighter materials can obviously make aircraft with greater lift and range. More powerful computers are invaluable in the design stage and of great utility in advanced avionics.

MEMS technology has been used to replace traditional large control structures on aircraft with large numbers of small MEMS controlled surfaces. This control system was used to operate a model airplane in a windtunnel. Nanotechnology should allow even finer control—finer control than exhibited by birds, some of which can hover in a light breeze with very little wing motion. Nanotechnology should also enable extremely small aircraft.

A reasonably advanced nanotechnology should be able to make simple atomically precise materials under software control. If the control is at the atomic level, then the full range of shapes possible with a given material should be achievable. Aircraft construction requires complex shapes to accommodate aerodynamic requirements. With molecular nanotechnology, strong complex-shaped components might be manufactured by general purpose machines under software control.

The aeronautics mission is responsible for launch vehicle development. Payload handling is an important function. Very efficient payload handling might be accomplished by a very advanced swarm. The sequence begins by placing each payload on a single large swarm located next to the shuttle orbiter. The swarm forms itself around the payloads and then moves them into the payload bay, arranging the payloads to optimize the center of gravity and other considerations. The swarm holds the payload in place during launch and may even damp out some launch vibrations. On orbit, satellites can be launched from the payload bay by having the swarm give them a gentle push. The swarm can then be left in orbit, perhaps at a space station, and used for orbital operations. This scenario requires a very advanced swarm that can operate in an atmosphere and on orbit in a vacuum. Besides the many and obvious difficulties of developing a swarm for a single environment, this provides additional challenges. Note that a simpler swarm might be used for aircraft payload handling.

Aerospace vehicles often require complex checkout procedures to insure safety and reliability. This is particularly true of reusable launch vehicles. A very advanced swarm with some special purpose appendages might be placed on a vehicle. It might then spread out over the vehicle and into all crevices to examine the state of the vehicle in great detail. Nanotechnology-enabled Earth-to-orbit transportation has the greatest potential to revolutionize human access to space by dropping the current $10,000 per pound cost of launch, but this was discussed above. Other less dramatic technologies include:

Space structures with a long design life (such as space station modules) need high-reliability materials that do not degrade. Active materials might help. The machines monitor structural integrity at the sub-micrometer scale. When a portion of the material becomes defective, it could be disassembled and then correctly reassembled. It should be noted that bone works somewhat along these lines. It is constantly being removed and added by specialized cells.

To effect timely repairs, space stations require a large store of spare parts and tools that are rarely used. A mature nanotechnology might create a "matter compiler," a machine that converts raw materials into a wide variety of products under software control. Contemporary examples of very limited matter compilers are numerically controlled machines and polypeptide sequencers. With a substantially more capable nanotechnology-based matter compiler, a space station crew could simply make spare parts and tools as needed. The programs could be stored on-board or on the ground. New tools invented on Earth could be transferred as software to the station for manufacture. Once used, unneeded tools and broken parts could be ionized in a solar furnace, transferred using controlled magnetic fields, and the constituent atoms stored for later manufacture into new products.

An advanced nanotechnology might be able to build filters that dynamically modify themselves to attract the contaminant molecules detected by the air and water quality sensors. Once attached to the filter, the filter could in principle move the offending molecules to a molecular laboratory for modifications to useful or at least inert products. A swarm might implement such an active filter if it was able to dynamically manufacture proteins that could bind contaminant molecules. The protein and bound contaminant might then be manipulated by the swarm for transportation. With a sufficiently advanced nanotechnology it might even be possible to directly generate food by non-biological means. Then agriculture waste in a self-sufficient space colony could be converted directly to useful nutrition. Making this food attractive will be a major challenge.

Sleeping crew members in the shuttle experience considerable pain and sleep disruption when the reaction control system fires and they collide with the cabin walls. If crew members were connected to the walls by a swarm, the swarm could absorb most or all of the force before the crew member struck the wall. The swarm could then gradually return the crew member to center (without the oscillations associated with bungee cords) in preparation for the next firing.

For resupply, spacecraft docking is a frequent necessity in space station operations. When two spacecraft are within a few meters of each other, a swarm could extend from each, meet in the middle, and form a stable connection before gradually drawing the spacecraft together. A swarm could support space-suited astronauts in simulated partial-g environments by holding them up appropriately. The swarm moves in response to the astronaut's motion providing the appropriate simulation of partial or 0 gravities. Tools and other objects are also manipulated by the swarm to simulate non-standard gravity.

Active nanotechnology materials (see active materials) might enable construction of a skin-tight space suit covering the entire body except the head (which is in a more conventional helmet). The material senses the astronaut's motions and changes shape to accommodate it. This should eliminate or substantially reduce the limitations current systems place on astronaut range of motion.

In situ resource utilization is undoubtedly necessary for large scale colonization of the solar system. Asteroids are particularly promising for orbital use since many are in near Earth orbits. Moving asteroids into low Earth orbit for utilization poses a safety problem should the asteroid get out of control and enter the atmosphere. Very small asteroids can cause significant destruction. The 1908 Tunguska explosion, which was calculated to be a 60 meter diameter stony asteroid, leveled 2,200 km^2 of forest. Hills calculated that 4 meter diameter iron asteroids are near the threshold for ground damage. Both these calculations assumed high collision speeds. At a density of 7.7 g/cm^3 Babadzhanov, a 3 meter diameter asteroid should have a mass of about 110 tons. Rabinowitz estimates that there are about one billion ten meter diameter near Earth asteroids and there should be far more smaller objects. For colonization applications one would ideally provide the same radiation protection available on Earth. Each square meter on Earth is protected by about 10 tons of atmosphere. Therefore, structures orbiting below the van Allen belts would like 10 tons/meter^2 surface area shielding mass. This would dominate the mass requirements of any system and require one small asteroid for each 11 meter^2 of colony exterior surface area. A 10,000 person cylindrical space colony such as Lewis One with a diameter of almost 500 meters and a length of nearly 2000 meters would require a minimum of about 90,000 retrieval missions to provide the shielding mass. The large number of missions required suggests that a fully automated, replicating nanotechnology may be essential to build large low Earth orbit colonies from small asteroids. A nanotechnology swarm along with an atomically precise lightsail is a promising small asteroid retrieval system. Lightsail propulsion insures that no mass will be lost as reaction mass. The swarm can control the lightsail by shifting mass. When a target asteroid is found, the swarm spreads out over the surface to form a bag. The interface to the sail must be active to account for the rotation of the asteroid—which is unlikely to have an axis-of-rotation in the proper direction to apply thrust for the return to Earth orbit. The active interface is simply swarm elements that transfer between each other to allow the sail to stay in the proper orientation. Of course, there are many other possibilities for nanotechnology based retrieval vehicles.

Extraterrestrial materials brought into orbit could be fed into a high-temperature solar furnace and partially ionized. Magnetic fields might then be used to separate the nuclei. These are fed in appropriate quantities to a matter-compiler to build the products desired.

Several authors, including Freitas have speculated that a sufficiently advanced nanotechnology could examine and repair cells at the molecular level. Should this capability become available—presumably driven by terrestrial applications—the small size and advanced capabilities of such systems could be of great utility on long duration space flights and on self-sufficient colonies.

Self-replicating systems permit efforts of great scope to be pursued economically. Adjusting the environment on another planet to suit the tastes of humans is one such undertaking. Heating and cooling can be achieved by (among many other methods) using space-based mirrors. Chemical modifications of the planetary surface and atmosphere can be achieved in relatively short periods by the use of self-replicating systems that absorb sunlight and raw materials, and convert them into the desired products. Much as plants changed the environment of the earth to what we see today, so self-replicating molecular manufacturing systems might more rapidly convert the environments of other planets.

As interstellar trips might last many years, the ability to conserve supplies by maintaining some crew members in a suspended state would be useful. An extremely advanced nanotechnology might use molecular manipulations of each cell to provide (a) better methods of slowing or suspending the metabolic activity of crew members and (b) better methods of restoring metabolic activity to a normal state when the destination is reached.

Molecular manufacturing should enable the creation of very precise mirrors. Unlike lightsail applications, telescope mirrors require a very precise and somewhat complicated shape. A swarm with special purpose appendages capable of bonding to the mirror might be able to achieve and maintain the desired shape.

A very advanced nanotechnology would be capable of imaging and then removing the surface atoms of an extra-terrestrial sample. By removing successive surface layers the location of each atom in the sample might be recorded, destroying the sample in the process. This data could then be sent to Earth. Besides requiring a very advanced nanotechnology, there is a more fundamental—but not necessarily fatal—problem: as the outside layer of atoms is removed the next layer may rearrange itself so the sample is not necessarily perfectly recorded.

Solar Power—Low Earth orbit spacecraft generally depend on solar cells and batteries for power. For energy collection, molecular manufacturing can be used to make solar photovoltaic cells at least as efficient as those made in the laboratory today. Efficiencies can therefore be > 30 per cent. In space applications, a reflective optical concentrator need consist of little more than a curved aluminum shell < 100 nanometers thick (photovoltaic cells operate with higher efficiency at high optical power densities). A metal fin with a thickness of 100 nanometers and a conduction path length of 100 microns can radiate thermal energy at a power density as high as 1000 W/m^2 with a temperature differential from base to tip of < 1 K. Accordingly, solar collectors can consist of arrays of photovoltaic cells several microns in thickness and diameter, each at the focus of a mirror of ~100 micron diameter, the back surface of which serves as a ~100 micron diameter radiator. If the mean thickness of this system is ~1 micron, the mass is ~10^{-3} kg/m^2 and the power per unit mass, at Earth's distance from the Sun, where the solar constant is ~1.4 kW/m^2, is > 10^5 W/kg." By comparison, the U.S. built Photovoltaic Panel Module solar cells currently used on the Mir Space Station and planned for use on the International Space Station generate about 118 W/kg.

Fuel Cells—A critical component in hydrogen/oxygen fuel cells is the PEM (Proton Exchange Membrane). This membrane must (a) permit the passage of protons while (b) blocking everything else. Present membranes do a rather poor job. One group at Ames is designing and computationally testing PEMs to study possible energy mechanisms in early life. While these studies are not meant to design optimal membranes for fuel cell use, the basic knowledge and approach may be of value. Another proposal is to design a diamond membrane a few nanometers thick with "proton pores." The pores might be lined with fluorine, oxygen and nitrogen to create a region with a high proton affinity. In addition, a positionally controlled platinum might be held at the mouth of the pore to verify that H_2 can be catalytically split into H+ and e–, and that the barrier for migration of the H+ into the pore is modest in size. Nanotechnology must provide precise control over the manufacturing process of the diamondoid PEM since the pores must be made very precisely.

Studies of H_2 absorption and packing in carbon nanotubes and nanoropes are in progress at NASA Ames and elsewhere. Nanotubes provide large pore sizes and nanoropes have different pore sizes depending on interstitial and other locations. Dillon estimated that the single walled nanotubes in their sample contained 5 to 10 per cent by weight of H_2. The nanotubes were about 0.1 to 0.2 per cent by weight of the total sample. Computational studies at Ames suggest that to store 7-10 per cent H_2 in single walled nanotubes at room temperature the H_2s must be stored inside the tubes, not merely adsorbed on the walls. This work suggests that carbon nanotubes might be developed into an excellent H_2 storage medium within 3-5 years. Calculations with oxygen suggest that a diamondoid sphere ~0.1 microns in diameter should easily hold oxygen at ~1,000 atmospheres. While higher pressures are feasible, they offer declining returns. At higher pressures, the pressure-volume relationship becomes severely non-linear and the density approaches a limiting value. Other gases might also be stored if diamondoid spheres can be built, but the analysis has not been done. High strength light-weight materials will allow greater efficiency of energy storage as angular momentum.

Many kinds of ultraminiature electromechanical devices have utility on a miniaturized space craft. It has been shown that manipulating carbon nanotubes changes their electrical properties. This might be exploited to build nanometer scale strain devices. This may be achievable within 3-5 years, and simulations along these lines are in progress. Similar results have been achieved experimentally with C_{60}. The electrical properties of a C_{60} molecule were changed by applying pressure to the molecule with an SPM tip.

Smaller, lighter spacecraft are cheaper to launch (current costs are about $10,000/lb) and generally cheaper to build. Diamondoid structural materials can radically reduce structural mass, miniaturized electronics can shrink the avionics and reduce power consumption, and atomically precise materials and components should shrink most other subsystems.

Thermal protection is crucial for atmospheric reentry and other tasks. The carbon nanotubes under investigation at NASA Ames and elsewhere may play a significant role. Most production processes for carbon nanotubes create a tangled mat of nanotubes that has a very low mass-to-volume ratio. Like graphite, the tubes should withstand high temperatures but the tangled mat should prevent them from ablating. This may lead to high temperature applications.

Many of the applications discussed here are speculative to say the least. However, they do not appear to violate the laws of physics. Something similar to these applications at these performance levels should be feasible if we can gain complete control of the three-dimensional structure of materials, processes and devices at the atomic scale. How to gain such control is a major, unresolved issue. However, it is clear that computation will play a major role regardless of which approach —positional control with replication, self-assembly, or some other means —is ultimately successful.

Computation has already played a major role in many advances in chemistry, SPM manipulation, and biochemistry. As we design and fabricate more complex atomically precise structures, modeling and computer aided design will inevitably play a critical role. Not only is computation critical to all paths to nanotechnology, but for the most part the same or similar computational chemistry software and expertise supports all roads to molecular nanotechnology. Thus, even if NASA's computational molecular nanotechnology efforts should pursue an unproductive path, the expertise and capabilities can be quickly refocused on more promising avenues as they become apparent. As nanotechnology progresses we may expect applications to become feasible at a slowly increasing rate. However, if and when a general purpose programmable assembler/ replicator can be built and operated, we may expect an explosion of applications. From this point, building new devices will become a matter of developing the software to instruct the assembler/replicators. Development of a practical swarm is another potential turning point. Once an operational swarm that can grow and divide has been built, a large number of applications become software projects.

REFERENCES

Altmann, J. & Gubrud, M.A.: 2004, 'Military, Arms Control, and Security Aspects of Nanotechnology', in: D. Baird, A. Nordmann & J. Schummer (eds.), *Discovering the Nanoscale*, Amsterdam: IOS Press, pp. 269-277.

Altmann, J.: 2004, 'Military Uses of Nanotechnology: Perspectives and Concerns', *Security Dialogue*, 35 (1): 61-79.

Attebery, B.: 2004, 'Dust, Lust, and Other Messages from the Quantum Wonderland', in: N.K. Hayles (ed.), *Nanoculture: Implications of the New Technoscience*, Bristol, UK: Intellect Books, pp. 161-172.

Averback, R.S., J. Bernholc, and D.L. Nelson. 1991. *MRS Symposium Proceedings* Vol. 206.

Baird, D. & Shew, A.: 2004, 'Probing the History of Scanning Tunneling Microscopy', in: D. Baird, A. Nordmann & J. Schummer (eds.), *Discovering the Nanoscale*, Amsterdam: IOS Press, pp. 145-156.

Baird, D.; Nordmann, A. & Schummer, J. (eds.): 2004, *Discovering the Nanoscale*, Amsterdam: IOS Press.

Baker, R.T.K. Synthesis, properties and applications of graphite nanofibers. In *R&D status and trends*, ed. Siegel et al.

Bate, R., Frazier, G., Frensley, W., Reed., M., "An Overview of Nanoelectronics," *Texas Instruments Technical Journal*, July-August 1989, pp. 13-20.

Beck, J.S., J.C. Vartuli, W.J. Roth, M.E. Leonowicz, C.T. Kresge, K.D. Schmitt, C.T.-W. Chu, D.H. Olsen, E.W. Shepard, S.B. McCullen, J.B. Higgins, and J.L. Schlenker. 1992. *J. Am. Chem. Soc.* 114:10834.

Bensaude-Vincent, B.: 2001, 'The Construction of a Discipline: Materials Science in the United States', *Historical Studies in the Physical and Biological Sciences*, 31, 223-248.

Bensaude-Vincent, B.: 2004, 'Two Cultures of Nanotechnology?', *Hyle: International Journal for Philosophy of Chemistry*, 10(2), 65-82.

Berube, D.M.: 2004, 'The Rhetoric of Nanotechnology', in: D. Baird, A. Nordmann & J. Schummer (eds.), *Discovering the Nanoscale*, Amsterdam: IOS Press, pp. 173-192.

Bowes, C.L., A. Malek, and G.A. Ozin. 1996. *Chem. Vap. Deposition* 2:97.

Braun, P.V., P. Osenar, and S.I. Stupp. 1996. *Nature*. 368: 2.

Braun, T.; Schubert, A. & Zsindely, S.: 1997, 'Nanoscience and nanotechnology on the balance', *Scientometrics*, 38, 321-325.

Brinker, C.J. 1996. *Curr. Opin. Solid State Mater. Sci.* 1:798.

Brown, N.: 2004, 'Needle on the Real: Technoscience and Poetry at the Limits of Fabrication', in: N.K. Hayles (ed.), *Nanoculture: Implications of the New Technoscience*, Bristol, UK: Intellect Books, pp. 173-190.

Bueno, O.: 2004, 'The Drexler-Smalley Debated on Nanotechnology: Incommensurability at Work?', *Hyle: International Journal for Philosophy of Chemistry*, 10(2), 83-98.

Bueno, O.: 2004, 'Von Neumann, Self-Reproduction and the Constitution of Nanophenomena', in: D. Baird, A. Nordmann & J. Schummer (eds.), *Discovering the Nanoscale*, Amsterdam: IOS Press, pp. 101-115.

Carroll, J.S.: 2001, 'Social Science Research Methods for Assessing Societal Implications of Nanotechnology', in: M.C. Roco and W.S. Bainbridge (eds.), *Societal Implications of Nanoscience and Nanotechnology*, Dordrecht: Kluwer, pp. 188-192.

Chen, C-Y., S.L. Burkett, H.-X. Li, and M.E. Davis. 1993. *Microporous Mater*. 2:27.

Chianelli, R.R. 1998. Synthesis, fundamental properties and applications of nanocrystals, sheets, and fullerenes based on layered transition metal chalcogenides. In *R&D status and trends,* ed. Siegel et al.

Chianelli, R.R., M. Daage, and M.J. Ledoux. 1994. *Advances in Catalysis* 40:177.

Claeson, T., and Likharev, K.," Single Electronics," *Scientific American,* June 1992, pp. 80-85.

Coenen, C.: 2004, 'Nanofuturismus: Anmerkungen zu seiner Relevanz, Analyse und Bewertung', *Technikfolgenabschätzung ' Theorie und Praxis*, 13 (2), 78-85.

Crandall, B.C. and Lewis, J., *Nanotechnology: Research and Perspectives*, MIT Press, Cambridge, 1992.

Crow, M. & Sarewitz, D.: 2001, 'Nanotechnology and Societal Transformation', in: M.C. Roco and W.S. Bainbridge (eds.), *Societal Implications of Nanoscience and Nanotechnology*, Dordrecht: Kluwer, pp. 45-54.

de Souza e Silva, A.: 2004, 'The Invisible Imaginary: Museum Spaces, Hybrid Reality and Nanotechnology', in: N.K. Hayles (ed.), *Nanoculture: Implications of the New Technoscience*, Bristol, UK: Intellect Books, pp. 27-82.

Decker, M.; Fiedeler, U.; Fleischer, Th.: 2004, 'Ich sehe was, was Du nicht siehst... zur Definition von Nanotechnologie', *Technikfolgenabschätzung ' Theorie und Praxis*, 13 (2), 10-16.

Dresselhaus, M., and G. Dresselhaus. 1995. *Ann. Rev. Mat. Sci*. 25:487.

Dresselhaus, M.S. 1998. Carbon-based nanostructures. In *R&D status and trends*, ed. Siegel et al.

Dresselhaus, M.S., G. Dresselhaus, and P. Eklund. 1996. *Science of fullerenes and carbon nanotubes*. San Diego: Academic Press.

Drexler, K.E.: 2004, 'Nanotechnology: From Feynman to Funding', *Bulletin of Science, Technology & Society*, 24, 21-27.

Dupuy, J.-P.: 2004, 'Complexity and Uncertainty: A Prudential Approach To Nanotechnology', in: European Commission (Community Health and Consumer Protection): *Nanotechnologies: A Preliminary Risk Analysis on the Basis of a Workshop, Brussels, 1-2 March 2004*, pp. 71-93 (www.europa.eu.int/comm/health/ ph_risk/documents/ev_20040301_en.pdf).

Dupuy, J.-P.: 2004, 'Pour une évaluation normative du programme nanotechnologique', *Annales des Mines*, (February 2004), 27-32.

Einsiedel, E.F.; Goldenberg, L.: 2004, 'Dwarfing the Social? Nanotechnology Lessons from the Biotechnology Front', *Bulletin of Science, Technology & Society*, 24, 28-33.

ETC: 2003a, *The Big Down: Atom Tech ' Technologies Converging at the Atomic Scale*. Winnipeg, Canada: Action Group on Erosion, Technology and Concentration [www.etcgroup.org/documents/TheBigDown.pdf].

ETC: 2003b, 'No Small Matter II: The Case for a Global Moratorium ' Size Matters!' *Occasional Paper Series* 7(1) [www.etcgroup.org/documents/Occ.Paper_Nanosafety.pdf].

Etzkowitz, H., 2001, 'Nano-Science and Society: Finding a Social Basis for Science Policy', in: M.C. Roco and W.S. Bainbridge (eds.), *Societal Implications of Nanoscience and Nanotechnology*, Dordrecht: Kluwer, pp. 121-128.

European Comission: *European Workshop on Social and Economic Research on Nanotechnologies and Nanosciences, Brussels, 14-15 April 2004* [www.stage-research.net/STAGE/PAGES/Nano.html].

European Commission (Community Health and Consumer Protection): 2004, *Nanotechnologies: A Preliminary Risk Analysis on theBasis of a Workshop, Brussels, 1-2 March 2004* [www.europa.eu.int/comm/health/ ph_risk/documents/ev_20040301_en.pdf].

European Commission: 2004, *Converging Technologies: Shaping the Future of European Societies (Report of the High Level Expert Group "Foresighting the New Technology Wave")*, Brussels: European Commission Research [http://europa.eu.int/comm/research/conferences/2004/ntw/index_en.html].

F. Buot, "Mesoscopic Physics and Nanoelectronics: Nanoscience and Nanotechnology," *Physics Reports*, pp.73-174, 1993.

F. Capasso and S. Datta, "Quantum Electron Devices," *Physics Today*, pp.74-82, February 1990.

Feynman, R., "There's Plenty of Room at the Bottom: An invitation to Enter a New Field of Physics," Talk at the Annual Meeting of the American Physical Society, 29 December 1959. Reprinted in Appendix B of Crandall and Lewis. See citation above.

Fielder, F.A.; Reynolds, G.H.: 1994, 'Legal Problems of Nanotechnology: An Overview', *Southern California Interdisciplinary Law Journal*, 3, 593-629.

Fleischer, Th.; Decker, M. & Fiedeler, U. (eds.): 2004, *Große Aufmerksamkeit für kleine Welten ' Nanotechnologie und ihre Folgen*, special issue of *Technikfolgenabschätzung ' Theorie und Praxis*, 13 (2), 5-85 [www.itas.fzk.de/tatup/042/inhalt.htm].

Fogelberg, H. & Glimell, H.: 2003, 'Molecular Matters: In Search of the Real Stuff', in: H. Fogelberg & H. Glimell, *Bringing Visibility to the Invisible: Towards A Social Understanding of Nanotechnology*, Göteborg: Göteborg University, pp. 5-32 [www.sts.gu.se/publications/STS_report_6.pdf].

Fogelberg, H. & Glimell, H.: 2003, *Bringing Visibility to the Invisible: Towards A Social Understanding of Nanotechnology*, Göteborg: Göteborg University [www.sts.gu.se/publications/STS_report_6.pdf].

Fogelberg, H.: 2003, 'The Grand Politics of Technoscience: Contextualizing Nanotechnology', in: H. Fogelberg & H. Glimell, *Bringing Visibility to the Invisible: Towards A Social Understanding of Nanotechnology*, Göteborg: Göteborg University, pp. 33-54 [www.sts.gu.se/publications/STS_report_6.pdf].

Fogelberg, H.: 2003, 'The Material Culture of Nanotechnology', in: H. Fogelberg & H. Glimell, *Bringing Visibility to the Invisible: Towards A Social Understanding of Nanotechnology*, Göteborg: Göteborg University, pp. 99-114.

Frazier, G., "An Ideology For Nanoelectronics," in *Concurrent Computations: Algorithms, Architecture, and Technology*, Plenum Press, New York, 1988.

Glimell, H.: 2001, 'Challenging Limits' Excerpts from an Emerging Ethnography of Nano Physicists', in: H. Glimell & O. Johlin (eds.), *The Social Production of Technology: On the Everyday Life with Things*, Göteburg, SE: BAS Publisher, chapter 7, pp. 111-131 (reprinted in: H. Fogelberg & H. Glimell, *Bringing Visibility to the Invisible: Towards A Social Understanding of Nanotechnology*, Göteborg: Göteborg University, pp. 115-139.

Glimell, H.: 2001, 'Dynamics of the Emerging Field of Nanoscience', in: M.C. Roco and W.S. Bainbridge (eds.), *Societal Implications of Nanoscience and Nanotechnology*, Dordrecht: Kluwer, pp. 156-160.

Glimell, H.: 2003, 'A Nano Narrative: The Mircopolitics of a New Generic Enabling Technology', in: H. Fogelberg & H. Glimell, *Bringing Visibility to the Invisible: Towards A Social Understanding of Nanotechnology*, Göteborg: Göteborg University, pp. 55-77 [www.sts.gu.se/publications/STS_report_6.pdf].

Glimell, H.: 2003, 'Dynamics of the Emerging Field of Nanoscience', in: H. Fogelberg & H. Glimell, *Bringing Visibility to the Invisible: Towards A Social Understanding of Nanotechnology*, Göteborg: Göteborg University, pp. 79-85 [www.sts.gu.se/publications/STS_report_6.pdf].

Glimell, H.: 2004, 'Grand Visions and Lilliput Politics: Staging the Exploration of the 'Endless Frontier", in: D. Baird, A. Nordmann & J. Schummer (eds.), *Discovering the Nanoscale*, Amsterdam: IOS Press, pp. 231-246.

Goldhaber-Gordon, D., Montemerlo, M.S., Love, J.C., Opiteck, G.J., and Ellenbogen, J.C., "Overview of Nanoelectronic Devices," submitted to the *Proceedings of the IEEE*, February 1997. For more information, please send e-mail to nanotech@mitre.org.

Gorman, M.E.; Groves, J.F. & Shrager, J.: 2004, 'Societal Dimensions of Nanotechnology as a Trading Zone: Results from a Pilot Project', in: D. Baird, A. Nordmann & J. Schummer (eds.), *Discovering the Nanoscale*, Amsterdam: IOS Press, pp. 63-73.

Grinbaum, A. & Dupuy, J.-P.: 2004, 'Living with Uncertainty: Toward the Ongoing Normative Assessment of Nanotechnology', *Techne: Research in Philosophy and Technology*, 8(3) (forthcoming).

Grinbaum, A.: 2004, 'La condition de l'homme moderne et les nanotechnologies', in: G. Nivat (ed.), *Les limites de l'humain. 39èmes Rencontres Internationales de Genève*, L'Age d'Homme, Genève, p. 141.

Grunwald, A.: 2004, 'Ethische Aspekte der Nanotechnologie. Eine Felderkundung', *Technikfolgenabschätzung ' Theorie und Praxis*, 13 (2), 71-78.

Gupta, V.K. & Pangannaya, N.B.: 2000, 'Carbon nanotubes: bibliometric analysis of patents', *World Patent Information*, 22, 185-189.

Hansson, S.O.: 2004, 'Great Uncertainty about Small Things', *Techne: Research in Philosophy and Technology*, 8(3) (forthcoming).

Haruta, M. 1997. *Catalyst surveys of Japan* 1:61 and references therein.

Hayles, N.K. (ed.): 2004, *Nanoculture: Implications of the New Technoscience*, Bristol, UK: Intellect Books.

Hayles, N.K.: 2004, 'Connecting the Quantum Dots: Nanotechscience and Culture', in: N.K. Hayles (ed.), *Nanoculture: Implications of the New Technoscience*, Bristol, UK: Intellect Books, pp. 11-26.

Hemerén, G.: 2004, 'Nano Ethics Primer', in: European Commission (Community Health and Consumer Protection): *Nanotechnologies: A Preliminary Risk Analysis on theBasis of a Workshop, Brussels, 1-2 March 2004* (www.europa.eu.int/comm/health/ ph_risk/documents/ ev_20040301_en.pdf), pp. 95-102.

Hennig, J.: 2004, 'Changes in the Design of Scanning Tunneling Microscopic Images from 1980 to 1990', *Techne: Research in Philosophy and Technology*, 8(3) (forthcoming).

Hessenbruch, A.: 2004, 'Nanotechnology and the Negotiation of Novelty', in: D. Baird, A. Nordmann & J. Schummer (eds.), *Discovering the Nanoscale*, Amsterdam: IOS Press, pp. 135-144.

Hullmann, A. & Meyer, M.: 2003, 'Publications and Patents in Nanotechnology. An overview of previous studies and the state of the art', *Scientometrics*, 58 (3) 507-527.

Huo, Q., D.I. Margolese, U. Ciesla, P. Feng, T.E. Gier, P. Sieger, R. Leon, P.M. Petroff, F. Schuth, and G.D. Stucky. 1994. *Nature* 368:317.

Jena, P., S.N. Khanna, and B.K. Rao. 1996. In *Science and technology of atomically engineered materials*, ed. P. Jena. River Edge, NJ: World Scientific.

Johansson, M.: 2003, 'Plenty of room at the bottom: Towards an anthropology of nanoscience', *Anthropology Today*, 19 (No. 6), 3-6.

Johnson, A.: 2004, 'The End of Pure Science: Science Policy from Bayh-Dole to the NNI', in: D. Baird, A. Nordmann & J. Schummer (eds.), *Discovering the Nanoscale*, Amsterdam: IOS Press, pp. 217-230.

Jones, C.W., and W.J. Koros. 1994. *Carbon* 32:1419.

Jounet, C., W.K. Maser, P. Bernier, A. Loiseau, M. Lamy de la Chapelle, S. Lefrant, P. Deniard, R. Lee, and J.E. Fischer. 1997. *Nature* 388:756.

K.E. Drexler, *Nanosystems: Molecular Machinery, Manufacturing, and Computation,* Wiley, New York, 1992.

Karger, J., and D.M. Ruthven. 1992. *Diffusion in zeolites*. New York: J. Wiley.

Keiper, A.: 2003, 'The Nanotechnology Revolution', *The New Atlantis*, 2, 17-34.

Khushf, G.: 2004, 'A Hierarchical Architecture for Nano-scale Science and Technology: Taking Stock of the Claims About Science Made By Advocates of NBIC Convergence', in: D. Baird, A. Nordmann & J. Schummer (eds.), *Discovering the Nanoscale*, Amsterdam: IOS Press, pp. 21-33.

Khushf, G.: 2004, 'Systems Theory and the Ethics of Human Enhancement: A Framework for NBIC Convergence', *Annals of the New York Academy of Sciences*, 1013, 124-149.

Krätschmer, W., L.D. Lamb, K. Fostiropoulos, and D.R. Huffman. 1990. *Nature* 347:354.

Kresge, C.T., M.E. Leonowicz, W.J. Roth, J.C. Vartuli, and J.S. Beck. 1992. *Nature* 359:710.

Kupperman, A., S. Nadimi, S. Oliver, G. Ozin, J. Garcés, and M. Olken. 1993. *Nature* 365:239.

Kuusi, O.; Meyer, M.: 2002, 'Technological generalizations and leitbilder ' the anticipation of technological opportunities', *Technological Forecasting & Social Change*, 69, 625-639.

Landon, B.: 2004, 'Less is More: Much Less is Much More: The Insistent Allure of Nanotechnology Narratives in Science Fiction', in: N.K. Hayles (ed.), *Nanoculture: Implications of the New Technoscience*, Bristol, UK: Intellect Books, pp. 131-146.

Laszlo, P.: 2004, 'Is There Life After Partington?', *Hyle: International Journal for Philosophy of Chemistry*, 10(2), 169-178.

Lenhard, J.: 2004, 'Nanoscience and the Janus-Faced Character of Simulations', in: D. Baird, A. Nordmann & J. Schummer (eds.), *Discovering the Nanoscale*, Amsterdam: IOS Press, pp. 93-100.

Lent, C.S., Tougaw, P.D., Porod, W., Bernstein, G.H., "Quantum Cellular Automata," *Nanotechnology*, Vol. 4, p. 49, 1993.

Lewak, S.E.: 2004, 'What's the Buzz? Tell Me What's A-Happening: Wonder, Nanotechnology, and Alice's Adventures in Wonderland', in: N.K. Hayles (ed.), *Nanoculture: Implications of the New Technoscience*, Bristol, UK: Intellect Books, pp. 201-.

Lide, D.R., ed. 1993-1994. *CRC Handbook of Chemistry and Physics*, 74th ed.

Lin-Easton, P.C.: 2001, 'It's Time for Environmentalists to Think Small ' Real Small: A Call for the Involvement of Environmental Lawyers in Developing Precautionary Policies Molecular Nanotechnology', *Georgetown International Law Review*, 14, 106-134.

López, J.: 2004, 'Bridging the Gaps: Science Fiction in Nanotechnology', *Hyle: International Journal for Philosophy of Chemistry*, 10(2), 129-152.

Lösch, A.: 2004, 'Nanomedicine and Space: Discursive Orders of Mediating Innovations', in: D. Baird, A. Nordmann & J. Schummer (eds.), *Discovering the Nanoscale*, Amsterdam: IOS Press, pp. 193-202.

Marshall, K.: 2004, 'Atomizing the Risk Technology', in: N.K. Hayles (ed.), *Nanoculture: Implications of the New Technoscience*, Bristol, UK: Intellect Books, pp. 147-160.

Martin, T.P., N. Malinowski, U. Zimmerman, U. Naher, and H. Schaber. 1993. *J. Chem. Phys.* 99:4210.

Martin, T.P., U. Naher, H. Schaber, U. Zimmerman. 1993. *Phys. Rev. Lett.* 70:3079.

Mayer, S.: 2002, 'From genetic modification to nanotechnology: the dangers of 'sound science", in: T. Gilland (ed.), *Science: Can We Trust the Experts?*, London: Hodder and Stoughton, pp. 1-15.

Mehta, M.D.: 2002, 'Nanoscience and Nanotechnology: Assessing the Nature of Innovation in These Fields', *Bulletin of Science, Technology & Society*, 22(4), 269-273.

Mehta, M.D.: 2004, 'From Biotechnology to Nanotechnology: What Can We Learn from Earlier Technologies?, *Bulletin of Science, Technology & Society*, 24, 34-39.

Meyer, M. & Kuusi, O.: 2004, 'Nanotechnology: Generalizations in an Interdisciplinary Field of Science and Technology', *Hyle: International Journal for Philosophy of Chemistry*, 10(2), 153-168.

Meyer, M.: 2000a, 'Does science push technology? Patents citing scientific literature', *Research Policy*, 29, 409-434.

Meyer, M.: 2000b, 'Patent citations in a novel field of technology: What can they tell about interactions of emerging communities of science and technology?', *Scientometrics*, 48, 151-178.

Meyer, M.: 2001a, 'Patent citations in a novel field of technology: An exploration of nano-science and nano-technology', *Scientometrics*, 51, 163-183.

Meyer, M.: 2001b, 'Socio-economic Research on Nanoscale Science and Technology: A European Overview and Illustration', in: M.C. Roco and W.S. Bainbridge (eds.), *Societal Implications of Nanoscience and Nanotechnology*, Dordrecht: Kluwer, pp. 217-241.

Meyer, M.; Persson, O.: 1998, 'Nanotechnology ' interdisciplinarity, patterns of collaboration and differences in application', *Scientometrics*, 47, 195'205.

Milburn, C.: 2002, 'Nanotechnology in the age of post-human engineering: science fiction as science', *Configurations*, 10, 261-295 [reprinted in: N.K. Hayles (ed.), *Nanoculture: Implications of the New Technoscience*, Bristol, UK: Intellect Books, 2004, pp. 109-130].

Milburn, C.: 2004, 'Nano/Splatter: Disintegrating the Postbiological Body', *New Literary History*, 35 (in print).

Mnyusiwalla, A.; Abdallah, S.D.; Singer, P.A.: 2003, 'Mind the gap: science and ethics in nanotechnology', *Nanotechnology*, 14, R9-R13.

Mody, C.C.M.: 2004, 'How Probe Microscopists Became Nanotechnologists', in: D. Baird, A. Nordmann & J. Schummer (eds.), *Discovering the Nanoscale*, Amsterdam: IOS Press, pp. 119-133.

Mody, C.C.M.: 2004, 'Instruments in Training: The Growth of American Probe Microscopy in the 1980s', in: D. Kaiser (ed.), *Pedagogy and the Practice of Science: Producing Physical Scientists, 1800-2000*, Cambridge, MA: MIT Press (forthcoming).

Mody, C.C.M.: 2004, 'Small, but Determined: Technological Determinism in Nanoscience', *Hyle: International Journal for Philosophy of Chemistry*, 10(2), 99-128.

Mody, C.C.M.: 2004, *Crafting the Tools of Knowledge: The Invention, Spread, and Commercialization of Probe Microscopy, 1960-2000*, Ph.D. dissertation, Cornell University.

Montemerlo, M.S., Love, J.C., Opiteck, G.J., Goldhaber, D. J., and Ellenbogen, J.C., "Technologies and Designs for Electronic Nanocomputers," MITRE Technical Report 96W0000044, The MITRE Corporation, McLean, VA, July 1996. For more information, send e-mail to nanotech@mitre.org.

Moor, J.H. & Weckert, J.: 2004, 'Nanoethics: Assessing the Nanoscale From an Ethical Point of View', in: D. Baird, A. Nordmann & J. Schummer (eds.), *Discovering the Nanoscale*, Amsterdam: IOS Press, pp. 301-310.

Munn Sanchez, E.: 2004, 'The Expert's Role in Nanoscience and Technology', in: D. Baird, A. Nordmann & J. Schummer (eds.), *Discovering the Nanoscale*, Amsterdam: IOS Press, pp. 257-266.

Nordmann, A.: 2003, 'Shaping the World Atom by Atom: Eine nanowissenschaftliche WeltBildanalyse', in: A. Grunwald (ed.), *Technikgestaitung zwischen Wunsch und Wirklichkeit*, Berlin: Springer, pp. 191-199.

Nordmann, A.: 2004, 'Molecular Disjunctions: Staking Claims at the Nanoscale', in: D. Baird, A. Nordmann & J. Schummer (eds.), *Discovering the Nanoscale*, Amsterdam: IOS Press, pp. 51-62.

Nordmann, A.: 2004, 'Nanotechnology: Convergence and Integration', Presentation at *EuroNanoForum, Trieste, December 10, 2003* (Proceedings in preparation).

Nordmann, A.: 2004, 'Nanotechnology's WorldView: New Space for Old Cosmologies', *IEEE Technology and Society Magazine*, 23 (forthcoming).

Nordmann, A.: 2004, 'Social Imagination for Nanotechnology', in: European Commission (Community Health and Consumer Protection): *Nanotechnologies: A Preliminary Risk Analysis on the Basis of a Workshop, Brussels, 1-2 March 2004*, pp. 111-113 [www.europa.eu.int/comm/health/ph_risk/documents/ev_20040301_en.pdf].

Nordmann, A.: 2004, 'Was ist TechnoWissenschaft? ' Zum Wandel der Wissenschaftskultur am Beispiel von Nanoforschung und Bionik', in: T. Rossmann & C. Tropea (eds.), *Bionik ' Neue Forschungsergebnisse aus Natur-, Ingenieur- und Geisteswissenschaften*, Berlin: Springer, 2004 (forthcoming)

Paschen, H.; Coenen, C.; Fleischer, T.; Grünwald, R.; Oertel, D. & Revermann, C.: 2004, *Nanotechnologie: Forschung, Entwicklung, Anwendung*, Berlin: Springer (*Nanotechnologie, TAB-Arbeitsbericht 92*, Berlin: Büro für Technikfolgen-Abschätzung beim Deutschen Bundestag).

Pitt, J.C.: 2004, 'The Epistemology of the Very Small', in: D. Baird, A. Nordmann & J. Schummer (eds.), *Discovering the Nanoscale*, Amsterdam: IOS Press, pp. 157-163.

Robinson, C.: 2004, 'Images in NanoScience/Technology', in: D. Baird, A. Nordmann & J. Schummer (eds.), *Discovering the Nanoscale*, Amsterdam: IOS Press, pp. 165-169.

Robison, W.L.: 2004, 'Nano-Ethics', in: D. Baird, A. Nordmann & J. Schummer (eds.), *Discovering the Nanoscale*, Amsterdam: IOS Press, pp. 285-300.

Roco, M.C. & Bainbridge, W.S. (eds.): 2001, *Societal implications of nanoscience and nanotechnology*, (Proceedings of a workshop organized by the National Science Foundation, September 28-29, 2000), Kluwer: Dordrecht [available online at http://itri.loyola.edu/nano/societalimpact/nanosi.pdf]

Roco, M.C. & Tomellini, R. (eds.): 2002, *Nanotechnology: Revolutionary Opportunities and Societal Implications, Workshop, Lecce (Italy), 31 January - 1 February 2002*, Luxemburg: Office for Official Publications of the European Communities, [200 pp.]

Roco, M.C.; Bainbridge, W.S. (eds.): 2002, *Converging Technologies for Improving Human Performance: Nanotechnology, Biotechnology, Information Technology and the Cognitive Science*, Arlington, VA: National Science Foundation.

Roher, H. 1993. *Jpn. J. Appl. Phys.* 32:1335.

Rohlfing, E.A., D.M. Cox, and A. Kaldor. 1984. *J. Chem. Phys.* 81:3846.

Rosen, A. 1998. A periodic table in three dimensions: A sightseeing tour in the nanometer world. In *Advances in quantum chemistry*. In press.

Ruthven, D.M., S. Farooq, K.S. Knaebel. 1994. *Pressure swing adsorption*. New York: VCH Publishers.

Sarewitz, D.; Woodhouse, E.: 2003, 'Small is Powerful', in: A. Lightman, D. Sarewitz & Chr. Desser, (eds.), *Living with the Genie: Essays on Technology and the Quest for Human Mastery*, Washington, DC: Island Press, pp. 63-83.

Schiemann, G.: 2004, 'Dissolution of the Nature-Technology Dichotomy? Perspectives on Nanotechnology from an Everyday Understanding of Nature', in: D. Baird, A. Nordmann & J. Schummer (eds.), *Discovering the Nanoscale*, Amsterdam: IOS Press, pp. 209-213.

Schmidt, J.C.: 2004, 'Unbounded Technologies: Working Through Technological Reductionism of

Nanotechnology', in: D. Baird, A. Nordmann & J. Schummer (eds.), *Discovering the Nanoscale*, Amsterdam: IOS Press, pp. 35-50.

Schuler, E.: 2004, 'Perception of Risks and Nanotechnology', in: D. Baird, A. Nordmann & J. Schummer (eds.), *Discovering the Nanoscale*, Amsterdam: IOS Press, pp. 279-284.

Schummer, J.: 2001, 'Ethics of Chemical Synthesis', *Hyle: International Journal for Philosophy of Chemistry*, 7, 103-124 [online: www.hyle.org/journal/issues/7/schummer.htm].

Schummer, J.: 2003, 'Aesthetics of Chemical Products: Materials, Molecules, and Molecular Models', *Hyle: International Journal for Philosophy of Chemistry*, 9, 73-104 [online: www.hyle.org/journal/issues/9-1/schummer.htm].

Schummer, J.: 2003, 'The Notion of Nature in Chemistry', *Studies in History and Philosophy of Science*, 34 (2003), 705-736.

Schummer, J.: 2004, 'Interdisciplinary Issues in Nanoscale Research', in: D. Baird, A. Nordmann & J. Schummer (eds.), *Discovering the Nanoscale*, Amsterdam: IOS Press, pp. 9-20.

Schummer, J.: 2004, 'Multidisciplinarity, Interdisciplinarity, and Patterns of Research Collaboration in Nanoscience and Nanotechnology', *Scientometrics*, 59, 425-465.

Schummer, J.: 2004, 'Naturverhältnisse in der modernen Wirkstoff-Forschung', in: K. Kornwachs (ed.), *Technik ' System ' Verantwortung*, Münster: LIT, pp. 629-638.

Schummer, J.: 2004, 'Why do Chemists Perform Experiments?', in: D. Sobczynska, P. Zeidler, E. Zielonacka-Lis (eds.), *Chemistry in the Philosophical Melting Pot*, Peter Lang (Frankfurt/M.), 2004, pp. 395-410.

Schummer, J.: 2005, 'Reading Nano: The Public Interest in Nanotechnology as Reflected in Book Purchase Patterns", *Public Understanding of Science*, 14 (2) (forthcoming).

Schummer, J.; Baird, D. (eds.): 2004-5, *Nanotech Challenges*, joint special issue of *Hyle: International Journal for Philosophy of Chemistry & Techne: Research in Philosophy and Technology* (forthcoming).

Schwarz, A.E.: 2004, 'Shrinking the 'Ecological Footprint' with NanoTechnoScience?', in: D. Baird, A. Nordmann & J. Schummer (eds.), *Discovering the Nanoscale*, Amsterdam: IOS Press, pp. 203-208.

Schwarz, R.B. 1998. Storage of hydrogen in powders with nanosized crystalline domains. In *R&D status and trends*, ed. Siegel et al.

Siegel, R.W., E. Hu, and M.C. Roco, eds. 1998. *R&D status and trends in nanoparticles, nanostructured materials, and nanodevices in the United States*. Baltimore: Loyola College, International Technology Research Institute. NTIS #PB98-117914.

Stix Gary., "Toward 'Point One'," *Scientific American*, February 1995, pp. 90-95.

Stuckless, J.T, D.E. Starr, D.J. Bald, C.T. Campbell. 1997. *J. Chem. Phys*. 107:5547.

Suchman, M.: 2001, 'Envisioning Life on the Nano-Frontier', in: M.C. Roco and W.S. Bainbridge (eds.), *Societal Implications of Nanoscience and Nanotechnology*, Dordrecht: Kluwer, pp. 211-216.

Suchman, M.: 2002, 'Social Science and Nanotechnologies', M. Roco & R. Tomellini (eds.), *Nanotechnology ' Revolutionary Opportunities and Societal Implications*, Luxembourg: European Communities, pp. 95-99.

Sweeney, A.E.; Seal, S. & Vaidyanathan, P.: 2003, 'The promises and perils of nanoscience and nanotechnology: Exploring emerging social and ethical issues', *Bulletin of Science, Technology & Society*, 23 (4), 236-245.

Tanev, P.T., M. Chibwe, and T.J. Pinnavaia. 1994. *Nature* 368:321.

Taylor, K.J., C.L. Pettiette-Hall, O. Cheshnovsky, and R.J. Smalley. 1992. *J. Chem. Phys*. 96:3319.

Tenner, E.: 2001, 'Nanotechnology and Unintended Consequences', in: M.C. Roco and W.S. Bainbridge (eds.), *Societal Implications of Nanoscience and Nanotechnology*, Dordrecht: Kluwer, pp. 241-246.

The Royal Society & The Royal Academy of Engineering: 2004, *Nanoscience and nanotechnologies: opportunities and uncertainties*, London [www.nanotec.org.uk/finalReport.htm].

Toumey, C.: 2004, 'Anticipating Public Reactions to Nanotechnology', *Techne: Research in Philosophy and Technology*, 8(3) (forthcoming).

Trudeau, M.L., and J.Y. Ying. 1996. *Nanostructured Materials* 7:245.

Tschope, A.S., W. Liu, M. Flyzani-Stephanopoulos, and J.Y. Ying 1995. *J. Catal*. 157:42.

Turton, R., *The Quantum Dot*, Oxford University Press, 1995.

Verbeek, A. & Callaert, J.: 2002, *Detailed analysis of the science-technology interaction in the field of Nanotechnology*, Luxemburg: Office for Official Publications of the European Communities (Linking Science to Technology. Bibliographic References in Patents, Vol. 9).

Vermaas, P.E.: 2004, 'Nanoscale Technology: A Two-Sided Challenge for Interpretations of Quantum Mechanics', in: D. Baird, A. Nordmann & J. Schummer (eds.), *Discovering the Nanoscale*, Amsterdam: IOS Press, pp. 77-91.

Wald, C.A.: 2004, 'Working Boundaries on the NANO Exhibition', in: N.K. Hayles (ed.), *Nanoculture: Implications of the New Technoscience*, Bristol, UK: Intellect Books, pp. 83-108.

Weckert, J.: 2001, 'The Control of Scientific Research: The Case of Nanotechnology', *Australian Journal of Professional and Applied Ethics*, 3, 29-44.

Weil, V.: 2001, 'Ethical Issues in Nanotechnology', in: M.C. Roco and W.S. Bainbridge (eds.), *Societal Implications of Nanoscience and Nanotechnology*, Dordrecht: Kluwer, pp. 193-198.

Whitesides, G.M., "Self-Assembling Materials

Wolfson, J.R.: 2003, 'Social and Ethical Issues in Nanotechnology: Lessons from Biotechnology and Other High Technologies', *Biotechnology Law Report*, 22 (4), 376-96.

Woods, S.; Jones, R. & Geldart, A.: 2003, *The Social and Economic Challenges of Nanotechnology*, Report to the Economic and Social Research Council (ESRC), Swindon, UK [www.esrc.ac.uk/esrccontent/DownloadDocs/Nanotechnology.pdf].

Yeo, Y.Y., C.E. Wartnaby, and D.A. King. 1995. *Science* 268:1731.

Zimmermann, U., N. Malinowski, A. Burkhardt, and T.P. Martin. 1995. *Carbon* 33:995.

4

Cruicial Issues in Nano Biotechnology

However broad the range of synthesis approaches, the critical control points fall into two categories:

1. control of the size and composition of the nanocluster components, whether they are aerosol particles, powders, semiconductor quantum dots, or other nanocomponents

2. control of the interfaces and distributions of the nanocomponents within the fully formed materials These two aspects of nanostructure formation are inextricably linked; nevertheless, it is important to understand how to exercise separate control over the nucleation of the nanostructure building blocks and the growth (for example, minimizing coagulation or agglomeration) of those components throughout the synthesis and assembly process. This latter issue is related to the importance of the following:

- the chemical, thermal, and temporal stability of such formed nanostructures
- the ability to scale-up synthesis and assembly strategies for low-cost, large-scale production of nanostructured materials, while at the same time maintaining control of critical feature size and quality of interfaces (economic viability is a compelling issue for any nanostructure technology) All researchers in this area are addressing these issues.

Outstanding Issues: Opportunities and Challenges

The broad explorations of this WTEC study have shown a great diversity of impressive work on nanostructured materials. One could argue that many aspects of the work on nanostructured materials have been long-established efforts, with well-developed techniques that have been brought forth into the manufacturing arena.

How do we then explain the current excitement and interest in nanostructured materials as a promising new endeavor? Part of the answer lies in the recognition of common scientific issues and common enabling technologies that link this group of researchers together. Ready availability of ever more sophisticated characterization methods that allow us to visualize and probe materials at the nanoscale has accelerated the pace of activities in this field; at the same time, recognition of the common critical issues of control over nanostructure size and placement motivates sharing of solutions over the boundaries of conventional disciplines. Thus, some of the most exciting findings of this study manifest the crossfertilization of techniques and ideas; for example, aerosol particles are being integrated with more traditionally fabricated electronic nanostructures, with placement achieved through the manipulation of STM tips, in order to explore ideas of electronic device enhancement at the nanostructure level (Junno et al. 1998). The ordered assembly of diblock copolymers define three-dimensional nanostructures, and those structures are transferred into electronic (semiconductor) substrates through high resolution pattern transfer processes (Möller 1998). Recognizing the need for effective utilization of ideas across disciplines brings the responsibility of establishing the educational infrastructure that will adequately train young scientists to more fully develop the concepts and applications of nanostructured materials in the future. Such an infrastructure will require researchers and educators who are familiar with the properties of a broad range of materials, including polymers, biomolecular materials, metals, ceramics, and semiconductors. It will benefit from an informed perspective on critical applications, and must provide access to a wide range of synthesis and assembly techniques and characterization methods that currently reside separately in the disciplines of physics, chemistry, electrical engineering, biology, etc. Finally, a critical enabler for the future of this field and for its educational infrastructure is further development of computational tools that encompass the full range of atomistic calculations to macroscopic materials properties. This will require a "systems approach" (Olson 1997) that will span a variety of different computational models, addressing the different critical length scales, starting with solutions of the Schrodinger and Poisson equations, solving interatomic forces, and scaling all the way up to simulations of macroscopic properties and behavior (Goddard 1998). Increased appreciation of and access to the diverse means of nanoparticle synthesis and assembly that have been developed within many different disciplines, and a common development of enabling tools and technologies, will enhance the pace of accomplishments in this new area of nanoscale synthesis and assembly.

Common Enabling Technologies

There has been steady technological progress in all fields of nanostructure synthesis and assembly, in no small part because of the more general availability of characterization tools having higher spatial, energy, and time resolution to clearly distinguish and trace the process of nanostructure formation. As transmission electron microscopy and X-ray diffraction techniques helped in an earlier period to relate the improved properties of "age-hardened" aluminum alloys to their nanostructure (Koch 1998), today's technological advances in materials characterization are providing new insights into the role of the nanostructure in determining macroscopic properties. The tightly-coupled iteration between characterizing the nanostructure, understanding the relationship between nanostructure and macroscopic material properties, and improved sophistication and control in determining nanostructure size and placement have accelerated the rate of progress and helped to define the critical components of this "new" field of nanostructure science and technology. Tightly focused (1-2 μm), high brightness synchrotron X-ray sources provide detailed structural information on colloids, polymers, alloys, and other material structures, highlighting the inhomogeneities of the material with suitable spatial resolution (Hellemans 1998).

Improved characterization - higher spatial resolution - higher sensitivity leads t o motivat es Better control of size and placement Understanding of - structure-property link - influence of interfaces stimulates.

Interactive cycle of characterization, understanding and enhanced control in the synthesis and assembly of nanostructures. Another important enabling technology has been the now widely available scanning probe technology, including scanning tunneling microscopy and atomic force microscopy. The power of these techniques has provided impetus for developing even higher performance scanning probe tips, fabricated through microfabrication techniques. Development of different tip structures in various materials has given rise to an entire family of powerful scanning probe techniques that encompass such a wide range of characterization capabilities that one can envision "a laboratory on a tip" (Berger et al. 1996). The development of a tip technology also impacts the synthesis and assembly processes themselves: scanning probe technologies have been used as the basis of materials patterning and processing at nanometer scales (Held et al. 1997; Snow et al. 1997; and Wilder et al. 1997) and have provided information on the mechanical and thermal properties of materials at the nanoscale (Nakabeppu et al. 1995; Tighe et al. 1997; and Zhang et al. 1996). More sophisticated in situ monitoring strategies have provided greater understanding and control in the synthesis of nanostructured building blocks, particularly those formed in vacuum environments. Molecular beam epitaxy (MBE) represents a physical vapor (gas phase) deposition technique where sub-monolayer control can be imposed on the formation of two-dimensional and, more recently, three-dimensional nanostructured materials (Leonard et al. 1993). A great deal of the understanding and control derives from the ability to carry out sensitive monitoring of the growth process in situ: reflection high energy electron diffraction (RHEED) details the nature of the surface and surface bonding, and oscillations of the RHEED intensity provide information on the growth rate (Neave et al. 1983). The improvements brought about by these advances in technology have been substantial, but perhaps of greater importance for this nascent field of nanostructure science and technology has been the development of strategies and technologies that have been formed across the former disciplines. More reliable means of controlling nanostructure size and placement, with an end view of being able to scale up the production of such materials while maintaining the control over the nanostructure, have given impetus to a common search for novel synthesis and assembly strategies. In that search, it is apparent that the naturally occurring synthesis and assembly of biological materials can provide us with some critical insights.

Major efforts in nanoparticle synthesis can be grouped into two broad areas: gas phase synthesis and sol-gel processing. Nanoparticles with diameters ranging from 1 to 10 nm with consistent crystal structure, surface derivatization, and a high degree of monodispersity have been processed by both gas-phase and sol-gel techniques. Typical size variances are about 20%; however, for measurable enhancement of the quantum effect, this must be reduced to less than 5% (Murray et al. 1993). Initial development of new crystalline materials was based on nanoparticles generated by evaporation and condensation (nucleation and growth) in a subatmospheric inert-gas environment (Gleiter 1989; Siegel 1991, 1994). Various aerosol processing techniques have been reported to improve the production yield of nanoparticles (Uyeda 1991, Friedlander 1998). These include synthesis by combustion flame (Zachariah 1994, Calcote and Keil 1997, Axelbaum 1997, Pratsinis 1997); plasma (Rao et al. 1997); laser ablation (Becker et al. 1997); chemical vapor condensation (Kear et al. 1997); spray pyrolysis (Messing et al. 1994); electrospray (de la Mora et al. 1994); and plasma spray (Berndt et al. 1997). Sol-gel processing is a wet chemical synthesis approach that can be used to generate nanoparticles by gelation, precipitation, and hydrothermal treatment (Kung and Ko 1996). Size distribution of semiconductor, metal, and metal oxide nanoparticles can be manipulated by either dopant introduction (Kyprianidou-Leodidou et al. 1994) or heat treatment (Wang et al. 1997). Better size and stability control of quantum-confined semiconductor nanoparticles can be achieved through the use of inverted micelles (Gacoin 1997), polymer matrix architecture based on block copolymers (Sankaran et al. 1993) or polymer blends (Yuan et al. 1992), porous glasses (Justus et al. 1992), and ex-situ particle-capping techniques (Majetich and Canter 1993; Olshavsky and Allcock 1997).

Additional nanoparticle synthesis techniques include sonochemical processing, cavitation processing, microemulsion processing, and highenergy ball milling. In sonochemistry, an acoustic cavitation process can generate a transient localized hot zone with extremely high temperature gradient and pressure (Suslick et al. 1996). Such sudden changes in temperature and pressure assist the destruction of the sonochemical precursor (e.g., organometallic solution) and the formation of nanoparticles. The technique can be used to produce a large volume of material for industrial applications. In hydrodynamic cavitation, nanoparticles are generated through creation and release of gas bubbles inside the sol-gel solution (Sunstrom et al. 1996). By rapidly pressurizing in a supercritical drying chamber and exposing to cavitational disturbance and high temperature heating, the sol-gel solution is mixed. The erupted hydrodynamic bubbles are responsible for nucleation, growth, and quenching of the nanoparticles. Particle size can be controlled by adjusting the pressure and the solution retention time in the cavitation chamber. Microemulsions have been used for synthesis of metallic (Kishida et al. 1995), semiconductor (Kortan et al. 1990; Pileni et al. 1992), silica (Arriagada and Osseo-Assave 1995), barium sulfate (Hopwood and Mann 1997), magnetic, and superconductor (Pillai et al. 1995) nanoparticles. By controlling the very low interfacial tension (~10-3 mN/m) through the addition of a cosurfactant (e.g., an alcohol of intermediate chain length), these microemulsions are produced spontaneously without the need for significant mechanical agitation. The technique is useful for large-scale production of nanoparticles using relatively simple and inexpensive hardware (Higgins 1997). Finally, high energy ball milling, the only top-down approach for nanoparticle synthesis, has been used for the generation of magnetic (Leslie- Pelecky and Reike 1996), catalytic (Ying and Sun 1997), and structural (Koch 1989) nanoparticles. The technique, which is already a commercial technology, has been considered dirty because of contamination problems from ball-milling processes. However, the availability of tungsten carbide components and the use of inert atmosphere and/or high vacuum processes have reduced impurities to acceptable levels for many industrial applications. Common drawbacks include the low surface area, the highly polydisperse size distributions, and the partially amorphous state of the asprepared powders.

Synthesis Issues Means to Achieve Monodispersity

One of the most challenging problems in synthesis is the controlled generation of monodispersed nanoparticles with size variance so small that size selection by centrifugal precipitation or mobility classification is not necessary. Among all the synthesis techniques discussed above, gas-phase synthesis is one of the best techniques with respect to size monodispersity, typically achieved by using a combination of rigorous control of nucleationcondensation growth and avoidance of coagulation by diffusion and turbulence as well as by the effective collection of nanoparticles and their handling afterwards. The stability of the collected nanoparticle powders against agglomeration, sintering, and compositional changes can be ensured by collecting the nanoparticles in liquid suspension. For semiconducting particles, stabilization of the liquid suspension has been demonstrated by the addition of polar solvent (Murray et al. 1993); surfactant molecules have been used to stabilize the liquid suspension of metallic nanoparticles.

Alternatively, inert silica encapsulation of nanoparticles by gas-phase reaction and by oxidation in colloidal solution has been shown to be effective for metallic nanoparticles (Andres et al. 1998). New approaches need to be developed for the generation of monodisperse nanoparticles that do not require the use of a size classification procedure. An example of this is a process developed in Japan where very monodispersed gold colloidal nanoparticles with diameters of about 1 nm have been prepared by reduction of metallic salt with UV irradiation in the presence of dendrimers (Esumi et al. 1998). Poly(amidoamine) dendrimers with surface amino groups of higher generations have spherical 3-D structures, which may have an effective protective action for the formation of gold nanoparticles. Although the specific role of dendrimers for the formation of monodispersed nanoparticles has yet to be defined, good monodispersity is thought to come from the complex reaction accompanying the decomposition of dendrimers, which eventually leads to the conversion of solution ions to gold nanoparticles.

Scaleup production is of great interest for nanoparticle synthesis. High energy ball milling, already a commercial high-volume process, as mentioned above, has been instrumental in generating nanoparticles for the preparation of magnetic, structural, and catalytic materials. However, the process produces polydispersed amorphous powder, which requires subsequent partial recrystallization before the powder is consolidated into nanostructured materials. Although gas-phase synthesis is generally a low production rate process (typically in the 100 milligrams per hour range) in research laboratories, higher rates of production (about 20 grams per hour) are being demonstrated at Ångström Laboratory at Uppsala University in Sweden. Even higher production rates (about 1 kg per hour) are now being achieved commercially. For sol-gel processing, the development of continuous processing techniques based on present knowledge of batch processing has yet to be addressed for economical scaleup production of nanoparticles. Other related sol-gel issues concern the cost of precursors and the recycling of solvent. Overall, sol-gel processing is attractive for commercial scale-up production.

A recent paradigm shift envisioned for optoelectronics and computational devices involves the assembly of molecular or quantum wires (Chidsey and Murray 1986). Large polymeric molecules have been used as nano building blocks for nanoporous molecular sieves, biocompatible materials, optical switching, data processing, and other nonlinear optical components. Chain aggregates of nanoparticles can be considered as polymer-like units with their primary particles composed of a few hundred to a few thousand molecules. Thus, these chain aggregates can be considered "heavy" quantum wires. In fact, nanoparticle chain aggregates have been studied extensively as magnetic materials (Zhang and Manthiram 1997), as reinforced elastomers (Pu et al. 1997), and as additives in concrete (Sabir 1997). These aggregates have been shown to have chemical and mechanical properties different from those of individual primary particles (Friedlander et al. 1998). Depending on the particle size and its compositional material, the bonding force responsible for holding the aggregates together varies from weak van der Waals force for micrometer particles to strong chemical bonds for nanometer particles to very strong magnetic dipolar bonds for nanosized magnetic particles. The mechanical, optical, and electronic transport properties can be varied by controlling the diameter and the monodispersity of the primary particles, the crystalline structure and morphology, aggregate length, interfacial properties, and material purity. These chain aggregates can be formed by allowing agglomeration of nanoparticles generated by any of the synthesis techniques discussed above, with the exception of high energy ball milling, which generates particles with low surface area and high anisotropic morphologies, both of which are detrimental for the formation of chain aggregates. Depending on the magnetic and electric charging properties of the nanoparticles, an external applied magnetic or electric field can be used to control the fractal dimension of aggregates. For optical applications of chain aggregates, lower fractal dimensions (i.e., relatively straight chain aggregates with few branches) are desirable.

Recent advances in the fabrication of nanometer fibers or tubes offer another form of building blocks for nanostructured materials. An effective way to generate nanometer fibers (or tubes) is based on the use of membrane-template techniques (Martin 1994). Membranes, with nanochannels generated by fission-fragment tracks or by electrochemical etching of aluminum metal, are used as templates for either chemical or electrochemical deposition of conductive polymers (Pathasarathy and Martin 1994), metal (van de Zande et al. 1997), semiconductor (Klein et al. 1993), and other materials for the generation of nanofibers or tubes. Since the nanochannels on membranes are very uniform in size, the diameter and the aspect ratio of the nanofibers (or tubes) synthesized by the membranetemplate technique can be precisely controlled. This greatly facilitates the interpretation of optical data and the processing of these fibers (or tubes) into 2-D nanostructured materials (de Heer et al. 1995). Single-crystal semiconductor nanofibers can also be grown catalytically by metalorganic vapor phase epitaxy and laser ablation vapor-liquid-solid techniques (Hiruma et al. 1995; Morales and Lieber 1998). The synthesis of these onedimensional structures with diameters in the range of 3 to 15 nm holds considerable technological promise for optoelectronic device applications, such as the p-n junctions for light emission at Hitachi Central Research Laboratory in Japan. The advent of carbon-based nanotubes has created yet another way to fabricate nanometer fibers and tubes. These nanotubes have been used as templates for the fabrication of carbide and oxide nanotubes (Dai et al. 1995; Kasuga et al. 1998). Synthesis of nanotubes based on BN, BC3 and BC2N have also been reported (Chopra et al. 1995; Miyamoto et al. 1994). These nanotubes potentially possess large third-order optical non-linearity and other unusual properties (Xie and Jiang 1998). Metallic nanofibers synthesized by carbon-nanotube-template techniques are useful in the design of infrared absorption materials. The carbon nanotubes can now be catalytically produced in large quantities and have been used for reinforcement of nanostructural composite materials and concrete (Peigney et al. 1997). The elegant complexity of biological materials represents the achievement of structural order over many length scales, with the full structure developed from the "nested levels of structural hierarchy" (Aksay et al. 1992), in which self-assembled organic materials can form templates or scaffolding for inorganic components. These notions of a multilevel material structure with strong interactions among levels and an interplay of perfection and imperfection forming the final material was discussed earlier by Cyril Stanley Smith (1981). Along with characteristic length scales, there are characteristic relaxation times of the material, bringing in the consideration of the temporal stability of the structured materials (Zener 1948). A more detailed discussion of the role of biological materials as both paradigm and tool for the fabrication of nanostructured materials is given. It is interesting how many of the synthesis and assembly approaches seek out and adapt two key features of biogenic fabrication: that of "self-assembly," and the use of natural nanoscaled templates.

Opportunities for Clusters and Nanocrystalline Materials

Clusters are groups of atoms or molecules that display properties different from both the smaller atoms or molecules and the larger bulk materials. Many techniques have been developed to produce clusters, beams of clusters, and clusters in a bottle for use in many different applications including thin film manufacture for advanced electronic or optical devices, production of nanoporous structures as thermal barrier coatings, and fabrication of thin membranes of nanoporous materials for filtration and separation. An apparatus developed at the University of Göteborg to measure cluster reactivity and sticking probability as a function of the number of metal atoms in the cluster. The unique properties of nanoparticles make them of interest. For example, nanocrystalline materials composed of crystallites in the 1-10 nm size range possess very high surface to volume ratios because of the fine grain size. These materials are characterized by a very high number of low coordination number atoms at edge and corner sites which can provide a large number of catalytically active sites. Such materials exhibit chemical and physical properties characteristic of neither the isolated atoms nor of the bulk material.

One of the key issues in applying such materials to industrial problems involves discovery of techniques to stabilize these small crystallites in the shape and size desired. This is an area of active fundamental research, and if successful on industrially interesting scales, is expected to lead to materials with novel properties, specific to the size or number of atoms in the crystallite. Schematic drawing of the experimental setup used in Göteborg for studies of chemical reactivity and/or sticking probability of various molecules with the clusters. The production of clusters is via laser vaporization of metal substrates and detection via photoionization time-of-flight mass spectrometry (A. Rosen, University of Göteborg, Sweden). A typical objective of nanoscale catalyst research is to produce a material with exceedingly high selectivity at high yield in the reaction product or product state, i.e., chemicals by design, with the option of altering the product or product state simply by changing the surface functionality, elemental composition, or number of atoms in the catalyst particle. For instance, new catalysts with increasing specificity are now being fabricated in which the stoichiometry may be altered due to size restrictions or in which only one or two spatial dimensions are of nanometer size. Five recent examples where nanocrystalline metallic and ceramic materials have been successfully investigated for catalytic applications are discussed below (Trudeau and Ying 1996).

Opportunities in Self-assembly

In self-assembly large molecular structures are obtained from the organization of a large number of molecules or atoms into a given shape, typically through specific interactions of the molecules among themselves and with a template. The interaction of the different bonding mechanisms is an area of strong fundamental research interest. Only two areas will be highlighted here: zeolites and carbon materials. Both of these materials exhibit characteristics of self-assembly, namely novel and reproducible structures that can be fabricated in industrially significant quantities.

Zeolitic Materials

Aluminosilicates (e.g., zeolites) are crystalline porous nanostructures with long range crystalline order with pore sizes which can be varied from about 4 Å to 15 Å in conventional zeolites. A 3-dimensional (e.g., MFI) zeolite cage structure together with a depiction of the straight and ziz-zag channels and a 2-dimensional zeolite with channels only in 2 directions. The vertices in the stick drawing denote position of the O atoms in the crystalline lattice. This particular zeolite has 10 atoms in the zeolite "window." The size of the window is determined by the number of oxygens in the ring. Typical zeolite structures together depicting the positions of the O atoms and two different zeolitic structures one (lower left) with a three dimensional structure and (lower right) a zeolite with a two dimensional channel structure.

For example normal hexane with a kinetic molecular diameter of about 5.1 Å can pass through a 10 ring or larger, whereas cyclohexane with a kinetic molecular diameter of 6.9 Å would be hard pressed to pass through a 10 ring. Thus all other things being equal, a 10-ring zeolite could be used to separate mixtures of normal hexane and cyclohexane. It is this property together with the ability to chemically modify the acidity of zeolitic materials that makes them extremely valuable as selective sorbants, as membranes and for use in selective catalytic reactions. In 1992, a new family of aluminosilicates (M41S) with pores sizes between 20 and 100 Å in diameter were reported by Mobil researchers (Beck et al. 1992; Kresge et al. 1992). One of particular interest is MCM-41, which consists of hexagonal arrays of uniform 2 to 10 nanometer-sized cylindrical pores. Not only can such materials be synthesized, but novel structures such as "tubules-within-a-tubule" have been fabricated as mesoporous molecular sieves in MCM-41 (Lin and Mou 1996).

Of particular interest is the possibility of expanding the so-called "liquid crystal templating" mechanism (Chen et al. 1993) to non-aluminum dopants within the silicate MCM-41 framework (Tanev et al. 1994) and to derive nonsiliceous MCM-41 type of materials (Braun et al. 1996). Another approach to synthesizing large pore and large single crystals of zeolytic materials is being pioneered by Geoffrey Ozin and his group at the University of Toronto, who have demonstrated that crystals as large as 5 mm can be synthesized (Kupperman et al. 1993). The ability to synthesize such large crystals has important implications for discovery of new sensors (selective chemical adsorbants) and membrane devices (selective transport of molecular species), since large single crystals can now be available to the laboratory researcher to carry out fundamental studies of adsorption and diffusion properties with such materials. These materials are expected to create new opportunities for applications in the fields of separations science, for use directly as molecular sieves or as new molecular sieving sorbant materials; in catalysis, as heterogeneous catalysts; and as supports for other catalytic materials as well as other novel applications (Bowes et al. 1996; Brinker 1996; Sayari 1996). The ability to synthesize zeolitic materials of precise pore size in the range between 4 and 100 Å continues to expand the possibilities for research and technological innovation in the catalytic, separations, and sorption technologies (Ruthven et al. 1994; Karger and Ruthven 1992).

Carbon Materials

The carbon-based materials of interest from a molecular self-assembly point of view include fullerenes and their relatives, including endohedral fullerenes and metal-coated fullerenes, carbon nanotubes, carbon nanoparticles, and porous carbons. Since 1990 with the discovery of techniques to produce soluble carbon in a bottle (for examples, see Krätschmer et al. 1990 and references therein), research on and with carbon materials has skyrocketed (Dresselhaus et al. 1996; Dresselhaus and Dresselhaus 1995). Not only can the molecular forms of carbon (the fullerenes and their derivatives) be synthesized, characterized, and studied for applications, but many other new carbon materials such as multi- and single-walled carbon nanotubes can now be produced in macroscopic quantities.

The broad variety of carbon nanotube structures whose properties are now being examined both theoretically and experimentally. A rich literature on these new carbon materials now exists. This report will only attempt to highlight a few important recent examples in the area of high surface area materials. Of particular interest for future catalytic applications is the recent report that not only can C60 be coated with metal atoms, but that the metal coating can consist of a precise number of metal atoms. For example, C60Li12 and C60Ca32 have been identified mass spectroscopically (Martin, Malinowski, et al. 1993; Martin, Naher, et al. 1993; Zimmerman et al. 1995). C60 has been coated with a variety of different metals, including Li, Ca, Sr, Ba, V, Ta and other transition metals. Interestingly, addition of more than 3 Ta atoms to C60 breaks the C60 cage. Replacement of one carbon atom in C60 by a transition metal atom such as Co or Ir is being studied for possible catalytic applications. The future technological challenge will be to discover techniques to fabricate large quantities of such materials, so that such catalyst materials can be put in a bottle and not just in molecular beams.

Examples of carbon nanotube structures, including multiwalled and metal-atomfilled nanotubes. Carbon nanotubes have the interesting property that they are predicted to be either semiconducting or conducting (metallic), depending on the chirality and diameter of the nanotube. Such materials are being studied as conductive additives to plastics and for use in electrochemical applications where the uniformity of the nanotube diameter and length is not overly critical (Dresselhaus 1998). Another approach is to use the carbon nanotube as a template for a nanotube of an inorganic oxide. Hollow nanotubes of zirconia and yttria-stablilized zirconia have been prepared by coating treated carbon nanotubes with a zirconium compound and then burning out the carbon template (Rao et al. 1997). Finally, large scale production of singlewalled nanotubes has recently been demonstrated, so one may anticipate a strong upsurge in the characterization and potential usage of single-walled carbon nanotubes in the future (Jounet et al. 1997). Porous carbons are of interest as molecular sieve materials, both as sorbants and as membranes, or as nanostraws for filtration.

One of the major research objectives is to develop materials or structures with exceedingly high storage capacity per unit volume and weight for gases such as H2 or CH4. H2 or CH4 could become an economic source of combustion fuel or a means to power fuel cells for ultralow-emission vehicles or for electric power generation. Microporous hollow carbon fibers have exhibited high permeance and high selectivity as hydrogen selective membranes, and development is now underway to scale up these membranes to commercial levels (Soffer et al. 1987; Jones and Koros 1994; Rao and Sircar 1993). Carbon fiber materials produced via catalytic decomposition of hydrocarbon vapors have also recently been reported to exhibit exceptionally high hydrogen adsorption capacity (Baker 1998). More mundane uses of nanotubes are as nanometer reinforcing rods in polymers or even in concrete. Incorporation of conducting carbon nanotubes in construction materials such as concrete or structural plastics opens opportunities for real time monitoring of material integrity and quality.

Microporous and Dense Ultrathin Films

Research and development of microporous thin films for use as molecular sieving membranes using inorganic crystalline materials such as zeolites or porous silica is another area of active research around the world. For molecular sieving membranes, one critical challenge rests on discovering ways to create large scale, thin, nearly defect-free membranes. One recent example is the fabrication of mesoporous conducting thin films grown from liquid crystal mixtures (Attard et al. 1997). Transmission electron microscopy (TEM) reveals an ordered array of 2.5 nm diameter cylindrical holes in a 300 nm thick Pt film. The hole diameter can be varied either by changing the chain length of the surfactant molecule or by adding an alkane to the plating solution. It is interesting that this technique produces a continuous thin film with nanoscale porosity in an electrically conducting material. Dense ultrathin films such as single monolayer films would be of significant importance in the semiconductor industry. Thin films of specialized coatings for corrosion, thermal, and/or chemical stability should be valuable for the chemical and aerospace industries. Novel chemical sensors may be anticipated through use of ultrathin films composed of specialized clusters. Typical techniques for production of thin films are physical vapor deposition, chemical vapor deposition, and Langmuir- Blodgett processes.

Social & Ethical Issues in Nano Biotechnology

At some point in the near future, nanomachines will produce consumer goods, rather than the bulky machines that are used today. NT works "...by piecing together one atom or molecule at a time..." For nanotechnology (NT) to work, small devices called assemblers must be created in order to combine atoms that will produce consumer goods. Replicators will also have to be made to produce enough assemblers that will be required to create goods.

Of the many positive features of NT, nanomachines will produce: stronger products than those produced by traditional methods; food and water will be able to be replicated, thus, eliminating hunger worldwide; computers could become dramatically more powerful without any increase in size; nanosurgeons could undertake delicate operations much more precise than current surgery is able to; and nanorobots could be used to alter one's physical features such as ears, nose, eye colour, etc.

Nanotechnology could also have many positive effects on the environment, including: reconstruction of the ozone layer; quickly cleaning up oil spills; purifying water; and eliminating the use of non-renewable resources (which could be made by nanomachines).

Brown, Doug "Nano Litterbugs? Experts See Potential Pollution Problems." *(From www.smalltimes.com).*

The U.S. Environmental Protection Agency (EPA) has begun research into the possible negative impacts associated with NT. Even as some products using NT have begun to be widely distributed, the environmental impact is still unknown.

A concern for Vicki Colvin, who is director of The Center for Biological and Environmental Nanotechnology, is that no research is being conducted into how to control nanomaterial if it was discovered to be harmful to humans or the environment.

Based on preliminary research on animals, it is known that nanomaterials do accumulate within the body. Also, nanomaterials can be overtaken by bacteria, which could mean nanomaterials could infiltrate the food chain. Bacteria enhanced by nanomaterials could produce more potent bacteria. While NT promises to have numerous benefits for society, many of the potential risks remain uninvestigated.

Joy, Bill "Why the future doesn't need us." (*In* Wired *magazine*. April, 2000).

Joy's concern over NT is amplified, due to the fact that engineered organisms can self-replicate. Such replication could lead to situations where things could quickly get out of hand. "A bomb is blown up only once—but one bot can become many, and quickly get out of control." An additional worry for Joy is, unlike weapons of mass destruction (WMD) – which require hard-to-get materials and highly specialized knowledge – genetics, nanotechnology, and robotics (GNR) are available to a wide range of people, thus making them all the more dangerous. While WMD are strictly controlled by the military, GNR's are corporate sponsored. Such a wide dispersion of NT increases the risk. Joy emphasizes prudence in our development of NT, because if we proceed too quickly, we could cause our own extinction. After all, we are still unable to control the spread of viruses and other undesirable components of life.

Joy draws a parallel between NT and the atomic bomb. The creation of the atomic bomb led to the development of an arms race that lasted nearly half a century and had many detrimental effects. Atomic and nuclear weapons have the ability to destroy the world many times over, as does NT. We should not repeat the same mistakes with NT that we did with WMD.

"The only realistic alternative I see is relinquishment: to limit development of the technologies that are too dangerous, by limiting our pursuit of certain kinds of knowledge." Limiting access to critical information involving NT must be enforced because it contains such a destructive force to humanity (e.g. exterminating the entire race). People will inevitably object to the infringement on their right to privacy and pursuit of intellectual property, but this is a necessary step if we are to avoid the dangers NT poses. Additionally, scientists will have to comply with strict ethical codes to ensure compliance with relinquishment rules.

While humanity's dream of immortality will likely continue forever, certain limits to the means creating this end must be imposed due to the inherent danger.

Mehta, Michael D. "On Nano-Panopticism: A Sociological Perspective."

This article deals with privacy issues associated with surveillance, and how surveillance will be changed with the development of NT.

It has been scientifically documented that observation of a subject will change the regular behaviour of that subject. For example, prisoners who are held in prison cells that are in open view of a prison guard behave better than those prisoners who are afforded more privacy. Therefore, if this concept is applied to society as a whole, one can see the ramifications increased surveillance might have on public behaviour. Increased surveillance with the use of NT may have many damaging effects that may express themselves through a decrease in public trust, less involvement in public life, and may result ultimately in an Orwellian state.

Mehta offers several suggestions on how to deal with the dangers related to nano-panopticism. They are: develop an independent agency to monitor and regulate progress in NT; introduce tough laws to ensure that privacy is protected; a review of a corporation's code of conducts to ensure nano-panopticism does not become a problem; public consultation of the appropriate uses of NT; and finally, increase the awareness of scientists so that they are better able to understand the impact new technologies have on society.

Mnyusiwalla, A. Abdallah S. Daar, & Peter A. Singer "'Mind the gap': science and ethics in nanotechnology." *(From* Nanotechnology, *February 13, 2003).*

As NT quickly develops, the ethical evaluation of such a development has yet to begin. Research into NT's impact on ethical, environmental, economic, legal, and social implications (NE^3LS) must try to keep pace with the technological progress that has been made. Otherwise, the technological progress will slow down.

K. Eric Drexler, author of *Engines of Creation*, and currently working for the Foresight Institute, has begun to develop guidelines for the use of NT, particularly concerning self-replication, but also issues such as wealth distribution and environmental protection.

While there is a call for increased research into the effects of NT on society, the environment, etc, and furthermore, that the funds for such research are available, very little research has been done. For example, The National Nanotechnology Initiative (NNI) spent less than half of its budget on studying the impacts NT will have on society. In addition, the National Science Foundation (NSF) did not fund any research into the societal implications of NT due to lack of research grant proposals.

The ethical issues associated with NT fall into a variety of categories including:

1. *Equity*: NT does not stand to help developed countries only, but also (and perhaps mainly) developing countries. For example, third world countries suffer most from things that can be improved upon from advancements made in NT. For instance, providing cleaner water, developing cheaper energy, and also the enormous health benefits to be reaped from NT are all aspects of NT that could have a dramatic impact on third world countries.
2. *Privacy and security*: While NT could dramatically improve surveillance systems and would undoubtedly have numerous military applications, some are left wondering how personal privacy would be affected. Questions concerning the regulation of this new technology are arising without many answers being offered.
3. *Environment*: The effects of NT materials on the environment are unknown. The EPA has increased funding to study the effects of NT on the environment.
4. *Human or machine?* How far are humans willing to go with replacement of human parts with replacement robot parts?

Earlier lessons from genomics and biotechnology (recommendations):

- A percentage of the research and development funds should go to study the societal impacts of NT. In the *Human Genome Project*, 3-5% of the total budget was dedicated to studying the societal impacts.
- Large-scale interdisciplinary research platforms should be avoided.
- The capacity for research should be increased. This would include research grants, career awards, increased funding, etc.

- There should be a cross-section of groups working on the implications of NT in order to have a variety of views expressed.
- A global opinion-leaders network for social and ethical implications ought to be established so that third world countries may be involved.
- Public engagement is crucial if NT is to have the impact it potentially holds. The public needs to have accurate information concerning the social and ethical implications of NT. Science museums could include exhibits on NT; as well, schools could discuss the issue in depth.

Moor, James & John Weckert "Nanoethics: Assessing the Nanoscale from an Ethical Point of View."

Weckert and Moor argue that nanoethics cannot develop fully or effectively until the technology develops further and its consequences become known. Weckert and Moor do acknowledge the potentially enormous impact NT will have on society, but are hesitant to develop strategies to deal with problems that do not yet exist. They recommend continued monitoring of NT.

Pagán, Sylvia "DNA nanoballs boost gene therapy." *(In* New Scientist*, May, 2002).*

NT is now beginning to enhance older technologies used to treat patients with cystic fibrosis. Nanoparticles made up of a single DNA molecule are much smaller, and therefore much better able to reach the nucleus of a cell, than were previous methods involving modified viruses. NT shows promising signs of contributing to the advancement of the cure for cystic fibrosis.

Phoenix, Chris & Mike Treder "Applying the Precautionary Principle to Nanotechnology." *(From* The Center for Responsible Nanotechnology*, January, 2003).*

There are two forms of the precautionary principle: (1) the strict form; and (2) the active form.

Strict form: "...requires inaction when action might pose a risk."

Active form: "...calls for choosing less risky alternatives when they are available, and for taking responsibility for potential risks."

One shortcoming of the strict form of the Precautionary Principle is that most actions carry some sort of risk. If we follow the strict form, much of the scientific research currently underway would have to be halted. Second, sometimes inaction can be more harmful than action. Sometimes the benefits to be gained by scientific research outweigh the minimal risks involved.

The Center for Responsible Nanotechnology (CRN) believes that a total ban on development of NT would not be practical; enforcement of such a ban would be nearly impossible. Not only is such a ban not practical, it may not be desired. There are many benefits that could be brought about by NT, including increased health, less environmental damage, cheap and efficient production of goods, etc. These factors must be considered when determining the fate of NT.

The CRN believes that the strict form of the Precautionary Principle is inadequate for dealing with NT. This is so because inaction carries with it certain risks:

1. No other solution may be found for certain pressing problems.
2. Inaction on the part of responsible people could simply lead to the development and use of MNT by less responsible people.
3. Lack of understanding of the technology will leave the world ill-equipped to deal with irresponsible use.

The CRN does, however, believe that the active form of the Precautionary

Principle is adequate in dealing with research into NT. Because the active form seeks to develop the least risky solution to existing problems, this is more desirable than the strict form.

The final recommendation of the CRN is to create "...one–*and only one*–molecular nanotechnology program, [which would permit] the widespread but restricted use of the resulting manufacturing capability.

Phoenix, Chris & Mike Treder "Safe Utilization of Advanced Nanotechnology." *(From* The Center for Responsible Nanotechnology, *January, 2003).*

There are many risks associated with NT. For instance, Drexler's "gray goo" problem–where self-replicating nanoassemblers multiply out of control; NT falling into the hands of rogue states; and decreased privacy due to the sizing-down of surveillance equipment, to name a few. Given these risks, it would appear that some regulation of NT is in order. If NT is regulated too strictly this could lead to the development of a black market where irresponsible users could do a great deal of damage. Phoenix and Treder propose a two-dimensional approach to the restriction of NT which will reduce risks associated with NT, yet allow the technology to further develop.

First, products that use NT do not necessarily pose any extraordinary risks, they are simply products that have been enhanced in some way by the use of NT; they will not be able to self-replicate whatsoever. However, the factories used to produce NT products could quite easily be very dangerous. NT factories, if controlled by the wrong people, could hold great danger for humankind with the potential to develop destructive weapons much greater than are available today. Phoenix and Treder stress that

"...research leading to advanced nanotechnology will have to be carefully monitored and controlled. However, the same *is not true of product research and development*."

If NT factories are to be safe, certain precautions must be taken. First, assemblers need not ever leave the factory, they can easily be fastened to the factory to avoid them escaping into the environment and getting out of control. Second, factories can be programmed to produce only specified products that are known to be safe. Third, one could go so far as to build in a tracking device to ensure that NT products were being used safely and for their intended purpose. Furthermore, nanofactories could be made tamper-proof so that they could shut down if attempt at alteration was made. Finally, a nanofactory could be connected to a central controller which would then determine if the desired product was safe. If the central controller determined the product to be unsafe, the controller could disable the factory. Additional advantages of this scenario would be that the controller could automatically provide updates to the nanofactories and also relay the location of the factory.

Rotman, David "Measuring the Risks of Nanotechnology." *(In* Technology Review, *April, 2003).*

Vicki Colvin, Director of The Center for Biological and Environmental Nanotechnology at Rice University, states that the possible dangers associated with NT are largely unknown at this point. Despite the fact that nanomaterials have been put to use in products such as sunscreen, the potential risks of these materials is unknown. It is believed that some nanoparticles can be transmitted from the blood stream to brain material. Colvin believes that the benefits of NT–especially its medical applications–far outweigh any risks that have been hypothesized.

Groups such as Greenpeace have begun to investigate NT for any risks it may pose.

"Handle Nano-Technology with Care," (In African Business. *March, 2003). [not available on-line]*

The action group on Erosion, Technology and Concentration (ETC) is worried that developments in NT are going unnoticed both by the public and government regulatory agencies. This is of great concern given that NT will have such a profound impact on society. One problem with regulating NT is that it has such a wide variety of applications and is therefore difficult to control under one regulatory agency. Furthermore, because many of the products that have incorporated NT merely use the same compounds that were previously used but on a lower scale, many people do not recognize any associated risks whatsoever. ETC is calling for a worldwide evaluation of the potential impacts of NT, and until such a process has begun, production of nanomaterials ought to cease.

Industry Organizations

- American Society for Cell Biology

http://www.ascb.org/

- American Society for Materials International

  http://www.asminternational.org/

- Cientifica

  http://www.cientifica.com/

- Electronic Device Failure Analysis Society/ISTFA

  http://www.asminternational.org/MSTemplate.cfm?Site=Electronic_Device_Failure_Analysis_Society

- European Microscopy Society

  http://www.eurmicsoc.org/

- IDEMA

  http://www.idema.org/

- International Federation of Societies for Microscopy

  http://www.ifsm.umn.edu/members.html

- Japanese Society of Microscopy

  http://wwwsoc.nii.ac.jp/jsm/index.html

- Materials Research Society

  http://www.mrs.org/

- NanoBusiness Alliance

  http://www.nanobusiness.org/

- SEMI

  http://www.semi.org/

Nanotechnology Links

- The Center for Nanotechnology at the University of Washington

  http://www.nano.washington.edu/index.asp

- nano tech 2006

  http://www.ics-inc.co.jp/nanotech/index_e.html

- The Nanotechnology Site

  http://inventors.about.com/library/inventors/blmicroscope.htm

- Nanotechnology and Cryonics

  http://www.futurescience.com/nanocryo.html

- Nanotechnology Now

http://www.nanotech-now.com/

- National Nanotechnology Initiative
  http://www.nano.gov/
- Open Encyclopedia Project: Nanotechnology
  http://open-site.org/Science/Technology/Nanotechnology/
- TNTlog: Global Nanotechnology Blog
  http://www.cientifica.com/archives/000031.html

Career Links

- FEI Company
  http://www.careerexchange.com/cejobs/templatefei.asp
- Career Resource Center for Materials Science & Engineering
  http://www.crc4mse.org/
- tinytechjobs
  http://www.tinytechjobs.com/

Electron Microscopy Links (General)

- History of the Microscope
  http://inventors.about.com/library/inventors/blmicroscope.htm
- Electron Microscope Web Course
  http://em-outreach.ucsd.edu/web-course/toc.html
- Electron Microscopy
  http://www.biologie.uni-hamburg.de/b-online/e03/03e.htm
- MicroElectronics—Atomic Scale Design
  http://www.atomicscaledesign.net/microelectronics/links/software/index.shtml

Publications

- Advanced Materials & Processing
  http://www.asminternational.org/
- European Semiconductor
  http://www.eurosemi.eu.com/
- Forbes/Wolfe Nanotech Report
  http://www.forbesinc.com/newsletters/nanotech/

- Materials Today
  http://www.materialstoday.com/
- Microscopy & Analysis
  http://www.microscopy-analysis.com/
- Microscopy Today
  http://www.microscopy-today.com/
- Micro
  http://www.micromagazine.com/
- MRS Bulletin
  http://www.mrs.org/publications/bulletin/
- News.Nano.Apex
  http://news.nanoapex.com/
- R&D Magazine
  http://www.mrs.org/publications/bulletin/
- Scanning
  http://www.scanning.org/
- Semicondcutor Manufacturing
  http://www.semimfg.com/
- Semiconductor International
  http://www.semiconductor.net/
- Small Times
  http://www.smalltimes.com/
- Solid State Technology
  http://www.solid-state.com/

Nanobiotechnology, Nanoscience, Nanocomposites and Nanotechnology

The definition of nanocomposite material has broadened significantly to encompass a large variety of systems such as one-dimensional, two-dimensional, three-dimensional and amorphous materials, made of distinctly dissimilar components and mixed at the nanometer scale. The general class of nanocomposite organic or inorganic materials is a fast growing area of research. Significant effort is focused on the ability to obtain control of the nanoscale structures via innovative synthetic approaches. The properties of nanocomposite materials depend not only on the properties of their individual parents but also on their morphology and interfacial characteristics. This rapidly expanding field is generating many exciting new materials with novel properties. The latter can derive by combining properties from the parent constituents into a single material. There is also the possibility of new properties which are unknown in the parent constituent materials. The inorganic components can be three-dimensional framework systems such as zeolites, two-dimensional layered materials such as clays, metal oxides, metal phosphates, chalcogenides, and even one-dimensional and zero-dimensional materials such as $(Mo_3Se_3\text{-})_n$ chains and clusters. Experimental work has generally shown that virtually all types and classes of nanocomposite materials lead to new and improved properties when compared to their macrocomposite counterparts. Therefore, nanocomposites promise new applications in many fields such as mechanically reinforced lightweight components, non-linear optics, battery cathodes and ionics, nanowires, sensors and other systems.

The general class of organic/inorganic nanocomposites may also be of relevance to issues of bio-ceramics and biomineralization in which *in-situ* growth and polymerization of biopolymer and inorganic matrix is occurring. Finally, lamellar nanocomposites represent an extreme case of a composite in which interface interactions between the two phases are maximized. Since the remarkable properties of conventional composites are mainly due to interface interactions, the materials dealt with here could provide good model systems in which such interactions can be studied in detail using conventional bulk sample (as opposed to surface) techniques. By judiciously engineering the polymer-host interactions, nanocomposites may be produced with a broad range of properties. Inorganic layered materials exist in great variety. They possess well defined, ordered intralamellar space potentially accessible by foreign species. This ability enables them to act as matrices or hosts for polymers, yielding interesting hybrid nano-composite materials. Lamellar nanocomposites can be divided into two distinct classes, intercalated and exfoliated. In the former, the polymer chains alternate with the inorganic layers in a fixed compositional ratio and have a well defined number of polymer layers in the intralamellar space. In exfoliated nano-composites the number of polymer chains between the layers is almost continuously variable and the layers stand >100 Å apart. The intercalated nanocomposites are also more compound-like because of the fixed polymer/layer ratio, and they are interesting for their electronic and charge transport properties. On the other hand, exfoliated nanocomposites are more interesting for their superior mechanical properties. A subclass of nanocomposites which is the lamellar class of intercalated organic/inorganic nanocomposites and namely those systems that exhibit electronic properties in at least one of the components offers the possibility of obtaining well ordered systems some of which may lead to unusual electrical and mechanical properties. Selected members of this class may be amenable to direct structural characterization by standard crystallographic methods. An important issue in this area is the few structural details that are available, therefore, any system which can be subjected to such analysis is of interest. Nanocomposites also offer the possibility to combine diverse properties which are impossible within a single material, e.g. flexible mechanical properties and superconducting properties. This work is now in its infancy and extensive investigations are underway. Another exciting aspect is the possibility of creating heterostructures composed of different kinds of inorganic layers, which could lead to new multifuntional materials. Several general synthetic routes have been developed for inserting polymer chains into host structures and many novel nanocomposites have been designed.

These include:

- *In-situ* intercalative polymerization (ISIP) of a monomer using the host itself as the oxidant. The rationale behind intercalative polymerization is that host matrices with high electron affinity can oxidatively polymerize appropriate monomers in their interior.
- Monomer intercalation followed by topotactic intralamellar solid state polymerization. This route creates conjugated polymers inside non-oxidizing hosts.
- Direct precipitative encapsulation of polymer chains by colloidally dispersed single layers of a host. This approach gives access to a large variety of nanocomposites with many kinds of polymers and hosts.

Materials with features on the scale of nanometers (10^{-9} meter) often have properties dramatically different from their bulk-scale counterparts. For example, nanocrystalline copper is five times harder than ordinary copper with its micrometer-sized crystalline structure. The development of such materials is currently a research area of great interest. Important among these nanoscale materials are nanocomposites, in which the constituents are mixed on a nanometer-length scale. They often have properties that are superior to conventional microscale composites and can be synthesized using surprisingly simple and inexpensive techniques.

A nanoscale dispersion of sheet-like inorganic silicate particles in a polymer matrix, for example, is superior to either constituent in such properties as optical clarity, strength, stiffness, thermal stability, reduced permeability, and flame retardancy. The key to the synthesis of nanocomposites is how the silicates are made to disperse in the polymer. Since the organic and inorganic components are typically immiscible (like water and oil), the silicate surface is modified by attaching surfactant molecules (like adding soap to water). Polymer-based nanocomposites are also being developed for electronics applications such as thin-film capacitors in integrated circuits and solid polymer electrolyes for batteries. Nature also makes fabulous nanocomposites, and scientists are trying to emulate such processes. The abalone shell, for example, has alternating layers of calcium carbonate and a rubbery biopolymer; it is twice as hard as and a thousand times tougher than its components. The study of nanocomposite materials requires a multidisciplinary approach, involving novel chemical techniques and an understanding of physics and surface science. It is a field of broad scientific interest with impressive technological promise.

Studies have revealed clearly the property advantages that nanomaterial additives can provide in comparison to both their conventional filler counterparts and base polymer. Properties which have been shown to undergo substantial improvements include:

- Mechanical properties e.g. strength, modulus and dimensional stability.
- Decreased permeability to gases, water and hydrocarbons.
- Thermal stability and heat distortion temperature.
- Flame retardancy and reduced smoke emissions.
- Chemical resistance.
- Surface appearance.
- Electrical conductivity.
- Optical clarity in comparison to conventionally filled polymers.

To date one of the few disadvantages associated with nanoparticle incorporation has concerned toughness and impact performance. Some of the data presented has suggested that nanoclay modification of polymers such as polyamides, could reduce impact performance. Clearly this is an issue which would require consideration for applications where impact loading events are likely. In addition, further research will be necessary to, for example, develop a better understanding of formulation/structure/property relationships, better routes to platelet exfoliation and dispersion etc.

In addition it is important to recognise that nanoparticulate/fibrous loading confers significant property improvements with very low loading levels, traditional microparticle additives requiring much higher loading levels to achieve similar performance. This in turn can result in significant weight reductions (of obvious importance for various military and aerospace applications) for similar performance, greater strength for similar structural dimensions and, for barrier applications, increased barrier performance for similar material thickness. Data provided by Hartmut Fischer of TNO in the Netherlands relating to polyamide – montmorillonite nanocomposites indicates tensile strength improvements of approximately 40 and 20 per cent at temperatures of 23°C and 120°C respectively and modulus improvements of 70 per cent and a very impressive 220 per cent at the same temperatures. In addition Heat Distortion Temperature was shown to increase from 65°C for the unmodified polyamide to 152°C for the nanoclay-modified material, all the above being achieved with just a 5 per cent loading of montmorillonite clay. Similar mechanical property improvements were presented for polymethyl methacrylate—clay hybrids.

Further data provided by Akkepeddi of Honeywell relating to polyamide-6 polymers confirms these property trends. In addition, the further benefits of short/long glass fibre incorporation, together with nanoclay incorporation, are clearly revealed.

REFERENCES

Adleman, L., "Molecular Computation of Solutions to Combinatorial Problems," Science, Vol. 266, 11 November 1994, pp. 1021-1023. *This is THE landmark paper that started the new field of DNA computing.*

Albus, J.S.: 2003, 'Engineering of Mind for Enhancing Human Productivity', in: M.C. Roco, W.S. Bainbridge (eds.), *Converging Technologies for Improving Human Performance*, Kluwer, Dordrecht, pp. 281-293.

Albus, J.S.; Bainbridge, W.S.; Banfield, M.; Dastoor, C.A.; Murray, K.; Carley,M.; Hirshbein, T.; Masciangioli, T.; Miller, R.; Norwood, R.; Price, P.; Rubin, J.; Sargent,G.; Wallace, W.A.: 2003, 'Enhancing Group and Societal Outcomes: Theme D Summary', in M.C. Roco, W.S. Bainbridge (eds.), *Converging Technologies for Improving Human Performance*, Kluwer, Dordrecht, pp. 275-277.

Alexandra Goho, "Protein Power: Solar cell produces electricity from spinach and bacterial proteins," Science News Online, week of June 5, 2004: Vol. 165, No. 2, p. 355.

Alibek, K. A., Testimony of Dr. Kenneth Alibek, Chief Scientist at Hadron, Inc., before the House Armed Services Committee, October 20, 1999. Ref: http://www.house.gov/hasc/testimony/106thcongress/99- 10-20alibek.htm.

Allara, D. L. 1995. "Critical Issues in Applications of Self-Assembled Monolayers", *Biosensors & Bioelectronics*, 10, 771-783.

Anne Eisberg, "Benign Viruses Shine on the Silicon Assembly Line," New York Times, February 12, 2004.

Antonio Regalado, "Nanotechnology Patents Surge as Companies Vie to Stake Claim," Wall Street Journal, June 18, 2004, p. 1.

Arap W., R. Pasqualini, E. Ruoslahti. 1998. "Cancer Treatment by Targeted Drug Delivery to Tumor Vasculature in a Mouse Model", *Science,* Jan 16; 279 (5349): pp. 377-380.

Arias, T.A. and Sohrab Ismail-Beigi. 2000. Novel algebraic formulation of density-functional calculation. *Computer Physics Communications*. In press.

Asimov, I., *A Choice of Catastrophes*, Simon and Schuster, NY (1979).

Atkins, P.: 2003 'The Future of Matter', in: J. Brockman (ed.), *The Next Fifty Years: Science in the First Half of the Twenty-First Century*, Vintage Books, New York, pp.194-205.

Babbie, E. *The Practice of Social Research*, 5th ed. Belmont, CA: Wadsworth, 1989. Bachand, G. and C.D. Montemagno. 2000. Constructing organic/inorganic NEMS devices powered by biomolecular motors. *Journal of Biomedical Microdevices*. 2, No.3, 179-184.

Bai, Q., K.D. Wise, and D.J. Anderson. 2000. A high-yield microassembly structure for three-dimensional microelectrode arrays. *IEEE Trans. Biomed Eng*. 47(3): p. 281-9.

Bainbridge, W.S.: 2002, 'Public Attitudes Toward Nanotechnology', *Journal of Nanoparticle Research*, 4, 561-570.

Barber, Benjamin. 1996. *Jihad vs. McWorld: How Globalism and Tribalism are Reshaping the World* (New York: Ballantine Books).

Baschnagel, J., K. Binder, K. Kremer et al. 2000. Bridging the gap between atomistic and coarse-grained models of polymers: status and perspectives. *Advances in Polymer Science*. Volume 152. Springer- Verlag.

Bass, T.A., "Gene Genie," Wired Magazine, August 1995, p. 114. *Well-informed and well-written survey of the technical developments and people in the new field of DNA computing.*

Ben-David, Joseph and Randall Collins, 1966. "Social Factors in the Origins of a New Science: the Case of Psychology." *American Sociological Review.* Vol. 31 pp 451-465.

Bijker, W. E. *Of Bicycles, Bakelites, and Bulbs: Toward a Theory of Sociotechnical Change.* Cambridge, MA: MIT Press, 1995.

Billings P.R., M.A. Kohn, M. de Cuevas, J. Beckwith, J.S. Alper, M.R. Natowicz. 1992. "Discrimination as a Consequence of Genetic Testing," *American Journal of Human Genetics*, Mar; 50 (3): pp. 476-82. Drews, J. 1996. "Genomic Sciences and the Medicine of Tomorrow," *Nature Biotechnology*, Nov., 14 (11), pp. 1516-1518.

Bimber, B., and Guston, D., eds., *Technology Assessment: The End of OTA*, special issue of *Technology Forecasting and Social Change*, Vol. 54, No. 2 and 3, 1997

Black, C.T.; Murray, C.B.; Sandstrom, R.L.; Sun, S.H. 2000 "Spin-dependent tunneling in self-assembled cobalt-nanocrystal superlattices", *Science*, 290, 1131-1134.

Blumenthal, D. 1992. "Academic-Industry Relationships: Extent, Consequences, and Management," *The Journal of the American Medical Association*, 268:3, December, pp. 3344 -3349.

Blumenthal, D.M., M. Gluck, K.S. Louis, M. Stoto, and D. Wise. 1986. "University-Industry Research Relationships in Biotechnology: Implications for the University," *Science* 232: pp. 1361-1366 (June 13).

BMBF. 1996. "Delphi-Bericht 1995 zur Entwicklung von Wissenschaft und Technik – Mini-Delphi." Bonn, author's translation.

Bockrath, M., D.H. Cobden, P.L. McEuen, N.G. Chopra, A. Zettl, A. Thess, and R.E. Smalley. 1997. "Single-Electron Transport in Ropes of Carbon Nanotubes." *Science*, 275.

Bonnemain, B. 1998. "Superparamagnetic agents in magnetic resonance imaging: Physicochemical characteristics and clinical applications - A review", *Journal of Drug Targeting*, 6, 167-174.

Brandt, A. 2000. *Multiscale Scientific Computation: 2000*. Weizmann Institute preprint. October.

Brave, Ralph. 2000. "Genome Hell is Revealed in Many Forms." *Baltimore Sun*, July 2, p. 1C.

Brown, J. S. and Duguid, P. Don't Count Society Out: A Response to Bill Joy. Paper prepared for the Societal Implications of Nanotechnology Conference, 2000.

Brown, J.S. , and Paul Duguid. 2001. Don't count society out. *Societal Implications of Nanoscience and Nanotechnology.* National Science Foundation. Arlington, VA. March.

Brown, J.S. and P. Daguid. 2001. "Don't Count Society Out: A Response to Bill Joy" in this volume: *Societal Implications of Nanoscience and Nanotechnology*. National Science Foundation, Arlington, VA.

Brown, Lawrence A. 1981. *Innovation Diffusion: A New Perspective*. New York: Methuen.

Brown, N.: 2003, 'Hope Against Hype – Accountability in Biopasts, Presents and Futures', *Science Studies*, 16(2), 3-21.

Bruchez, M.; Moronne, M.; Gin, P.; Weiss, S.; Alivisatos, A.P. 1998. "Semiconductor nanocrystals as fluorescent biological labels", *Science*, 281, 2013-2016.

Bush, Vannevar. 1945. *Science, the Endless Frontier*. Washington D.C. U.S. Government Printing Office.

Byl, J., Physica D 34,295-299 (1989). *Self-Reproduction in Small Cellular Automata.*

Caflisch, R.E., W. E, M.F. Gyure, B. Merriman, and C. Ratsch. 1999. Kinetic model for a step edge in epitaxial growth. *Phys. Rev. E.* Vol 59. 6879.

Callon, M.: 1986, 'Some Elements of a Sociology of Translation', in J. Law (ed.), *Power, Action and Belief: A New Sociology of Knowledge?*, Routledge, London, pp. 196-223.

Carroll, J.S.: 2001, 'Social Science Research Methods for Assessing Societal Implications of Nanotechnology', in: M.C. Roco, W.S. Bainbridge (eds.), *Societal Implications of Nanoscience and Nanotechnology*, National Science Foundation, pp. 188-192.

Casey, Jr., H.C., and M.B. Panish. 1978. *Heterostructure Lasers, Part A: Fundamental Principles.* (Academic Press. New York). pp. 1-9.

Cerf, Christopher, and Victor Navasky. 1984. *The Experts Speak.* New York: Pantheon Books.

Cetron, M., and Davies, O. 1997. *Probable Tomorrows: How Science and Technology Will Transform Our Lives in the Next Twenty Years* (New York: St. Martin's Press).

Chan, W.C.W. and S. Nie. 1998. Quantum dot bioconjugates for ultrasensitive nonisotopic detection. *Science.* 281, 2016 (25 September).

Chen, C.-Y., K. Ishihara, N. Nakabayashi, E. Tamiya, and I. Karube. 1999. Multifunctional biocompatible membrane and its application to fabricate a miniaturized glucose sensor with potential for use *in vivo. Journal of Biomedical Microdevices.* 1, No.2, 155-166.

Chondroudis, K. and D.B. Mitzi. 1999. Electroluminescence from an organic-inorganic perovskite incorporating a quaterthiophene dye within lead halide perovskite layers. *Chemistry of Materials.* Vol 11. p. 3028.

Christensen, Clayton M. *The Innovator's Dilemma: When New Technologies Cause Great Firms to Fail.* New York: Harperbusiness, 2000.

Cipra, Barry A. 2000. The best of the 20th century: editors name top 10 algorithms. *SIAM News.* vol. 33(4).

Clifford Lau of the US Defense Department to Barnaby Feder, "Frontier of Military Technology is the Size of a Molecule," New York Times, April 8, 2003, p. C2.

Collier, C.P., E.W. Wong, M. Belohradsky, F.M. Raymo, J.F. Stoddart, P.J. Kuekes, R.S. Williams, and J.R. Heath. 1999. "Electronically Configurable Molecular-Based Logic Gates." *Science*, 285.

Cook, T. K. and Campbell, D. T. *Quasi-Experimentation.* Rand-McNally, 1979 Fischer, C. S. *America Calling: A Social History of the Telephone to 1940.* Berkeley: U. of California, 1992.

Crommie, M.F.; Lutz, C.P.; Eigler, D.M.; Heller, E.J. 1996. "Quantum interference in 2D atomic-scale structures," *Surface Science*, 362, 864-869.

Crow, M. & Sarewitz, D.: 2001, 'Nanotechnology and Societal Transformation', in: M.C. Roco, W.S. Bainbridge (eds.), *Societal Implications of Nanoscience and Nanotechnology*, National Science Foundation, pp. 45-54.

Crowe, W. J., Statement of Admiral William J. Crowe, Jr., before the House Committee on Government Reform, October 12, 1999. Ref: http://www.house.gov/reform/hearings/healthcare/99.10.12/ crowe.htm.

Csicsery-Ronay, I. Jr.: 1991, 'The SF of Theory: Baudrillard and Haraway', *Science Fiction Studies*, 18, 387-404.

Csicsery-Ronay, I. Jr.: 1996, 'The Seven Beauties of Science Fiction', *Science Fiction Studies,* 23, 385-392.

Darden, T., L. Perera, L. Li, L. Pedersen, 1999, New tricks for modelers from the crystallization toolkit: The particle mesh Ewald algorithm and its use in nucleic acid simulations, *Structure.* Vol. 7, R55-R60.

David, Paul. 1991. "Computer and Dynamo: The Modern Productivity Paradox in a Not-too-Distant Mirror," *Technology and Productivity: The Challenge for Economic Policy* (Paris: OECD).

de Gennes, P.G. 1974. *The Physics of Liquid Crystals.* Oxford University Press.

Delaney, S.R.: 1984, *Starboard Wine: More Notes on the Language of Science Fiction*, Dragon Press, New York.

Delaney, S.R.: 1994, *Silent Interviews*, Univ. Press of New England, Hanover, NH.

DeSimone, J.M., E.T. Samulski, et al. 2000. Critical micelle density for the self-assembly of block copolymer surfactants in supercritical carbon dioxide. *Langmuir*. Vol.16. 416-421.

Di Paolo, E. A., *Artificial Life Bibliography of On-line Publications*, June 28, 2000.

Dimitrakopoulos, C.D., S. Purushothaman, J. Kymissis, A. Callegari, and J.M. Shaw. 1999. Low-voltage organic transistors on plastic comprising high dielectric constant gate insulators. *Science.* Vol. 283, 5 February. pp. 822 - 824.

DiVincenzo, D. P., "Quantum Computation," Science, vol. 270, 13 October 1995, pp. 255-261

Doering, R.R. and P.K. Chatterjee. 1998. "The Future of Microelectronics." *Proceedings of the IEEE*, 86, 1.

Doering, Robert and Yoshio Nishi. 2000. "Limits of Integrated-Circuit Manufacturing." *Proceedings of the IEEE*. September.

Doi, M. and S.F. Edwards. 1986. *The Theory of Polymer Dynamics*. Oxford University Press (Clarendon), London-New York.

Donald, A.M. and A.H. Windle. 1992. Liquid crystalline polymers. *Cambridge Solid State Science Series*. Cambridge University Press.

Drell, S. D., A. D. Sofaaer, and G. D. Wilson, Hoover Digest 2000, No. 1, *The Present Threat.*

Drell, S. D., A. D. Sofaaer, and G. D. Wilson, *The New Terror, Facing the Threat of Biological and Chemical Weapons*, Hoover Institution Press, Stanford University (1999).

Drexler, E.: 1990 [1986], *Engines of Creation*, Anchor Books, New York.

Drexler, E.; Smalley, R.E.: 2003, 'Point-Counterpoint: Nanotechnology', *Chemical and Engineering News*, 81(48), 37-42.

Drexler, K. Eric. 1986. *Engines of Creation.* Garden City, N.Y.: Anchor Press/Doubleday.

Ehrlich, Paul R., Anne H. Ehrlich, and John P. Holdren. 1977. *Ecoscience: Population, Resources, Environment*. San Francisco: W.H. Freeman.

Elkins, C.: 1979, 'Science Fiction versus Futurology: Dramatic Versus Rational Models', *Science Fiction Studies*, 6, 20-31.

Ellul, J. 1964. *The Technological Society*. New York: Alfred Knopf. (First published 1954).

Empedocles, S.; Bawendi, M. 1999. "Spectroscopy of single CdSe nanocrystallites," *Accounts of Chemical Research*, 32, 389-396.

Esant. 1999. "Economic and Social Aspects of Nanotechnology." First draft report of working group 6 of the Euroconferences on Nanoscience for Nanotechnology. Also at http://www.nano.org.uk/ESANT99.htm.

Etzkowitz, H., M. Gulbrandsen and J. Levitt. 2000. *Public Venture Capital: Government Funding Sources for Technology Entrepreneurs*. New York: Harcourt.

Etzkowitz, Henry 2001. "The Endless Transition: Finding a Social Basis for Science Policy." *Scipolicy: The Journal of Science and Health Policy*. Vol. 1. No.2, Spring 2001.

Etzkowitz, Henry, Carol Kemelgor and Brian Uzzi. 2000. *Athena Unbound: The Advancement of Women in Science and Technology*. Cambridge: Cambridge University Press.

Evans, J.H.: 2002, *Playing God? Human Genetic Engineering and the Rationalization of Public Bioethical Debate*, Chicago University Press, Chicago.

Fantel, Hans. 1987. "Portable CD Players Advance." *New York Times*, May 17, Section 2, p. 33, col. 1.

Fermi, E., J. Pasta, and S. Ulam. 1974. Studies of nonlinear problems, I. *Los Alamos Report*. LA1940. 1955.

Feynman, R.P. 1961. "There is plenty of room at the bottom." In *Miniaturization*. New York: Reinhold.

Fisher, D. E. and Fisher, M. J. *Tube: The Invention of Television*. Harcourt Brace, 1997.

FitzGerald, Frances. 2000. *Way Out There in the Blue: Reagan, Star Wars, and the End of the Cold War*. New York: Simon & Schuster.

Fogelberg, H.: 2003, '"Science" and "Technology" as Knowledge Cultures', in: H. Fogelberg, H. Glimell, *Bringing Visibility to the Invisible: Towards a Social Understanding of Nanotechnology*, Göteborg University, Göteborg, pp. 77-85.

Fogelberg, H.; Glimell, H.: 2003, *Bringing Visibility to the Invisible: Towards a Social Understanding of Nanotechnology*, Göteborg University, Göteborg.

Forest, G. 2000. Comment at Conference on Societal Implications of Nanoscience and Nanotechnology, September 29, 2000, NSF, Arlington, VA.

Forest, M.G., Q. Wang, and H. Zhou. 2000a. Exact banded patterns from a Doi-Marrucci-Greco model of nematic liquid crystal polymers. *Physical Review E*. Vol. 61(6). 6665-6672.

Forest, M.G., Q. Wang, and H. Zhou. 2000b. Homogeneous pattern selection and director instabilities of nematic liquid crystal polymers induced by elongational flows. *Physics of Fluids*. Vol. 12(3). 490-498.

Forum Proceedings of Sigma Xi, the Scientific Research Society, *Ethics, Values, and the Promise of Science*, Feb. 25-26, 1993.

Foucault, M. 1977. *Power/Knowledge: Selected Interviews and Other Writings 1972-1977*. New York, Pantheon Books.

Fourier, J. 1878. *The Analytical Theory of Heat*. translated by A. Freeman. Cambridge University Press. London and New York.

Frankel, M. 1996. "Perception, Reality, and the Political Context of Conflict of Interest in University-Industry Relationships," *Academic Medicine*, 71:12, December, pp. 1297-1304.

Franklin, N.R.; Dai, H.J. 2000. "An enhanced CVD approach to extensive nanotube networks with directionality", *Advanced Materiais*, 12, 890-894.

Freitas, L. *Nanomedicine*. Bioscience, 1999.

Frenken, J.W.M. 1998. "Scanning Tunneling Microscopy", section 5.2, A. ten Wolde (ed.), *Nanotechnology: Towards a molecular construction kit*. STT Report #60, The Hague, 289-299.

Gates, B.; Park, S.H.; Xia, Y.N. 2000. "Tuning the photonic bandgap properties of crystalline arrays of polystyrene beads by annealing at elevated temperatures", *Advanced Materials*, 12, 653-656.

Geneva Convention, *Protocol for the Prohibition of the Use in War of Asphyxiating, Poisonous or Other Gases, and of Bacteriological Methods of Warfare*, signed at Geneva on June 17,1925.

Gergen, K. J. Social psychology as history. *Journal of Personality and Social Psychology*, 1973, 26, 309- 20.

Gerlach, N.; Hamilton, S.: 2000, 'Telling the Future, Managing the Present: Business Restructuring Literature as SF', *Science Fiction Studies*, 27, 461-477.

Goldbeck-Wood, G. and A.H. Windle. 1999. Lattice modeling of nematodynamics. *Rheol. Acta*. Vol. 38. 548-561.

Graham, P.: 2002, 'Hypercapitalism: Language, New Media and Social Perceptions of Value', *Discourse and Society*, 13(2), 227-249.

Gref, Ruxandra, Yoshiharu Minamitake, Maria Teresa Peracchia, Vladimir Trubetskoy, Vladimir Torchilin, Robert Langer. 1994. "Biodegradable Long Circulating Polymeric Nanospheres", *Science*, Vol. 263, 18 Mar, pp 1600-1603.

Griliches, Zvi. 1957. "Hybrid Corn: An Exploration in the Economics of Technological Change," *Econometrica*, 25 (4), pp 501-522.

Griliches, Zvi. 1960. "Hybrid Corn and the Economics of Innovation," *Science*, 132, pp 275-280. Mass, William. 1989. "Mechanical and Organizational Innovation: The Case of the Draper Loom," *Business History Review*, 63 (Winter).

Gröndahl B., W. Puppe, A. Hoppe, I. Kühne, J.A. Weigl, H.J. Schmitt. 1999. "Rapid Identification of Nine Microorganisms Causing Acute Respiratory Tract Infections by Single-Tube Multiplex Reverse Transcription-PCR: Feasibility Study." *Journal of Clinical Microbiology*, Jan; 37 (1): pp. 1-7.

Grosberg, A.Y. and A.R. Khokhlov. 1994. *Statistical Physics of Macromolecules*, AIP Series in Polymers and Complex Materials, AIP Press, New York.

Gubrud, M. A., *Nanotechnology and International Security*, Fifth Foresight Conference on Molecular Nanotechnology, November 5-8, 1997. Ref: http://www.foresight.org/Conferences/MNT05/Papers/Gubrud/.

Gurley, Bill. 2000. Digital music: the real law is Moore's law. *Fortune*. 2 October. page 268. Joy, Bill. 2000. Why the future doesn't need us. *Wired*. April. pp. 238 - 262.

Hadjiconstantinou, N.G. 1999. Combining atomistic and continuum simulations of contact-line motion. *Physical Review E*. Vol. 59(2).2, 2475-2478.

Hamilton, S.: 2003, 'Biotechnology, SF and the Media', *Science Fiction Studies*, 30, 267-282.

Haraway, D.J.: 1997, *Modest Witness@Second Millenium.FemaleManã Meets OncoMouseä*, Routledge, London and New York.

Hayes, B., "The Square Root of NOT," *American Scientist*, July-August 1995, pp. 304-308.

Hayward, R.C.; Saville, D. A.; Aksay, I.A. 2000. "Electrophoretic assembly of colloidal crystals with optically tunable micropatterns", *Nature*, 404, 56-59.

Hendrickson, C.T., A. Horvath, S. Joshi, and L.B. Lave. 1998. Economic input-output models for environmental life-cycle assessment. *Environmental Science & Technology*. p. 32.

Hughes, P. M., *Global Threats and Challenges: The Decades Ahead*, Lt. Gen. Patrick M. Hughes, Director of the Defense Intelligence Agency, in Statement for the Senate Committee on Intelligence, January 28, 1998. Ref:http://www.fas.org/irp/congress/1998_hr/s980128h.htm.

Huijer, M.: 2003, 'Reconsidering Democracy: History of the Human Genome Project', *Science Communication*, 24(4), 479-502.

Huntington, Samuel P. 1996. *The Clash of Civilizations and the Remaking of the World Order* (New York: Simon and Schuster).

Jacobs, K.M., et al. 2000. Postlesional epilepsy: the ultimate brain plasticity. *Epilepsia*. 41(Suppl 6): p. S153-61.

James, E.: 1994, *Science Fiction in the 20th Century*, Oxford University Press, Oxford.

Jameson, F.: 1982, 'Progress Versus Utopia; or, Can We Imagine the Future', *Science Fiction Studies*, 9, 147-158.

Janet Raloff, "Nano Hazards: Exposure to minute particles harms lungs, circulatory system," Science News Online, Week of March 19, 2005; Vol. 167, No. 12.

Johnson, D.G. and H. Nissenbaum (eds). 1995. *Computers, Ethics and Social Values*. Prentice Hall:Englewood Cliffs, NJ. pp. 262-393.

Johnson, P.C.: 2003, 'Implications of the Continuum of Bioinformatics', in: M.C. Roco; W.S. Bainbridge (eds.), *Converging Technologies for Improving Human Performance*, Kluwer, Dordrecht, pp. 207-212.

Joy, Bill. "Why the Future Doesn't Need Us." *Wired*, April 8, 2000.

Joy, W. 2000. Why the Future Doesn't Need Us. *Wired*, April.

Judd, C., Smith, E. and Kidder, L. *Research Methods in Social Relations*, (6th Edition). Holt, Rinehart, Winston, 1991.

Kalman B., F.D. Lublin. 1999. "The Genetics of Multiple Sclerosis. A Review." *Biomedicine and Pharmacotherapy*, Sep; 53 (8): pp. 358-370.

Kay, L.: 2000, *Who Wrote the Book of Life? A History of the Genetic Code*, Stanford University Press, Stanford.

Kennedy, P., *The Rise and Fall of the Great Powers, Economic Change and Military Conflict from 1500 to 2000*,Random House (1987).

Kennedy, Paul. *The Rise and Fall of the Great Powers, Economic Change and Military Conflict from 1500 to 2000*. New York: Random House,1987.

Keyssar, Alexander. 1986. *Out of Work: The First Century of Unemployment in Massachusetts* (Cambridge: Cambridge University Press).

Kolata, G., "A Vat of DNA May Become a Vast Computer of the Future," N.Y. Times, 11 April 1995, p. C1. *Large, much cited spread at the top of the front page of the Tuesday "Science Times" Section that first brought the Adelman discovery to wide public attention. Beautiful illustrations that explain how it works.*

Kruskal, M., and N. Zabusky. 1965. Interaction of solitons in a collisionless plasma and the recurrence of initial states, *Phys. Rev. Lett.* Vol. 15. 240-243.

Kupferman, R., M. Kawaguchi, and M.M. Denn. 2000. Emergence of structure in models of liquid crystalline polymers with elastic coupling. *J. Non-Newtonian Fluid Mech.* Vol. 91. 255-271.

Kurzweil, R. 1999. *The Age of Spiritual Machines*. Penguin Books.

Kuusi, O. 1999b. *Expertise in the future use of generic technologies*. VATT Research Reports #59, Helsinki.

Kuusi, O. and M. Meyer. 2001. "Technology Generalizations and Leitbilds – The Anticipation of Technological Opportunities." Manuscript submitted to *Technological Forecasting and Social Change*.

Kuusi, O., 1999. "Epistemic value analysis – a tool to evaluate technological options." Personal communication.

Laibinis, P.E.; Hickman, J.J.; Wrighton, M.S.; Whitesides, G.M. 1989. "Orthogonal Self-Assembled Monolayers - Alkanethiols On Gold and Alkane Carboxylic-Acids On Alumina", *Science*, 245, 845- 847.

Lambright, W. Henry. 1995. *Powering Apollo: James E. Webb of NASA*. Baltimore: Johns Hopkins University Press.

Landon, B.: 2004, 'Less is More: Much Less is Much More: The Insistent Allure of Nanotechnology Narratives in Science Fiction', in: K. Hayles (ed.), *NanoCulture*, Intellect Books, Bristol, pp. 59-78.

Lapham, E.V., C. Kozma, J.O. Weiss. 1996. "Genetic Discrimination: Perspectives of Consumers." *Science*, Oct 25; 274 (5287): pp. 621-624.

Larson, R.G. 1999. *The Structure and Rheology of Complex Fluids*. Oxford University Press.

Latour, B. & Woolgar, S.: 1986, *Laboratory Life: The Social Construction of Scientific Facts*, 2nd edn., Princeton University Press, Princeton.

Latour, B.: 1986, 'Visualisation and Cognition: Thinking With Eyes and Hands', *Knowledge and Society*, 6, 1-40.

Latour, B.: 2002, 'Morality and Technology: The End of the Means', *Theory, Culture and Society*, 19(5/6), 247-260.

Lave, L.B. & H.L. MacLean. 2000. *An Environmental-Economic Evaluation of Hybrid Electric Vehicles: Toyota's Prius vs. Its Conventional Internal Combustion Engine Corolla*. working paper. August.

Lave, L.B., A. Russell, F.C. McMichael, and C.T. Hendrickson. 1996. Environmental implications of battery powered vehicles as a transitional technology: ozone vs. lead. *Environmental Science & Technology*. p. 30.

Lave, L.B., C.T. Hendrickson, and F. C. McMichael. 1995b. Environmental implications of electric cars. *Science.* vol. 268: 992-5. May 19.

Lave, L.B., E. Cobas-Flores, C.T. Hendrickson, and F.C. McMichael. 1995a. Using input-output analysis to estimate economy wide discharges. *Environmental Science & Technology*, 29(9).

Lax, Peter D. 1989. The flowering of applied mathematics in America. in *A Century of Mathematics in America-Part II*. edited by Peter Duren et al. American Mathematical Society. 455-466.

Leah Eisenberg, "Medicating Death Row Inmates So They Qualify for Execution," AMA Case in Health Law, Vol. 6, No. 9, September 2004.

Ledley, Fred D. 1995. "Nonviral Gene Therapy: The promise of Genes as Pharmaceutical Products," *Human Gene Therapy*, Vol. 6, Sep, pp 1129-1144.

Lee, M.S. and M. Head-Gordon. 2000. Absolute and relative energies from polarized atomic orbital selfconsistent field calculations and a second order convergence with size and composition of the secondary basis. *Computers and Chemistry*. Vol. 24(3,4). 295-301.

Lee, S.C. 1998. The nanobiological strategy for construction of nanodevices. *Biological molecules in nanotechnology: the convergence of biotechnology, polymer chemistry and materials science*, edited by S.C. Lee and L. Savage (IBC Press, Southborough, MA).

Lemann, Nicholas. 1991. *The Promised Land: The Great Black Migration and How it Changed America* (New York: Random House).

Lipson, H. and J. B. Pollack, Nature, 406, 974-978, (2000). *Automatic design and manufacture of robotic life forms.* See, also, several additional articles on this subject in contemporary issues of Nature.

Lloyd, S., "A Potentially Realizable Quantum Computer," *Science*, 17 September 1993, pp. 1569-1571.

López, J.: 2003, *Society and Its Metaphors: Language Social Theory and Social Structure*, Continuum, New York and London.

López, J.: 2004, 'How Sociology Can Save Bioethics...Maybe', *Sociology of Health and Illness*, 26(7), 1-22.

Lu, H.P.; Xun, L.Y.; Xie, X.S. 1998. "Single-molecule enzymatic dynamics", *Science*, 282, 1877-1882.

Lynn, David M., Robert Langer. 2000. "Degradable Poly(b-amino esters): Synthesis, Characterization, and Self-Assembly with Plasmid DNA," *Journal of the American Chemical Society*, Vol. 122.

M.C. Roco, "National Nanotechnology Initiative: Overview," September 20, 2004.

Mahmood U., C.H. Tung, A. Bogdanov Jr., R. Weissleder. 1999. "Near-Infrared Optical Imaging of Protease Activity for Tumor Detection." *Radiology*, Dec; 213 (3): pp. 866-870.

Malinow, R., Z.F. Mainen, and Y. Hayashi. 2000. LTP mechanisms: from silence to four-lane traffic. *Curr. Opin. Neurobiol.* 10(3): p. 352-7.

Mark, H., Statement of The Honorable Hans Mark, Director, Defense Research and Engineering, before the House Armed Service Committee, October 20, 1999. Ref: http://www.house.gov/hasc/testimony/ 106thcongress/99-10-20mark.htm.

Marshall, K.: 2004, 'Atomizing Risk Technology', in: K. Hayles (ed.), *NanoCulture*, Intellect Books, Bristol, pp. 79-102.

Matsuki, H., K. Ofuji, N. Chubachi, and S. Nitta. 1996. Signal transmission for implantable medical devices using figure-of-eight coils. *IEEE Trans. on Magnetics.*, vol. 32. No.5, 5121. September.

Matsuki, H., Y. Yamakata, N. Chubachi, S. Nitta, and H. Hashimoto. 1996. Transcutaneous DC-DC converter for totally implantable artificial heart using synchronous rectifier. *IEEE Trans. On Magnetics*. 32, No.5, 5118 (Sept.).

McCain, L.: 2002, 'Informing Technology Policy Decisions: the US Human Genome Project's Ethical, Legal and Social Implications Programs as a Critical Case', *Technology and Society*, 24, 111-132.

McConnell, S.A., ed. 1998. *Pharmaceuticals*. Dun and Bradstreet Industry Reference Handbooks, Gale, Detroit, pp. 30-34.

McHale, B.: 1987, *Postmodern Fiction*, Methuen, New York & London.

Merz, J.L. 1986. "The Optoelectronics Joint Research Laboratory: Light Shed on Cooperative Research in Japan." *Scientific Bulletin* (ONR Far East Office). 11 (4), 1-30 (Oct.-Dec.).

Meyer, M. 2000. "Patent citations in a novel field of technology: What can they tell about interactions of emerging communities of science and technology?" *Scientometrics*, XXXXIIX (2), 151-178.

Meyer, M. 2000b. "Hurdles on the way to growth. Commercializing novel technologies." Institute of Strategy and International Business, Working Papers Series 2000/1, Helsinki University of Technology, Espoo.

Meyer, M. and O. Persson. 1998. "Nanotechnology - interdisciplinarity, patterns of collaboration and differences in application," *Scientometrics*, XXXXII (2), 195-205.

Michalet, X. and D. Bensimon. 1988. Vesicles of toroidal topology: observed morphology and shape transformations. *J. Phys. II*. France. Vol. 5. 598.

Mihail Roco and William Sims Bainbridge, eds., Converging Technologies for Improving Human Performance, NSF/DOC Report, June 2002.

Miksanek, T.: 2001, 'Microscopic Doctors and Molecular Black Bag: Science Fiction's Prescription for Nanotechnology and Medicine', *Literature and Medicine*, 21(1), 55-70.

Milburn, C.: 2002 'Nanotechnology in the Age of Posthuman Engineering: Science Fiction as Science', *Configurations*, 10, 261-295.

Mirsky, Steve. 2000. "Tantalizing Tubes." *Scientific American*. June.

Mizrach, S., *Should there be a limit placed on the integration of humans and computers and electronic technology?* Ref: http://www.limmat.ch/koni/texte/cyborg-ethics.html.

Mnyusiwalla, A.; Daar, A.S.; Singer, P.A.: 2003, "'Mind the Gap": Science and Ethics in Nanotechnology, *Nanotechnology*, 14, R9-R13.

Moronne, M.M.; Ben Dahan, M.; Bruchez, M.; Hamamoto, D.J.; Weiss, S.; Alivisatos, A.P. 1999. "Nanocrystal phosphors as biological probes for fluorescence microscopy", *Biophysical Journal*, 76, A450-A450.

Mumford, Lewis. 1934. *Technics and Civilization* (New York: Harcourt Brace Jovanovich, Inc.). National Science and Technology Council (NSTC). 1999. *Nanotechnology: Shaping the World Atom by Atom* (Washington, DC: September; available at http://www.nano.gov).

Murray, C.B.; Kagan, C.R.; Bawendi, M.G. 2000. "Synthesis and characterization of monodisperse nanocrystals and close-packed nanocrystal assemblies", *Annual Review of Materials Science*, 30, 545- 610.

Nardi, B. 2001. "Cultural Ecology of Nanotechnology" in this volume: *Societal Implications of Nanoscience and Nanotechnology*. National Science Foundation, Arlington, VA.

Nardi, B., O'Day, V. 1999. *Information Ecologies: Using Technology with Heart*. MIT Press.

National Coordination Office for Information Technology, *Ethical, Legal, and Societal Implications of Information Technology Research and Development*. Available at: http://www.itrd.gov, October 1999.

National Human Genome Research Institute, *Ethical, Legal, and Social Implications of Human Genetics Research of the National Institutes of Health*. Available at:

National Science and Technology Council (NSTC). 2000. *National Nanotechnology Initiative: Leading to the Next Industrial Revolution* (Supplement to the President's FY 2001 Budget) (Washington, DC: February; available at http://www.nano.gov).

National Science and Technology Council (NSTC). 2000. *National Nanotechnology Initiative: Leading to the Next Industrial Revolution*. February. (http://www.nano.gov/nni.pdf).

Naughton, G. 1999. The Advanced Tissue Sciences story. *Sci. Am*. 280(4): p. 84-5.

Nauta, K. and R.E. Miller. 1999. Non-equilibrium self-assembly of long chains of polar molecules in superfluid helium. *Science*. Vol. 283. 1895-1897.

Nicholas Wade, "A DNA Success Raises Bioterror Concern," New York Times, January 12, 2005.

Nowotny, H.; Scott, P.; Gibbons, M.: 2001, *Re-thinking Science*, Polity, Cambridge.

NSET. 2000. *National Nanotechnology Initiative: Leading to the Next Industrial Revolution*. 2000. A report by the Interagency Working Group on Nanoscience, Engineering and Technology (NSET), of the National Science and Technology Council's Committee on Technology. Washington, D.C., February.

NSTC (National Science and Technology Council): 1999, *Nanotechnology: Shaping the World Atom by Atom* [www.wtec.org/loyola/nano/IWGN.Public.Brochure/IWGN.Nanotechnology.Brochure.pdf, accessed: 15 Nov. 2003].

Parikh, A.N.; Schivley, M.A.; Koo, E.; Seshadri, K.; Aurentz, D.; Mueller, K.; Allara, D.L. 1997. "nalkylsiloxanes: From single monolayers to layered crystals. The formation of crystalline polymers from the hydrolysis of n- octadecyltrichlorosilane", *Journal of the American Chemical Society*, 119, 3135-3143.

Parr, R.G. and W. Yang. 1989. *Density-Functional Theory of Atoms and Molecules*. Oxford University Press. Oxford.

Peng, X.G.; Manna, L.; Yang, W.D.; Wickham, J.; Scher, E.; Kadavanich, A.; Alivisatos, A.P. 2000. "Shape control of CdSe nanocrystals", *Nature*, 404, 59-61.

Perkal M., C. Marks, M.I. Lorber, W.H. Marks. 1992. "A three-year experience with serum anodal trypsinogen as a biochemical marker for rejection in pancreatic allografts. False positives, tissue biopsy, comparison with other markers, and diagnostic strategies." *Transplantation*, Feb; 53 (2): pp. 415-419.

Philip Ball, "Synthetic Biology: Starting from Scratch," Nature, 431, pp. 624-626, 7 October 2004.

Philipse, A.P. 1998. "Colloidal Dispersions," section 3.4, A. ten Wolde (ed.), *Nanotechnology: Towards a molecular construction kit*. STT Report #60, The Hague, 171-178.

Phoenix, C.; Drexler, E.: 2004, 'Safe Exponential Manufacturing', *Nanotechnology*, 15, 869-872.

Pierce, N.A. and M.B. Giles. 2000. Adjoint reconstruction of superconvergent functionals from PDE approximations. *SIAM Rev*. Vol. 42(2). 247-264.

Piveteau, Laurent-Dominique, Anthe S. Zandvliet and Robert Langer. 2001. "Nitroxide and Galactose Grafted Dendrimer: A Specifically Targeted Contrast Agent for MRI and EPR Imaging of the Liver," *in preparation*.

Polanyi, Karl. 1944. *The Great Transformation: The Political and Economic Origins of Our Time* (Boston: Beacon Press, 1957; orig. ed. 1944).

Poldrack, R.A. 2000. Imaging brain plasticity: conceptual and methodological issues — a theoretical review. *Neuroimage*. 12(1): p. 1-13.

Porod, W., C.S. Lent, G.H. Bernstein, A.O. Orlov, I. Amlani, G.L. Snider, and J.L. Merz. 1999. Quantumdot cellular automata: computing with coupled quantum dots. *Int. Journal of Electronics* 86, No.5, 549-590.

Pritchard, M.S. 1996. "Conflicts of Interest: Conceptual and Normative Issues," *Academic Medicine*, 71:12, December, pp. 1305-1313.

Putnam, D., Gentry, C.A., Pack, D.W., Langer, R. 2001. "Polymer-based Gene Delivery with Low Cytotoxicity by a Unique Balance of Side-chain Termini." *Proceedings of the National Academy of Sciences*, U.S.A., Volume 98, Number 3, pages 1200-1205.

Ramos, L.; Lubensky, T.C.; Dan, N.; Nelson, P.; Weitz, D.A. 1999. "Surfactant-mediated two-dimensional crystallization of colloidal crystals", *Science*, 286, 2325-2328.

RAND Corporation. *The Potential of Nanotechnology for Molecular Manufacturing*. 1995.

Reed, Mark A. and James M. Tour. 2000. "Computing with Molecules." *Scientific American*. June.. SIA (Semiconductor Industry Association). 1999. "1999 International Technology Roadmap for Semiconductors."

Rinzler, A.G.; Liu, J.; Dai, H.; Nikolaev, P.; Huffman, C.B.; Rodriguez-Macias, F.J.; Boul, P.J.; Lu, A.H.; Heymann, D.; Colbert, D.T.; Lee, R.S.; Fischer, J.E.; Rao, A.M.; Eklund, P.C.; Smalley, R.E. 1998.

Roco, M. C., R. S. Williams, and P. Alivisatos, *Nanotechnology Research Directions: Vision for Nanotechnology in the Next Decade*, Kluwer Academic Publishers (2000).

Roco, M.C. & Bainbridge, W.S., (eds.): 2001, *Societal Implications of Nanoscience and Nanotechnology*, National Science Foundation [online: www.wtec.org/loyola/nano/societalimpact/nanosi.pdf , accessed: 15 November 2003].

Roco, M.C. & Bainbridge, W.S., (eds.): 2003, *Converging Technologies for Improving Human Performance: Nanotechnology, Biotechnology, Information Technology and Cognitive Science*, Kluwer, Dordrecht.

Roco, M.C., R.S. Williams, and P. Alivisatos (eds). 1999. *Nanotechnology Research Directions: IWGN Workshop Report, Vision for Nanotechnology R&D in the Next Decade*. (Kluwer Academic Publishers.

Roco, M.C.: 2003a, 'Broader Societal Issues of Nanotechnology', *Journal of Nanoparticle Research*, 5, 181-189.

Roco, M.C.: 2003b, 'Coherence and Divergence in Megatrends in Science and Engineering', in: M.C. Roco, W.S. Bainbridge (eds.), *Converging Technologies for Improving Human Performance*, Kluwer, Dordrecht, pp. 79-96.

Roco, M.C.: 2003c, 'Fundamentally New Manufacturing Processes and Products', in: Roco, M.C. & Bainbridge, W.S., (eds.), *Converging Technologies for Improving Human Performance*, Kluwer, Dordrecht, pp. 300-302.

Roco, M.C.: 2004, 'The Emergence and Policy Implications of Converging Technologies' [www.nsf.gov/home/crssprgm/nano/nbic_roco_04_0422_@aaas_57sl.pdf, accessed May 2004].

Rodney Brooks, "The Merger of Flesh and Machines," The Next Fifty Years: Science in the First Half of the Twenty-First Century, ed. by John Brockman, 2002, p. 191.

Rosenberg, Nathan. 1994. *Exploring the Black Box: Technology, Economics, and History*. New York: Cambridge University Press.

Rothman, B.K.: 1998, *Genetic Maps and Human Imagination*, W.W. Norton, New York & London,

San Francisco Chronicle, *Scientist Is Fearful of Computer Mutiny; Sun Micro co-founder says replicating robots could replace humans*, March 13, 2000.

Santini, John T., Jr., Michael J. Cima and Robert Langer. 1999. "A Controlled-release Microchip" *Nature*, Vol 397, 28 Jan, p 335-338.

Schummer, J.: 2004, 'Interdisciplinary Issues in Nanoscale Research', in: D. Baird, A. Nordmann, J. Schummer (eds.), *Discovering the Nanoscale*, IOS Press, Amsterdam, pp. 9-20.

Schumpeter, Joseph Alois. 1942. *Capitalism, Socialism, and Democracy*. New York: Harper.

Senturia, S., N. Aluru, and J. White. 1997. Simulating the behavior of MEMS devices: computational methods and needs. *IEEE Computational Science and Engineering*. Vol. 4(1). 30-43.

Shor, P., *Proc. 35th Ann. Symp. Foundations of Computer Science*, 1994.

Shultz, W.: 2000, 'Nanotechnology: The Next Big Thing', *Chemical and Engineering News*, 78(18), 41-47.

Simon, Julian L. (ed). 1995. *The State of Humanity*, Oxford: Blackwell.

Simos, P.G., et al. 2000. Insights into brain function and neural plasticity using magnetic source imaging. *J.*

Smalley, R.E. 2000. Hearing of the Senate Science and Technology Caucus on Nanotechnology. April 5, (unpublished).

Smalley, R.E.: 2001, 'Of Chemistry, Love and Nanobots', *Scientific American*, 285(3), September, 76-77.

Smith, A., *The Wealth of Nations: An Inquiry into the Nature and Causes*, Random House, New York, 1994.

Smith, D., H. Babcock, and S. Chu. 1999. Single-polymer dynamics in steady shear flow. *Science*. Vol. 283. 1724-1727. March.

Snider, G.L.; Orlov, A.O.; Amlani, I.; Zuo, X.; Bernstein, G.H.; Lent, C.S.; Merz, J.L.; Porod, W. 1999. "Quantum-dot cellular automata: Review and recent experiments (invited)", *Journal of Applied Physics*, 85, 4283-4285.

Soong, R.K.; Bachand, G.D.; Neves, H.P.; Olkhovets, A.G.; Craighead, H.G.; Montemagno, C.D. 2000. "Powering an inorganic nanodevice with a biomolecular motor", *Science*, 290, 1555-1558.

Steelman, John R. 1947. *Science and Public Policy* (Washington, DC: US Government Printing Office). Stephenson, Neal. 1995. *The Diamond Age, or, Young Lady's Illustrated Primer* (New York: Bantam Books).

Sterling, B., *When Robots Act Like Rabbits*, The Wall Street Journal, September 6, 2000.

Sterman, J. D. (1989b). Modeling Managerial Behavior: Misperceptions of Feedback in a Dynamic Decision Making Experiment. *Management Science*, 35(3), 321-339.

Steve, Jurvetson, "Transcending Moore's Law with Molecular Electronics," Nanotechnology Law & Business Journal, Vol. 1, No. 1, article 9, p. 9.

Stipe, B.C.; Rezaei, M.A.; Ho, W. 1998. "Single-molecule vibrational spectroscopy and microscopy", *Science*, 280, 1732-1735.

Stix, G.: 1996, 'Trends in Nanotechnology: Waiting for Breakthroughs', *Scientific American*, 274(4), 94-99.

Stokes, Donald E. 1997. *Pasteur's Quadrant: Basic Science and Technological Innovation*. Washington, D.C.: Brookings Institution Press.

Storrs, J., Hall, *Flying Saucers?* http://www.nanocentral.com/NanoWorld/Perspectives/JoshHall/askjosh_aug96.html.

Stranick, S.J.; Atre, S.V.; Parikh, A.N.; Wood, M.C.; Allara, D.L.; Winograd, N.; Weiss, P.S. 1996. "Nanometer-scale phase separation in mixed composition self-assembled monolayers", *Nanotechnology*, 7, 438-442.

Sun, S., C.B. Murray, D. Weller, L. Folks, and A. Moser. 2000. Monodisperse fept nanoparticles and ferromagnetic fept nanocrystal superlattices. *Science.* Vol. 287, 17 March. pp. 1989 - 1992.

Sun, S.H.; Murray, C.B. 1999. "Synthesis of monodisperse cobalt nanocrystals and their assembly into magnetic superlattices", *Journal of Applied Physics*, 85, 4325-4330.

Sun, S.H.; Murray, C.B.; Weller, D.; Folks, L.; Moser, A. 2000. "Monodisperse FePt nanoparticles and ferromagnetic FePt nanocrystal superlattices", *Science*, 287, 1989-1992.

Suvin, D.: 1979, *Metamorphoses of Science Fiction*, Yale University Press, New Haven.

Tadmor, E., R. Phillips, and M. Ortiz. 2000. Hierarchical modeling in the mechanics of materials. *Int. J. Solids Struct*. Vol. 37. 379.

Thacker, E.: 2000, 'SF, Technoscience, Net.art: The Politics of Extrapolation (science fiction)', *Art Journal*, 59(3), 64-73.

Theis, T.: 2001, 'Information Technology Based on a Mature Nanotechnology', in: M.C. Roco, W.S. Bainbridge (eds.), *Societal Implications of Nanoscience and Nanotechnology*, National Science Foundation, pp. 60-67.

Thomas, E.L., D.M. Anderson, C.S. Henkee, and D. Hoffman. 1988. Periodic area-minimizing surfaces in block copolymers. *Nature.* Vol. 334. 598.

Thompson, P.B.: 2001, 'Social Acceptance of Nanotechnology', in: M.C. Roco, W.S. Bainbridge (eds.), *Societal Implications of Nanoscience and Nanotechnology*, online: National Science Foundation, pp. 198-202.

Tilton, John E. (ed). 1991. *World Metal Demand*, Washington, DC: Resources for the Future.

Tobio, M., R. Gref, A. Sanchez, R. Langer and M.J. Alonso. 1998. "Stealth PLA-PEG Nanoparticles as Protein Carriers for Nasal Administration", *Pharmaceutical Research*, Vol. 15, No 2, pp 270-275.

Tramontin, A.D. and E.A. Brenowitz. 2000. Seasonal plasticity in the adult brain. *Trends Neurosci.* 23(6): p. 251-8.

Tsuji, T. and A. Rey. 1998. Orientation mode selection mechanisms for sheared nematic liquid crystalline materials. *Phys. Rev. E.* Vol. 57(5). 5609-5625.

Turnbull, D.: 2000, *Masons, Tricksters and Cartographers*, Harwood, Amsterdam.

Tushman, Michael and Philip Anderson (1996), Managing Strategic Innovation and Change: A Collection of Readings. London: Oxford University Press.

Vaia, R., D. Tomlin, M. Schulte, and T. Bunning. 2000. Two-phase nanoscale morphology of polymer/LC composites. *Polymer*. In press.

Vale, R.D.; Milligan, R.A. 2000. "The way things move: Looking under the hood of molecular motor proteins", *Science*, 288, 88-95.

Valenti. Plaza y Valdes, S.A. de C.V. Instituto de Investigaciones Sociales. Universidad Nacional Autonoma de Mexico. Universidad Autonoma Metropolitana, pp.81-93.

Van Maanen, J. *Tales of the Field*. Chicago: U. Chicago, 1988.

Vettiger, P.; Brugger, J.; Despont, M.; Drechsler, U.; Durig, U.; Haberle, W.; Lutwyche, M.; Rothuizen, H.; Stutz, R.; Widmer, R.; Binnig, G. 1999. "Ultrahigh density, high-data-rate NEMS-based AFM data storage system", *Microelectronic Engineering*, 46, 11-17.

Vettiger, P.; Despont, M.; Drechsler, U.; Durig, U.; Haberle, W.; Lutwyche, M.I.; Rothuizen, H.E.; Stutz, R.; Widmer, R.; Binnig, G.K. 2000. "The "Millipede" - More than one thousand tips for future AFM data storage", *IBM Journal of Research and Development*, 44, 323-340.

Voter, A.F. 1997. Hyperdynamics: accelerated molecular dynamics of infrequent events. *Phys. Rev. Lett.* Vol. 78. 3908.

Warner, M. and E. M. Terentjev. 1996. Nematic elastomers: a new state of matter? *Progress in Polymer Science*. Vol. 21(5). 853.

Warner, M., E. M. Terentjev, R. B. Meyer, and Y. Mao. 2000. Untwisting of a cholesteric elastomer by a mechanical field. *Phys. Rev. Lett.* Vol. 85(11). 2320.

Washington, London, and Moscow Convention, Convention on the prohibition of the development, production and stockpiling of bacteriological (biological) and toxin weapons and on their destruction, signed Washington, London, and Moscow April 10,1972, entered into force March 26, 1975.

Webb, E. J., Campbell, D. T., Schwartz, R. D., Sechrest, L., and Grove, J. B. *Unobtrusive Measures: Nonreactive Research in the social Sciences* (2nd ed.). Boston: Houghton Mifflin, 1981.

Webster, A. and H. Etzkowitz. 1991. "Academic-industry Relations: The Second Academic Revolution?" Science Policy Support Group, London, 1991.

Weil, V. 1988. "Afterword," *Biotechnology: Professional Issues and Social Concerns*. (AAAS Publication) co-edited with Paul DeForest, Mark Frankel, and Jeanne Poindexter.

Weil, V. 1996. "Biotechnology: Societal Impact and Quandaries," in *Biotechnology and Ethics: A Blueprint for the Future, Report of NSF Workshop*, Northwestern University, Center for Biotechnology.

Weil, V. 2000. "Como Pueden Convivir Los Valores Universitarios Y Las Normes Empresariales?" in *Dos Ejes en La Vinculacion de Las Universidades a la Produccion*. ed. by Rosalba Casas and Giovanna

Weiss, S. 1999. "Fluorescence spectroscopy of single biomolecules", *Science*, 283, 1676-1683.

White, Lynn Jr. 1974. Technology assessment from the stance of a medieval historian. *The American Historical Review*. 79(1): 1-13.

White, Lynn T. 1962. *Medieval Technology and Social Change* (London: Oxford University Press).

Whitesides, G.M.; Mathias, J.P.; Seto, C.T. 1991. "Molecular Self-Assembly and Nanochemistry - a Chemical Strategy For the Synthesis of Nanostructures", *Science*, 254, 1312-1319.

Wilbur, J.L.; Whitesides, G.M. 1999. *Self-Assembly and Self-Assembled Monolayers in Micro- and Nanofabrication*; Timp, G., Ed.; Springer-Verlag New York, Inc.: New York, pp 331-370.

William-Jones, B.; Corrigan, O.: 2003, 'Rhetoric and Hype: Where's the Ethics in Pharmacogenomics?', *American Journal of Pharmacogenomics*, 3(6), 375-383.

Winner, Langdon, ed., *Democracy in a Technological Society, in Philosophy and Technology*, Vol. 9, Kluwer Academic Publishers, Dordrecht, The Netherlands, 1992.

Wong, H.-S.P., D.J. Frank, and P.M. Solomon. 1998. "Device Design Considerations for Double-Gate, Ground-Plane, and Single-Gated Ultra-Thin SOI MOSFETs at the 25-nm Channel-Length Generation." *Technical Digest of the 1998 International Electron Devices Meeting.*

Wood, S., Jones, R.; Geldart, A.: 2003, *The Social and Economic Challenges of Nanotechnology*, UK Economic and Social Research Council [www.esrc.ac.uk/esrccontent/DownloadDocs/Nanotechnology.pdf, accessed: 3 Feb. 2004].

Wynne, B. 1991. "Knowledges in Context." *Science, Technology, and Human Values*, 16:1, Winter 1991, pp. 111-121.

Zachary, G. Pascal. 1997. *Endless Frontier: Vannevar Bush, Engineer of the American Century.* New York: Free Press.

Zinchenko, V. 1996. Developing Activity Theory. In B. Nardi, Ed. *Context and Consciousness: Activity Theory and Human-Computer Interaction.* MIT Press.

5

Scientific Bibliography on Nano Biotechnology

Nanotechnology-based Drug Delivery for Cancer. Jain KK. Technol Cancer Res Treat. 2005 Aug;4(4):407-16.

Nanobiotechnologies have been applied to improve drug delivery and to overcome some of the problems of drug delivery in cancer. These can be classified into many categories that include use of various nanoparticles, nanoencapsulation, targeted delivery to tumors of various organs, and combination with other methods of treatment of cancer such as radiotherapy. Nanoparticles are also used for gene therapy for cancer. Some of the technologies enable combination of diagnostics with therapeutics which will be important for the personalized management of cancer. Some of the limitations of these technologies and prospects for future development are discussed.

Editorial: targeted drug delivery for cancer. Jain KK. Technol Cancer Res Treat. 2005 Aug;4(4):311-4.

Cancer drug delivery is no longer simply wrapping the drug in new formulations for different routes of delivery. Knowledge and experience from other technologies such as nanotechnology, advanced polymer chemistry, and electronic engineering, are being brought together in developing novel methods of drug delivery. Advances in our knowledge of molecular biology of cancer and pathways involved in malignant transformation of cells are revolutionizing the approach to cancer treatment with a focus is on targeted cancer therapy. The newer approaches to cancer treatment not only supplement the conventional chemotherapy and radiotherapy but also aim to prevent damage to the normal tissues and overcome drug resistance. Innovative methods of cancer treatment require new concepts of drug delivery in cancer.

Selective tumor targeting by enhanced permeability and retention effect. Synthesis and antitumor activity of polyphosphazene-platinum (II) conjugates. Jun YJ, Kim JI, Jun MJ, Sohn YS. J Inorg Biochem. 2005 Aug;99(8):1593-601

Nanosized polyphosphazene-platinum (II) conjugates with a wide range of molecular weight from 24,000 to 115,000 were synthesized to study their tumor selectivity by enhanced permeability and retention (EPR) effect and their antitumor activity. It has been found from biodistribution study that the present polyphosphazene-Pt(II) conjugates exhibit high tumor selectivity by EPR effect with the tumor to tissue ratio (TTR) from 3.6 to 13 depending on the molecular size. These polymer conjugates have shown excellent in vivo antitumor activity against both murine and human cancer cell lines. In particular, xenograft trials of the conjugates have shown outstanding tumor inhibition effect on the stomach cancer cell line, YCC-3, which is one of the least responsive to the anticancer agents currently in clinical use, although the reason is not clearly explainable yet. The high in vivo activity seems to be attributed to the controlled-release of the antitumor active platinum (II) moiety, [GlyGluPt(dach] (dach=trans-(+/-)-1,2-diaminocyclohexane) from the phosphazene backbone by degradation in aqueous solution.

Preparation and in vitro evaluation of chitosan nanoparticles containing a caspase inhibitor. Aktas Y, Andrieux K, Alonso MJ, Calvo P, Gursoy RN, Couvreur P, Capan Y. Int J Pharm. 2005 Jul 25;298(2):378-83.

The aim of this work was to develop a formulation for Z-DEVD-FMK, a peptide which is a caspase inhibitor and has been used in experimental animal studies for a decade. Peptide loaded chitosan nanoparticles were obtained by ionotropic gelation process and Z-DEVD-FMK was quantified by an HPLC method. The influence of the initial peptide concentration on the nanoparticle characteristics and release behavior was evaluated. The CS nanoparticles have a particle diameter (Z-average) ranging from approximately 313-412 nm and a positive zeta potential (20-28 mV). The formulation with the initial peptide concentration of 400 ng/ml provided the highest loading capacity (0.46%) and the highest extent of release (65% at 24 h) suggesting the possibility to achieve a therapeutic dose. According to the data obtained, this chitosan-based nanotechnology opens new and interesting perspectives for anticaspase activity.

A methodology to study intracellular distribution of nanoparticles in brain endothelial cells. Garcia-Garcia E, Andrieux K, Gil S, Kim HR, Le Doan T, Desmaele D, d'Angelo J, Taran F, Georgin D, Couvreur P. Int J Pharm. 2005 Jul 25;298(2):310-4.

Cell internalisation and intracellular distribution of PEG-coated polyhexadecylcyanoacrylate (PEG-PHDCA) nanoparticles in rat brain endothelial cells (RBEC) have been investigated. A cell fractionation method has been developed based on the selective permeabilisation of RBEC plasma membrane by digitonin. By interacting with membrane cholesterol, digitonin creates pores allowing the release of soluble and diffusible species outside the cell. The selectivity of plasma membrane permeabilisation was controlled by using compartment markers such as lactate dehydrogenase (LDH) for cytoplasm and cathepsin B for lysosomes. An optimal digitonin concentration of 0.003% (w/v) has been identified to induce a pattern of membrane permeabilisation corresponding to the extraction of 72% LDH and less than 15% of Cathepsin B.

Membrane permeabilisation at this digitonin concentration allows one to distinguish between the cell cytoplasm and its endo/lysosomal fraction. This methodology was applied to investigate the intracellular distribution of the nanoparticles after their incubation with the RBEC. The results showed that PEG-PHDCA nanoparticles were able to be internalised to a higher extent than PHDCA nanoparticles (after 20 min incubation). Additionally, these nanoparticles displayed different patterns of intracellular capture, depending on their specific surface composition: PEG-PHDCA nanoparticles were 48% in the plasma membrane, 24% in the cytoplasm, 20% in vesicular compartments and 8% associated with the fraction of the nucleus, the cytoskeleton and caveolae suggesting that PEG-PHDCA nanoparticle uptake by RBEC is specific and presumably due to endocytosis. Confocal microscopy studies confirmed the cellular uptake of PEG-PHDCA nanoparticles.

Preparation and in vitro anticancer activity of wheat germ agglutinin (WGA)-conjugated PLGA nanoparticles loaded with paclitaxel and isopropyl myristate. Mo Y, Lim LY. J Control Release. 2005 Jul 25; [Epub ahead of print].

The purpose of this study was to develop a novel lectin-conjugated isopropyl myristate (IPM)-incorporated PLGA nanoparticle system (NP) for the local delivery of paclitaxel to the lungs. Wheat germ agglutinin (WGA) was conjugated onto preformed IPM- and paclitaxel-loaded PLGA NPs by a two-step carbodiimide method following comparative uptake studies of Concanavalin A, Ricinus communis-120 and WGA on A549, H1299 and CCL-186 cells. WIT-NP with mean diameter of 331 nm and zeta potential of -4.3 mV were prepared with yield of 66% and paclitaxel encapsulation efficiency of 61%. Particle size was expanded by surface conjugation with WGA, while zeta potential was reduced by the addition of IPM and WGA. In vitro paclitaxel release profile was not affected by WGA but initial drug release was enhanced by adding IPM into the formulation. The WIT-NP showed a burst-release of about 32% of the paclitaxel load within the first 5 h followed by a slow zero-order release of another 7% of the drug load in the next 115 h. Compared with the clinical paclitaxel formulation, paclitaxel-loaded nanoparticles without IPM or WGA, or paclitaxel-loaded nanoparticles with only IPM or WGA, the WIT-NP had superior in vitro cytotoxicity against A549 and H1299 cells. IC50 for WIT-NP after 5 and 24 h incubation with A549 cells were not significantly different (15.5 and 15 muM, respectively) whereas the clinical formulation was not cytotoxic after 5 h but had IC50 of 14 muM after 24 h incubation. WIT-NP exhibited stronger cell-killing effect because of more efficient cellular uptake via WGA-receptor-mediated endocytosis and IPM-facilitated release of paclitaxel from the NPs.

Fusogenic liposome delivers encapsulated nanoparticles for cytosolic controlled gene release. Kunisawa J, Masuda T, Katayama K, Yoshikawa T, Tsutsumi Y, Akashi M, Mayumi T, Nakagawa S. J Control Release. 2005 Jul 20;105(3):344-53.

Therapeutic agents based on DNA or RNA oligonucleotides (e.g., antisense DNA oligonucleotide, small interfering RNA) require a regulation of their kinetics in cytoplasm to maintain an optimal concentration during the treatment period. In this respect, delivery of functional nanoparticles containing these drugs into cytoplasm has been thought to have a potential for the cytosolic controlled gene release. In this study, we establish a protocol for the encapsulation of nanoparticles into liposome, which is further fused with ultra violet-inactivated Sendai virus to compose fusogenic liposomes. When nanoparticles were encapsulated in conventional liposomes, endocytosis-mediated uptake of nanoparticles was observed. In contrast, numerous amounts of nanoparticles were delivered into the cytoplasm without any cytotoxicity when the particles were encapsulated in fusogenic liposomes. Additionally, fusogenic liposome showed a high ability to deliver nanoparticles containing DNA oligonucleotides into cytoplasm. These results indicate that this combinatorial nanotechnology using fusogenic liposome and nanoparticle is a valuable system for regulating the intracellular pharmacokinetics of gene-based drugs.

Targeting ECM-Integrin Interaction with Liposome-Encapsulated Small Interfering RNAs Inhibits the Growth of Human Prostate Cancer in a Bone Xenograft Imaging Model. Bisanz K, Yu J, Edlund M, Spohn B, Hung MC, Chung LW, Hsieh CL. Mol Ther. 2005 Jul 19; [Epub ahead of print].

The intricate intracellular communication between stromal and epithelial cells, which involves cell-cell-, cell-insoluble extracellular matrix- (ECM), and cell-soluble factor-mediated signaling processes, is an attractive target for therapeutic intervention in hormone-refractory and bone-metastatic prostate cancer. In the present study we demonstrated that androgen-independent PC3 prostate cancer cells adhered to and migrated on vitronectin (VN), a major noncollagenous ECM in mature bone, through the expression of alphav-containing integrin receptors alphavbeta1 and alphavbeta5 on the cell surface, as determined by antibody function blocking assay and flow cytometry analysis. Small interfering RNAs (siRNAs) targeting human integrin alphav markedly reduced their respective mRNA and protein expression in cells, resulting in nearly complete reduction in VN-mediated cancer progression in vitro. In vivo quantitative bioluminescence analysis of human prostate cancer bone xenografts demonstrated for the first time that intratumoral administration of liposome-encapsulated human alphav-siRNAs significantly inhibits the growth of luciferase-tagged PC3 tumors in skeleton, which was associated with decreased integrin alphav expression and increased apoptosis in tumor cells. This integrin-based gene therapy is particularly suitable for the treatment of prostate cancer bone metastasis.

Nanoparticles of poly(lactide) / vitamin E TPGS copolymer for cancer chemotherapy: Synthesis, formulation, characterization and in vitro drug release. Zhang Z, Feng SS. Biomaterials. 2005 Jul 15; [Epub ahead of print].

Paclitaxel is one of the best anticancer drugs, which has excellent therapeutic effects against a wide spectrum of cancers. The formulation of paclitaxel used in its currently clinical administration includes Cremophor EL, which has been found to cause serious side effects. Nanoparticle formulation of paclitaxel may provide an ideal solution for this problem and achieve a sustained chemotherapy. A novel copolymer, poly(lactide)-vitamin E TPGS (PLA-TPGS), was synthesized from lactide and d-alpha-tocopheryl polyethylene glycol 1000 succinate by bulk polymerization for nanoparticle formulation of anticancer drugs. (1)H NMR, FTIR and GPC were used to detect molecular structure of the copolymer. Paclitaxel-loaded PLA-TPGS nanoparticles were fabricated by a modified solvent extraction/evaporation technique with or without emulsifier involved, which were characterized by laser light scattering for size and size distribution; field emission scanning electron microscopy for surface morphology; zeta potential for surface charge; X-ray photoelectron spectroscopy for surface chemistry. The drug encapsulation efficiency and the in vitro drug release kinetics were measured by high-performance liquid chromatography. Formulation optimization was pursued. The particles were found of around 300nm in size and narrow size distribution. Of all, 89% drug encapsulation efficiency has been achieved for nanoparticles of 5% drug loading. The drug release from PLA-TPGS nanoparticles was found to be biphasic with an initial burst of 17% in the first day, followed by a sustained pattern with 51% release after 31 days.

Radiosensitivity of human colon cancer cell enhanced by immunoliposomal docetaxel. Wang QW, Lu HL, Song CC, Liu H, Xu CG. World J Gastroenterol. 2005 Jul 14;11(26):4003-7.

AIM: To enhance the radiosensitivity of human colon cancer cells by docetaxel. METHODS: Immunoliposomal docetaxel was prepared by coupling monoclonal antibody against carcinoembryonic antigen to cyanuric chloride at the PEG terminus of liposome. LoVo adenocarcinoma cell line was treated with immunoliposomal docetaxel or/and irradiation. MTT colorimetric assay was used to estimate cytotoxicity of immunoliposomal docetaxel and radiotoxicity. Cell cycle redistribution and apoptosis were determined with flow cytometry. Survivin expression in LoVo cells was verified by immunohistochemistry. D801 morphologic analysis system was used to semi-quantify immunohistochemical staining of survivin. RESULTS: Cytotoxicity was induced by immunoliposomal docetaxel alone in a dose-dependent manner. Immunoli-posomal docetaxel yielded a cytotoxicity effect at a low dose of 2 nmol/L. With a single dose irradiation, the relative surviving fraction of LoVo cells showed a dose-dependent response, but there were no significant changes as radiation delivered from 4 to 8 Gy. Compared with liposomal docetaxel or single dose irradiation, strongly radiopotentiating effects of immunoliposomal docetaxel on LoVo cells were observed. A low dose of immunoliposomal docetaxel could yield sufficient radiosensitivity. Immunoliposomal docetaxel were achieved both specificity of the conjugated antibody and drug radiosensitization. Combined with radiation, immunoliposomal docetaxel significantly increased the percentage of G(2)/M cells and induced apoptosis, but significantly decreased the percentage of cells in G(2)/G(1) and S phase by comparison with liposomal docetaxel. Immunohistochemical analysis showed that the brown stained survivin was mainly in cytoplasm of LoVo cells. Semi-quantitative analysis of the survivin immunostaining showed that the expression of survivin in LoVo cells under irradiation with immunoliposomal docetaxel was significantly decreased. CONCLUSION: Immunoliposomal docetaxel is strongly effective for target radiosensitation in LoVo colon carcinoma cells, and may offer the potential to improve local radiotherapy.

Novel Functionalized Biodegradable Polymers for Nanoparticle Drug Delivery Systems. Kallinteri P, Higgins S, Hutcheon GA, St Pourcain CB, Garnett MC. Biomacromolecules. 2005 Jul 11;6(4):1885-1894.

We have prepared and screened a library of novel functionalized polymers for development of nanoparticle drug delivery systems. The polymer backbone consisting of two ester-linked, nontoxic, biological monomers, glycerol and adipic acid, was prepared using a hydrolytic enzyme. The specificity of the chosen enzyme yields a linear polymer with one free pendant hydroxyl group per repeat unit, which can be further functionalized. This protocol gives control over the backbone polymer molecular weight, together with the ability to incorporate various amounts of different fatty acyl substituents. These functionalized polymers are able to self-assemble into well-defined small particles of high homogeneity with a very low toxicity. They are able to incorporate a water soluble drug, dexamethasone phosphate, with a high efficiency and drug loading which varies with the polymer specification. The above characteristics strongly suggest that these polymers could be developed into useful nanoparticulate drug delivery systems.

Phase I Study of Liposome-DNA Complexes Encoding the Interleukin-2 Gene in Dogs with Osteosarcoma Lung Metastases. Dow S, Elmslie R, Kurzman I, Macewen G, Pericle F, Liggitt D. Hum Gene Ther. 2005 Jul 7; [Epub ahead of print].

Systemic gene delivery using cationic liposome-DNA complexes (LDCs) has been shown to elicit potent antitumor activity in mice with tumor metastases to the lungs. However, intravenous gene delivery for treatment of established cancer has not been evaluated previously in a spontaneous, large animal model. We therefore evaluated the safety, toxicity, and efficacy of intravenous gene delivery, using LDCs in dogs with established tumor metastases. Twenty dogs with chemotherapy-resistant osteosarcoma metastases to the lungs received a series of intravenous infusions of cationic liposomes and plasmid DNA encoding the canine interleukin-2 (IL-2) cDNA. Effects of intravenous gene delivery on immune activation, clinical and hematologic parameters, tumor responses, and survival times were assessed. We found that slow intravenous administration of IL-2 LDCs resulted in detectable IL-2 transgene expression in lung tissues of dogs. Repeated intravenous infusions of LDCs were well tolerated by dogs with lung tumor metastases and elicited systemic immune activation, as reflected by fever, leukogram changes, monocyte activation, and increased natural killer cell activity. Three of 20 dogs experienced partial or complete regression of lung metastases after infusion of IL-2 LDCs. Overall survival times were significantly increased in treated dogs compared with historical control animals with the same stage of disease. We conclude that repeated intravenous infusion of LDCs in cancerbearing dogs is safe and well tolerated at low doses and may be capable of eliciting antitumor activity in some animals with advanced tumor metastases.

Encapsulation and Controlled Release of a Hydrophobic Drug Using a Novel Nanoparticle-Forming Hyperbranched Polyester. Zou J, Shi W, Wang J, Bo J. Macromol Biosci. 2005 Jul 7;5(7):662-668.

An amphiphilic, hyperbranched polymer suitable for use in controlled drug delivery is reported. This polymer was obtained by modification of the hyperbranched aliphatic polyester Boltorntrade mark H20 (H20) with succinic anhydride and then glycidyl methacrylate, and formed nanoparticles in aqueous solution. The critical association concentration was 7.4 x 10(-3) g. L(-1), as determined by fluorescence spectroscopy using pyrene as a molecular probe. A static/dynamic laser light scattering (LLS) study revealed that the average particle size was 39.4 nm with a low particle size distribution (PDI = 0.04), and that each particle was composed of about 350 amphiphilic molecules. Daidzein, a hydrophobic traditional Chinese medicine, was encapsulated during particle formation and the release properties were determined. The optimal feeding concentration of daidzein to hyperbranched polyester was 4.9 x 10(-5) g. mL(-1) to 5.0 x 10(-3) g. mL(-1) with a loading efficiency of 76.1%. In the presence of the enzyme Lipase PS, the drug loaded nanoparticles degraded in a random one-by-one manner and released the drug over a few days. This system is therefore a novel controlled drug release system based on nanoparticles formed of hyperbranched polyester.Encapsulation of daidzein by hyperbranched polyester particles.

Applied nanotechnology for the management of breakthrough cancer pain. Sprintz M, Benedetti C, Ferrari M. Minerva Anestesiol. 2005 Jul-Aug;71(7-8):419-23.

Pain, often considered the 5th vital sign, plays a significant role in the pathophysiology of cancer. Often addressed as an afterthought, untreated or under-treated cancer-related pain can have deleterious effects on a patient's physical and psychological well-being. Additionally, patients with breakthrough cancer pain tend to have more intense and more frequent background pain than patients without breakthrough pain. Currently, only oral transmucosal fentanyl citrate (OTFC) has reached the market for the treatment of cancer-related breakthrough pain. OTFC is an excellent first step in addressing the unmet need for symptomatic relief of breakthrough cancer pain; however, there is much room for improvement. Nanoscale science and engineering advancements have the long-term potential to bring revolutionary changes in society and across virtually all physical, biological and engineering disciplines, particularly medicine. Nanotechnology offers the potential to address multiple, major unmet problems in the diagnosis, treatment and symptom management of a large variety of diseases and conditions, including cancer. Nanotechnology can engender transformational progress in crucial aspects of the fight against cancer, spanning the continuum that ranges from prevention, to early detection, screening and monitoring, to innovative diagnostics and therapeutic modalities in the era of patient-centered, molecular medicine. Specifically, the authors will discuss their current research in the field of biomedical nanotechnology and its application to the management of breakthrough cancer pain.

Cationic lipid-protamine-DNA (LPD) complexes for delivery of antisense c-myc oligonucleotides. Junghans M, Loitsch SM, Steiniger SC, Kreuter J, Zimmer A. Eur J Pharm Biopharm. 2005 Jul;60(2):287-94.

In the present study, cationic lipid-peptide-DNA-complexes (LPDs) consisting of AH-Chol-liposomes and protamine-phosphodiester-oligonucleotide-particles (proticles) were introduced as carriers for antisense therapy. The LPDs were physically characterized, and a possible mechanism for adsorption of oligonucleotides (ODNs) was suggested. An increase in stability of ODNs against DNase I and serum nuclease digestion by these carriers was demonstrated. The hydrodynamic diameter increased after incubation with FCS which could be attributed to a protein coating of the particle surface. However, in cell culture medium lower particle sizes of the complexes occurred. In an antisense c-myc in vitro model, the effect of LPDs was tested using U937 cells. The C-MYC level was reduced after treatment of these antisense ODN carrier complexes. Furthermore, no changes in target mRNA concentration of the treated cells was found by reverse transcription and competitive multiplex-PCR.

Phase I trial of liposomal encapsulated doxorubicin (MyocetTM; D-99) and weekly docetaxel in advanced breast cancer patients. Mrozek E, Rhoades CA, Allen J, Hade EM, Shapiro CL. Ann Oncol. 2005 Jul;16(7):1087-93.

BACKGROUND: : We conducted a phase I trial to determine the safety and maximum tolerated dose (MTD) of non-pegylated liposome-encapsulated doxorubicin (Myocettrade mark; D-99) administered with weekly docetaxel in metastatic breast cancer (MBC) patients. PATIENTS AND METHODS: : Twenty-one patients with no prior chemotherapy for MBC received D-99 (60 or 50 mg/m(2)) intravenously (i.v.) on day 1 and escalating doses of docetaxel (25, 30 and 35 mg/m(2)) i.v. on days 1 and 8 in cohorts of three to six patients. Treatment cycles were repeated every 21 days for a maximum of six cycles. RESULTS: : The maximum tolerated dose (MTD) was 50 mg/m(2) of D-99 in combination with 25 mg/m(2) of weekly docetaxel. The most common grade 4 toxicity was neutropenia that occurred in 42 (41%) of treatment cycles, with 10 hospitalizations for febrile neutropenia. Serious protocol-defined cardiac events occurred in three (14%) patients, with two (10%; 95% confidence interval [CI] 1% to 30%) developing congestive heart failure (CHF) after a total cumulative anthracycline dose (adjuvant doxorubicin + D-99) of 540 mg/m(2). CONCLUSIONS: : D-99 in combination with weekly docetaxel, at the doses and schedule as administered in this trial, is not recommended for phase II testing. Additional trials, using different doses and schedules, are required to evaluate the potential side-effects and efficacy of D-99 and docetaxel.

Substantial increases in idarubicin plasma concentration by liposome encapsulation mediates improved antitumor activity. Dos Santos N, Waterhouse D, Masin D, Tardi PG, Karlsson G, Edwards K, Bally MB. J Control Release. 2005 Jun 20;105(1-2):89-105.

Idarubicin has been successfully encapsulated in cholesterol-free liposomes, however, little is known about how the rate of drug release from circulating liposomes influences therapeutic activity. The studies described herein assess the attributes of a liposome formulation required to significantly increase the plasma levels of idarubicin and further establish whether increases in the circulation longevity of the drug mediate improved antitumor activity. Pharmacokinetic assessments of 6 different (3)[H]-labelled liposome formulations were compared to free idarubicin. The highest idarubicin plasma concentrations were observed with DSPC/DSPE-PEG(2000) liposomes formulated with 2 mol% DSPE-PEG(2000) and 150 mM (iso-osmotic) internal citrate concentration. It was shown that increased levels of PEG-lipid incorporation augmented IDA release and the optimal liposomal formulation needed to be prepared under iso-osmotic conditions. For efficacy studies in a murine leukemia model, groups of 12-14 mice were treated i.v. with saline or equivalent doses (1, 2, 3 mg/kg) of free or liposomal IDA. Liposomal treatment groups exhibited a higher % increase in life span (ILS) as compared to equivalent doses of free drug. Efficacy studies completed in two drug resistant models, P388/ADR and MDA435LCC6/MDR1, demonstrated that neither the free nor liposomal formulation of idarubicin was therapeutically active. Encapsulation of IDA in liposomes increased antitumor activity in an IDA sensitive model, however, the significant increase in plasma drug levels was not sufficient to overcome multidrug resistance.

Controlled release of paclitaxel from microemulsion containing PLGA and evaluation of anti-tumor activity in vitro and in vivo. Kang BK, Chon SK, Kim SH, Jeong SY, Kim MS, Cho SH, Lee HB, Khang G. Int J Pharm. 2004 Nov 22;286(1-2):147-56.

The main objective of this study was to develop an optimal paclitaxel microemulsion prepared by self-microemulsifying drug delivery system (SMEDDS) which is a mixture of paclitaxel, tetraglycol, Cremophor ELP, and Labrafil 1944 and a paclitaxel microemulsion containing poly(D,L-lactide-co-glycolide) (PLGA) in order to offer controlled release of paclitaxel. To achieve this goal, paclitaxel and PLGA were dissolved by solubilizer like tetraglycol. There was not observed any change in molecular weight of PLGA after being solubilized by tetraglycol. The droplet size for all of the formulation of microemulsion was found in the range of 45-270nm by dynamic light scattering (DLS). It was observed that the droplet size of microemulsion without PLGA was smaller than that of microemulsion containing PLGA by transmission electron microscopy (TEM). The droplet of microemulsion containing PLGA was almost of spherical shape with smooth surface and there was no aggregation or adhesion among droplet of microemulsion by atomic force microscopy (AFM). The release behaviour of paclitaxel from microemulsion containing PLGA having various molecular weights (8K, 33K, and 90K) exhibited a biphasic pattern characterized by a fast initial release during the first 48h, followed by a slower and continuous release for 144h, in contrast that the release of paclitaxel from microemulsion without PLGA was finished during 24h. This result was identical with the result of anti-tumor activity in vitro of paclitaxel from microemulsion containing PLGA against human breast cancer cell line MCF7 and this formulation enhanced anti-tumor activity in vivo compared with microemulsion without PLGA against SKOV-3 human ovarian cancer cells bearing nude mice model.

Efficacy of transferrin-conjugated paclitaxel-loaded nanoparticles in a murine model of prostate cancer. Sahoo SK, Ma W, Labhasetwar V. Int J Cancer. 2004 Nov 1;112(2):335-40

Chemotherapy remains the preferred choice of treatment for prostate cancer but modest drug response and significant toxicity by conventional methods of administration limit their efficacy. In our study, we determined the efficacy of paclitaxel (Tx)-loaded biodegradable nanoparticles (NPs) on tumor inhibition. We hypothesized that NPs following conjugation to transferrin (Tf) ligand (NPs-Tf) would enhance the therapeutic efficacy of the encapsulated drug. The antiproliferative activity of NPs was determined in human prostate cancer cell line (PC3) and their effect on tumor inhibition in a murine model of prostate cancer. NPs (approximately 220 nm in diameter, 5.4% w/w drug loading) under in vitro conditions exhibited sustained release of the encapsulated drug (60% release in 60 days). The IC50 (concentration of drug for 50% inhibition of cell growth) of the drug with Tf-conjugated NPs (Tx-NPs-Tf) was about 5-fold lower than that with unconjugated NPs (Tx-NPs) or drug in solution. Animals that received a single-dose intratumoral injection of Tx-NPs-Tf (Tx dose= 4 mg/kg) demonstrated complete tumor regression and greater survival rate than those that received either Tx-NPs or Tx-Cremophor EL formulation. In conclusion, sustained release NPs demonstrated greater antitumor activity following their conjugation to Tf ligand.

Size-dependent immunogenicity: therapeutic and protective properties of nano-vaccines against tumors. Fifis T, Gamvrellis A, Crimeen-Irwin B, Pietersz GA, Li J, Mottram PL, McKenzie IF, Plebanski M. J Immunol. 2004 Sep 1;173(5):3148-54.

Infection can protect against subsequent disease by induction of both humoral and cellular immunity, but inert protein-based vaccines are not as effective. In this study, we present a new vaccine design, with Ag covalently conjugated to solid core nano-beads of narrowly defined size (0.04-0.05 microm) that localize to dendritic cells (DEC205(+) CD40(+), CD86(+)) in draining lymph nodes, inducing high levels of IFN-gamma production (CD8 T cells: precursor frequencies 1/5000 to 1/1000) and high Ab titers in mice. Conjugation of Ag to these nano-beads induced responses that were significantly higher (2- to 10-fold) than those elicited by other bead sizes, and higher than a range of currently used adjuvants (alum, QuilA, monophosphoryl lipid A). Responses were comparable to CFA/IFA immunization for Abs and ex vivo peptide-pulsed dendritic cell immunization for CD8 T cells. A single dose of Ag-conjugated beads protected mice from tumors in two different model challenges and caused rapid clearance of established tumors in mice. Thus, a range of Ags conjugated to nano-beads was effective as immunogens in both therapeutic and prophylactic scenarios.

Body distribution of nanoparticle-containing adriamycin injected into the hepatic artery of hepatoma-bearing rats. Chen JH, Wang L, Ling R, Li Y, Wang Z, Yao Q, Ma Z. Dig Dis Sci. 2004 Aug;49(7-8):1170-3.

The aim of the study was to investigate the body distribution of nanoparticle-containing adriamycin (NADR) injected into the hepatic artery of hepatoma-bearing rats. Thirty Walker-256 hepatoma-bearing rats were divided into two groups at random, with 15 rats in each. NADR and free adriamycin (FADR) were injected into the hepatic artery of animals on the seventh day after tumor implantation. At 1, 5, and 15 hr, after administration, five animals in each group were sacrificed and the ADR concentrations in the plasma, liver, heart, spleen, lungs, kidneys, and tumor were determined. The results showed that NADR substantially increased the ADR concentrations in liver, spleen, and tumor of rats compared to FADR, whereas the concentrations in plasma, heart, and lungs were significantly decreased. In conclusion, the body distribution of ADR can be modified by its encapsulation into nanoparticles and administration via the hepatic artery.

Preparation of magnetic polybutylcyanoacrylate nanospheres encapsulated with aclacinomycin A and its effect on gastric tumor. Gao H, Wang JY, Shen XZ, Deng YH, Zhang W. World J Gastroenterol.2004 Jul 15;10(14):2010-3.

AIM: To evaluate the effect of aclacinomycin A-loaded magnetic polybutylcyanoacrylate nanoparticles on gastric tumor growth in vivo and in vitro. METHODS: Magnetic polybutylcyanoacrylate (PBCA) nanospheres encapsulated with aclacinomycin A (MPNS-ACM) were prepared by interfacial polymerization. Particle size, shape and drug content were examined. Female BABL/c nude mice were implanted with MKN-45 gastric carcinoma tissues subcutaneously to establish human gastric carcinoma model. The mice were randomly divided into 5 groups of 6 each: ACM group (8 mg/kg bm); group of high dosage of MPNS-ACM (8 mg/kg bm); group of low dosage of MPNS-ACM (1.6 mg/kg bm); group of magnetic PBCA nanosphere (MPNS) and control group (normal saline). Magnets (2.5 T) were implanted into the tumor masses in all of the mice one day before the therapy. Above-mentioned drugs were administered intravenously to the mice of every group on the first day and sixth day. When the mice were sacrificed, tumor weight was measured, and the assay of granulocyte- macrophage colony forming-unit (CFU-GM) was performed on semi-solid culture. White blood cell, alanine aminotransferase and creatine were examined. 3-(4-dimethylthiazol-2-yl)-2,5-diphenyltetrazolium bromide (MTT) was used to examine the viability of MKN-45 cells after incubation with different concentrations of ACM, MPNS and MPNS-ACM suspension respectively for 48 h. RESULTS: Content of ACM in MPNS-ACM was 12.0% and the average diameter of the particles was 210 nm. The inhibitory rates of ACM (8 mg/kg bm), high dosage of MPNS-ACM (8 mg/kg bm), low dosage of MPNS-ACM (1.6 mg/kg bm) and MPNS on human gastric carcinoma in nude mice were 22.63%, 52.55%, 30.66% and 10.22%, respectively. There was a significant decrease in the number of

CFU-GM of bone marrow in ACM group compared with control group, whereas no obvious change was observed in that of the nanosphere groups. The values of 50% inhibition concentration (IC50) of ACM, MPNS and MPNS-ACM were 0.09, 97.78 and 1.07 microg/mL, respectively. CONCLUSION: The tumor inhibitory rate of MPNS-ACM was much higher than that of ACM under magnetic field and the inhibition on bone marrow was alleviated significantly compared with ACM group.

Enhanced antitumor efficacy on hepatoma-bearing rats with adriamycin-loaded nanoparticles administered into hepatic artery. Chen JH, Ling R, Yao Q, Wang L, Ma Z, Li Y, Wang Z, Xu H. World J Gastroenterol. 2004 Jul 1;10(13):1989-91.

AIM: To investigate the antitumor activity of adriamycin (ADR) encapsulated in nanoparticles (NADR) and injected into the hepatic artery of hepatoma-bearing rats. METHODS: NADR was prepared by the interfacial polymerization method. Walker-256 carcinosarcomas were surgically implanted into the left liver lobes of 60 male Wistar rats, which were divided into 4 groups at random (15 rats per group). On the 7th day after implantation, normal saline (NS), free ADR (FADR), NADR, or ADR mixed with unloaded nanoparticles (ADR+NP) was respectively injected via the hepatic artery (i.a.) of rats in different groups. The dose of ADR in each formulation was 2.0 mg/kg body weight and the concentration was 1.0 mg/mL. Survival time, tumor enlargement ratio, and tumor necrosis degree were compared between each group. RESULTS: Compared with the rats that received NS i.a., the rats that received FADR or ADR+NP acquired apparent inhibition on tumor growth, as well as prolonged their life span. Further significant anticancer efficacy was observed in rats that received i.a. administration of NADR. Statistics indicated that NADR brought on a more significant tumor inhibition and more extensive tumor necrosis, as compared to FADR or ADR+NP. The mean tumor enlargement ratio on the 7th day after NADR i.a. was 1.106. The mean tumor-bearing survival time was 39.50 days. Prolonged life span ratio was 109.22% as compared with rats that accepted NS. CONCLUSION: Therapeutic effect of ADR on liver malignancy can be significantly enhanced by its nanopaticle formulation and administration via hepatic artery.

Photo-thermal tumor ablation in mice using near infrared-absorbing nanoparticles. O'Neal DP, Hirsch LR, Halas NJ, Payne JD, West JL. Cancer Lett. 2004 Jun 25;209(2):171-6.

The following study examines the feasibility of nanoshell-assisted photo-thermal therapy (NAPT). This technique takes advantage of the strong near infrared (NIR) absorption of nanoshells, a new class of gold nanoparticles with tunable optical absorptivities that can undergo passive extravasation from the abnormal tumor vasculature due to their nanoscale size. Tumors were grown in immune-competent mice by subcutaneous injection of murine colon carcinoma cells (CT26.WT). Polyethylene glycol (PEG) coated nanoshells (approximately 130 nm diameter) with peak optical absorption in the NIR were intravenously injected and allowed to circulate for 6 h.

Tumors were then illuminated with a diode laser (808 nm, 4 W/cm2, 3 min). All such treated tumors abated and treated mice appeared healthy and tumor free >90 days later. Control animals and additional sham-treatment animals (laser treatment without nanoshell injection) were euthanized when tumors grew to a predetermined size, which occurred 6-19 days post-treatment. This simple, non-invasive procedure shows great promise as a technique for selective photo-thermal tumor ablation.

Characterization of poly(D,L-lactic-co-glycolic acid) based nanoparticulate system for enhanced delivery of antigens to dendritic cells. Elamanchili P, Diwan M, Cao M, Samuel J. Vaccine.2004 Jun 23;22(19):2406-12.

Biodegradable nanoparticles made of poly(D,L-lactic acid-co-glycolic acid) (PLGA) copolymer were characterized for enhanced delivery of antigens to murine bone marrow derived dendritic cells (DCs) in vitro. PLGA nanoparticles were efficiently phagocytosed by the DCs (CD11c+, MHC class II+, CD86+) in culture, resulting in their intracellular localization. The efficiency of the uptake was influenced by the incubation time and nanoparticle concentration. DCs pulsed with PLGA nanoparticles containing an immunomodulator, monophosphoryl lipid A (MPLA), showed upregulation of surface expression of MHC class II and CD86 molecules. Delivery of a cancer-associated antigen (MUC1 mucin peptide: BLP25) and MPLA in PLGA nanoparticles was shown to be superior to their delivery in the soluble form for activation of naive T cells of normal and MUC1-transgenic mice. These results strongly suggest that PLGA nanoparticles provide an efficient vaccine delivery system for targeting DCs and the development of DC based cellular vaccines.

Chemotherapy of glioblastoma in rats using doxorubicin-loaded nanoparticles. Steiniger SC, Kreuter J, Khalansky AS, Skidan IN, Bobruskin AI, Smirnova ZS, Severin SE, Uhl R, Kock M, Geiger KD, Gelperina SE. Int J Cancer. 2004 May 1;109(5):759-67.

Glioblastomas belong to the most aggressive human cancers with short survival times. Due to the blood-brain barrier, they are mostly inaccessible to traditional chemotherapy. We have recently shown that doxorubicin bound to polysorbate-coated nanoparticles crossed the intact blood-brain barrier, thus reaching therapeutic concentrations in the brain. Here, we investigated the therapeutic potential of this formulation of doxorubicin in vivo using an animal model created by implantation of 101/8 glioblastoma tumor in rat brains. Groups of 5-8 glioblastoma-bearing rats (total n = 151) were subjected to 3 cycles of 1.5-2.5 mg/kg body weight of doxorubicin in different formulations, including doxorubicin bound to polysorbate-coated nanoparticles. The animals were analyzed for survival (% median increase of survival time, Kaplan-Meier). Preliminary histology including immunocytochemistry (glial fibrillary acidic protein, ezrin, proliferation and apoptosis) was also performed. Rats treated with doxorubicin bound to polysorbate-coated nanoparticles had significantly higher survival times compared with all other groups. Over 20% of the animals in this group showed a long-term remission. Preliminary histology confirmed lower tumor sizes and lower values for proliferation and apoptosis in this group. All groups of animals treated with polysorbate-containing formulations also had a slight inflammatory reaction to the tumor. There was no indication of neurotoxicity. Additionally, binding to nanoparticles may reduce the systemic toxicity of doxorubicin. This study showed that therapy with doxorubicin bound to nanoparticles offers a therapeutic potential for the treatment of human glioblastoma.

Efficient cancer therapy with a nanobody-based conjugate. Cortez-Retamozo V, Backmann N, Senter PD, Wernery U, De Baetselier P, Muyldermans S, Revets H. Cancer Res. 2004 Apr 15;64(8):2853-7.

Nanobodies are the smallest fragments of naturally occurring single-domain antibodies that have evolved to be fully functional in the absence of a light chain. Nanobodies are strictly monomeric, very stable, and highly soluble entities. We identified a nanobody with subnanomolar affinity for the human tumor-associated carcinoembryonic antigen. This nanobody was conjugated to Enterobacter cloacae beta-lactamase, and its site-selective anticancer prodrug activation capacity was evaluated. The conjugate was readily purified in high yields without aggregation or loss of functionality of the constituents. In vitro experiments showed that the nanobody-enzyme conjugate effectively activated the release of phenylenediamine mustard from the cephalosporin nitrogen mustard prodrug 7-(4-carboxybutanamido) cephalosporin mustard at the surface of carcinoembryonic antigen-expressing LS174T cancer cells. In vivo studies demonstrated that the conjugate had an excellent biodistribution profile and induced regressions and cures of established tumor xenografts. The easy generation and manufacturing yield of nanobody-based conjugates together with their potent antitumor activity make nanobodies promising vehicles for new generation cancer therapeutics.

Enhanced cellular uptake of a triplex-forming oligonucleotide by nanoparticle formation in the presence of polypropylenimine dendrimers. Santhakumaran LM, Thomas T, Thomas TJ. Nucleic Acids Res.2004 Apr 15;32(7):2102-12. Print 2004.

We used polypropylenimine dendrimers for delivering a 31 nt triplex-forming oligonucleotide (ODN) in breast, prostate and ovarian cancer cell lines, using 32P-labeled ODN. Dendrimers enhanced the uptake of ODN by approximately 14-fold in MDA-MB-231 breast cancer cells, compared with control ODN uptake. Dendrimers exerted their effect in a concentration- and molecular weight-dependent manner, with generation 4 (G-4) dendrimer having maximum efficacy. A similar increase in ODN uptake was found with MCF-7 and SK-BR-3 (breast), LNCaP (prostate) and SK-OV-3 (ovarian) cancer cells. The dendrimers had no significant effect on cell viability at concentrations at which maximum ODN uptake occurred. [3H]Thymidine incorporation showed that complexing the ODN with G-4 significantly increased the growth-inhibitory effect of the ODN. Western blot analysis showed a significant 65% reduction of c-myc protein level in ODN-G-4 treated cells compared with that of ODN-treated/control cells. Gel electrophoretic analysis showed that ODN remained intact in cells even after

48 h of treatment. The hydrodynamic radii of nanoparticles formed from ODN in the presence of the dendrimers were in the range of 130-280 nm, as determined by dynamic laser light scattering. Taken together, our results indicate that polypropylenimine dendrimers might be useful vehicles for delivering therapeutic oligonucleotides in cancer cells.

A novel antiangiogenesis therapy using an integrin antagonist or anti-Flk-1 antibody coated 90Y-labeled nanoparticles. Li L, Wartchow CA, Danthi SN, Shen Z, Dechene N, Pease J, Choi HS, Doede T, Chu P, Ning S, Lee DY, Bednarski MD, Knox SJ. Int J Radiat Oncol Biol Phys. 2004 Mar 15;58(4):1215-27.

PURPOSE: Integrin alpha(v)beta(3) and vascular endothelial growth factor receptor 2 (Flk-1) have been shown to be involved in tumor-induced angiogenesis. Selective targeting of upregulated alpha(v)beta(3) and Flk-1 on the neovasculature of tumors is a novel antiangiogenesis strategy for treating a wide variety of solid tumors. In the studies described here, we investigated the potential therapeutic efficacy of two three-component treatment regimens using two murine tumor models. METHODS AND MATERIALS: The treatment regimens used nanoparticle (NP) based targeting agents radiolabeled with (90)Y. The small molecule integrin antagonist (IA) 4-[2-(3,4,5,6-tetrahydropyrimidin-2-ylamino)ethoxy]-benzoyl-2-(5)-aminoeth ylsulfonylamino-beta-alanine, which binds to the integrin alpha(v)beta(3), and a monoclonal antibody against murine Flk-1 (anti-Flk-1 MAb) were used to target the NPs. Murine tumor models K1735-M2 (melanoma) and CT-26 (colon adenocarcinoma) were used to evaluate the treatment efficacy. RESULTS: In K1735-M2 and CT-26 tumors, a single treatment with IA-NP-(90)Y (14.2 microg/g IA, 5 or 6 microCi/g (90)Y) caused a significant tumor growth delay compared to untreated control tumors, as well as tumors treated with IA, IA-NP, and NP-(90)Y, respectively ($p < 0.025$, Wilcoxon test). In K1735-M2 tumors, a single treatment with anti-Flk-1 MAb-NP-(90)Y (0.36 microg/g anti-Flk-1 MAb, 5 microCi/g (90)Y) also caused a significant tumor growth delay ($p < 0.05$, Wilcoxon test) compared to untreated tumors, as well as tumors treated with anti-Flk-1 MAb, anti-Flk-1 MAb-NP, and conventional radioimmunotherapy with (90)Y-labeled anti-Flk mAb. Anti-CD31 staining showed a marked decrease in vessel density in tumors treated with anti-Flk-1 MAb-NP-(90)Y), which was associated with a high level of apoptotic death in these tumors, as shown by terminal deoxynucleotidyl transferase-mediated dUTP nick-end labeling staining. CONCLUSIONS: The present studies provide proof of principle that targeted radiotherapy works using different targeting agents on a nanoparticle, to target both the integrin alpha(v)beta(3) and the vascular endothelial growth factor receptor. These encouraging results demonstrate the potential therapeutic efficacy of the IA-NP-(90)Y and anti-Flk-1 MAb-NP-(90)Y complexes as novel therapeutic agents for the treatment of a variety of tumor types.

Gold glyconanoparticles as new tools in antiadhesive therapy. Rojo J, Diaz V, de la Fuente JM, Segura I, Barrientos AG, Riese HH, Bernad A, Penades S. Chembiochem. 2004 Mar 5;5(3):291-7.

Gold glyconanoparticles (GNPs) have been prepared as new multivalent tools that mimic glycosphingolipids on the cell surface. GNPs are highly soluble under physiological conditions, stable against enzymatic degradation and nontoxic. Thereby GNPs open up a novel promising multivalent platform for biological applications. It has recently been demonstrated that specific tumor-associated carbohydrate antigens (glycosphingolipids and glycoproteins) are involved in the initial step of tumor spreading. A mouse melanoma model was selected to test glyconanoparticles as possible inhibitors of experimental lung metastasis. A carbohydrate-carbohydrate interaction is proposed as the first recognition step for this process. Glyconanoparticles presenting lactose (lacto-GNPs) have been used successfully to significantly reduce the progression of experimental metastasis. This result shows for the first time a clear biological effect of lacto-GNPs, demonstrating the potential application of this glyconanotechnology in biological processes.

Potential of gene therapy for the treatment of pituitary tumors. Goya RG, Sarkar DK, Brown OA; Herenu CB. Curr Gene Ther. 2004 Mar;4(1):79-87.

Pituitary adenomas constitute the most frequent neuroendocrine pathology, comprising up to 15% of primary intracranial tumors. Current therapies for pituitary tumors include surgery and radiotherapy, as well as pharmacological approaches for some types. Although all of these approaches have shown a significant degree of success, they are not devoid of unwanted side effects, and in most cases do not offer a permanent cure. Gene therapy-the transfer of genetic material for therapeutic purposes-has undergone an explosive development in the last few years. Within this context, the development of gene therapy approaches for the treatment of pituitary tumors emerges as a promising area of research. We begin by presenting a brief account of the genesis of prolactinomas, with particular emphasis on how estradiol induces prolactinomas in animals. In so doing, we discuss the role of each of the recently discovered growth inhibitory and growth stimulatory substances and their interactions in estrogen action. We also evaluate the cell-cell communication that may govern these growth factor interactions and subsequently promote the growth and survival of prolactinomas. Current research efforts to implement gene therapy in pituitary tumors include the treatment of experimental prolactinomas or somatomammotropic tumors with adenoviral vector-mediated transfer of the suicide gene for the herpes simplex type 1 (HSV1) thymidine kinase, which converts the prodrug ganciclovir into a toxic metabolite. In some cases, the suicide transgene has been placed under the control of pituitary cell-type specific promoters, like the human prolactin or human growth hormone promoters. Also, regulatable adenoviral vector systems are being assessed in gene therapy approaches for experimental pituitary tumors. In a different type of approach, an adenoviral vector, encoding the human retinoblastoma suppressor oncogene, has been successfully used to rescue the phenotype of spontaneous pituitary tumors of the pars intermedia in mice. We close the article by discussing the future of molecular therapies. We point out that although, gene therapy represents a key step in the development of molecular medicine, it has inherent limitations. As a consequence, it is our view that at some point, genetic therapies will have to move from exogenous gene transfer (i.e. gene therapy) to endogenous gene repair. This approach will call for radically new technologies, such as nanotechnology, whose present state of development is outlined.

Efficacy of lytic peptide-bound magnetite nanoparticles in destroying breast cancer cells. Kumar CS, Leuschner C, Doomes EE, Henry L, Juban M, Hormes J. J Nanosci Nanotechnol. 2004 Mar;4(3):245-9.

A 23-amino-acid synthetic lytic peptide (Hecate) was covalently linked to magnetite nanoparticles and the lytic peptide-bound nanoparticles were characterized by X-ray absorption near-edge structure spectroscopy, transmission electron microscopy, and electron diffraction. Investigation of magnetic properties with a superconducting quantum interference device (SQUID) magnetometer has shown a reduction in the saturation magnetization (Ms) of magnetite nanoparticles after binding with lytic peptide. An in vitro cell culture assay with breast cancer cell lines MDA-MB-435S revealed that the lytic peptide-bound magnetite nanoparticles were therapeutically active.

Nanoparticles of biodegradable polymers for clinical administration of paclitaxel. Feng SS, Mu L, Win KY, Huang G. Curr Med Chem. 2004 Feb;11(4):413-24.

Paclitaxel is one of the best antineoplastic drugs found from nature in the past decades, which has been found effective against a wide spectrum of cancers including ovarian cancer, breast cancer, small and non small cell lung cancer, colon cancer, head and neck cancer, multiple myeloma, melanoma, and Kaposi's sarcoma. Like many other anticancer drugs, it has difficulties in clinical administration due to its poor solubility in water and most pharmaceutical reagents. In its current clinical application, an adjuvant called Cremophor EL has to be employed, which has been found to be responsible for many serious side effects. Nanoparticles of biodegradable polymers can provide an ideal solution to such an adjuvant problem and realize a controlled and targeted delivery of the drug with better efficacy and less side effects. With further development, such as particle size optimization and surface coating, nanoparticle formulation of paclitaxel can promote a new concept of chemotherapy to realize its full efficacy and to improve quality of life of the patients, which includes personalized chemotherapy, local chemotherapy, sustained chemotherapy, oral chemotherapy, chemotherapy across the blood-brain barrier, chemotherapy across the microcirculation barrier, etc. The present research proposes a novel formulation for fabrication of nanoparticles of poly(lactic-co-glycolic acid) (PLGA) by a modified solvent extraction/evaporation technique, in which natural emulsifiers, such as phospholipids, cholesterol and vitamin E TPGS are creatively applied to achieve high drug encapsulation efficiency, desired drug released kinetics, high cell uptake and high cytotoxicity. The nanoparticles composed of various recipes and manufactured under various conditions were characterized by laser light scattering (LLS) for size and size distribution, scanning electron microscopy (SEM) and atomic force microscopy (AFM) for morphological properties, X-ray photoelectron spectroscopy (XPS) and Fourier Transformation Infrared Spectroscopy (FTIR) for surface chemistry, zeta-potential for surface charge, and differential scanning calorimetry (DSC) for the thermogram properties. The drug encapsulation efficiency and the drug release kinetics under in vitro conditions were measured by high performance liquid chromatography (HPLC). It was found that these natural emulsifiers have great advantages for nanoparticle formulation of paclitaxel over the traditional macromolecular emulsifiers, such as polyvinyl alcohol (PVA). Nanoparticles of desired small size and narrow size distribution can be obtained. The drug encapsulation efficiency can be achieved as high as 100 %. The released kinetics can be made under control. The HT-29 cancer cell line experiment showed that after 24 hours of incubation, the cell mortality caused by the drug administered by such nanoparticle formulation could be more than 13 times higher than that caused by the free drug under similar conditions.

Critical determinants in PLGA / PLA nanoparticle-mediated gene expression. Prabha S, Labhasetwar V. Pharm Res. 2004 Feb;21(2):354-64.

PURPOSE: The aim of the study was to determine the critical determinants in nanoparticle-mediated gene transfection. It was hypothesized that different formulation parameters could affect the nanoparticle characteristics and hence its gene transfection. METHODS: Nanoparticles encapsulating plasmid DNA encoding for firefly luciferase were formulated using polylactide (PLA) and poly (D,L-lactide-co-glycolide) (PLGA) polymers of different compositions and molecular weights. A multiple-emulsion solvent-evaporation method with polyvinyl alcohol (PVA) as an emulsifier was used to formulate DNA-loaded nanoparticles. Gene expression of nanoparticles was determined in breast cancer (MCF-7) and prostate cancer (PC-3) cell lines. RESULTS: Nanoparticles formulated using PLGA polymer demonstrated greater gene transfection than those formulated using PLA polymer, and this was attributed to the higher DNA release from PLGA nanoparticles. Higher-molecular-weight PLGA resulted in the formation of nanoparticles with higher DNA loading, which demonstrated higher gene expression than those formulated with lower-molecular-weight PLGA. In addition, the nanoparticles with lower amount of surface-associated PVA demonstrated higher gene transfection in both the cell lines. Higher gene transfection with these nanoparticles was attributed to their higher intracellular uptake and cytoplasmic levels. Further study demonstrated that the molecular weight and the degree of hydrolyzation of PVA used as an emulsifier also affect the gene expression of nanoparticles. CONCLUSIONS: Results thus demonstrate that the DNA loading in nanoparticles and its release, and the surface-associated PVA influencing the intracellular uptake and endolysosomal escape of nanoparticles, are some of the critical determinants in nanoparticle-mediated gene transfection.

Nanoshell-mediated near-infrared thermal therapy of tumors under magnetic resonance guidance. Hirsch LR, Stafford RJ, Bankson JA, Sershen SR, Rivera B, Price RE, Hazle JD, Halas NJ, West JL. Proc Natl Acad Sci U S A. 2003 Nov 11;100(23):13549-54. Epub 2003 Nov 03.

Metal nanoshells are a class of nanoparticles with tunable optical resonances. In this article, an application of this technology to thermal ablative therapy for cancer is described. By tuning the nanoshells to strongly absorb light in the near infrared, where optical transmission through tissue is optimal, a distribution of nanoshells at depth in tissue can be used to deliver a therapeutic dose of heat by using moderately low exposures of extracorporeally applied near-infrared (NIR) light. Human breast carcinoma cells incubated with nanoshells in vitro were found to have undergone photothermally induced morbidity on exposure to NIR light (820 nm, 35 W/cm2), as determined by using a fluorescent viability stain. Cells without nanoshells displayed no loss in viability after the same periods and conditions of NIR illumination. Likewise, in vivo studies under magnetic resonance guidance revealed that exposure to low doses of NIR light (820 nm, 4 W/cm2) in solid tumors treated with metal nanoshells reached average maximum temperatures capable of inducing irreversible tissue damage (DeltaT = 37.4 +/- 6.6 degrees C) within 4-6 min. Controls treated without nanoshells demonstrated significantly lower average temperatures on exposure to NIR light (DeltaT < 10 degrees C). These findings demonstrated good correlation with histological findings. Tissues heated above the thermal damage threshold displayed coagulation, cell shrinkage, and loss of nuclear staining, which are indicators of irreversible thermal damage. Control tissues appeared undamaged.

Heat shock protein 70 gene therapy combined with hyperthermia using magnetic nanoparticles. Ito A, Matsuoka F, Honda H, Kobayashi T. Cancer Gene Ther. 2003 Dec;10(12):918-25.

Heat shock proteins (HSPs) are recognized as significant participants in immune reactions. We previously reported that expression of HSP70 in response to hyperthermia, produced using our original magnetite cationic liposomes (MCLs), induces antitumor immunity. In the present study, we examine whether the antitumor immunity induced by hyperthermia is enhanced by hsp70 gene transfer. A human hsp70 gene mediated by cationic liposomes was injected into a B16 melanoma nodule in C57BL/6 mice in situ. At 24 hours after the injection of the hsp70 gene, MCLs were injected into melanoma nodules in C57BL/6 mice, which were subjected to an alternating magnetic field for 30 minutes. The temperature at the tumor reached 43 degrees C and was maintained by controlling the magnetic field intensity. The combined treatment strongly arrested tumor growth over a 30-day period, and complete regression of tumors was observed in 30% (3/10) of mice. Systemic antitumor immunity was induced in the cured mice. This study demonstrates that this novel therapeutic strategy combining the use of hsp70 gene therapy and hyperthermia using MCLs may be applicable to patients with advanced malignancies.

Nanoparticles for the delivery of genes and drugs to human hepatocytes. Yamada T, Iwasaki Y, Tada H, Iwabuki H, Chuah MK, VandenDriessche T, Fukuda H, Kondo A, Ueda M, Seno M, Tanizawa K, Kuroda S. Nat Biotechnol.2003 Aug;21(8):885-90.

Hepatitis B virus envelope L particles form hollow nanoparticles displaying a peptide that is indispensable for liver-specific infection by hepatitis B virus in humans. Here we demonstrate the use of L particles for the efficient and specific transfer of a gene or drug into human hepatocytes both in culture and in a mouse xenograft model. In this model, intravenous injection of L particles carrying the gene for green fluorescent protein (GFP) or a fluorescent dye resulted in observable fluorescence only in human hepatocellular carcinomas but not in other human carcinomas or in mouse tissues. When the gene encoding human clotting factor IX was transferred into the xenograft model using L particles, factor IX was produced at levels relevant to the treatment of hemophilia B. The yeast-derived L particle is free of viral genomes, highly specific to human liver cells and able to accommodate drugs as well as genes. These advantages should facilitate targeted delivery of genes and drugs to the human liver.

Colloidal anticancer drugs bioavailabilities in oral administration models. Moutardier V, Tosini F, Vlieghe P, Cara L, Delpero JR, Clerc T. Int J Pharm. 2003 Jul 9;260(1):23-38.

Liposomes have been prepared with a polymerised core. Drug release and gastrointestinal (GI) degradation of liposomes with this polymerised core was slightly less important than those of classical liposomes. Vincristine, 5-fluorouracil (5-FU), and methotrexate (MTX) have been incorporated into the liposomes, and studies carried out using the differentiated cell lines Caco-2 and TC7, and with 150 histologically normal sections of human colon. Encapsulation of the drugs in liposomes had variable effects, depending on the test system and the drug used. For 5-FU and MTX calculated to be in a therapeutic range, liposomal formulation enhanced drug permeation, but not for the other drugs tested. In the excised human colon model, the treatment history of the patients can affect bioavailability: pre-operative radiation increased the drug tissue uptake. Transmucosal transport of ions was modified by prior chemotherapy. These results should be taken into account in the design of oral anticancer treatments both at the level of nutritional and pharmacological considerations.

Ceramic-based nanoparticles entrapping water-insoluble photosensitizing anticancer drugs: a novel drug-carrier system for photodynamic therapy. Roy I, Ohulchanskyy TY, Pudavar HE, Bergey EJ, Oseroff AR, Morgan J, Dougherty TJ, Prasad PN. J Am Chem Soc. 2003 Jul 2;125(26):7860-5.

A novel nanoparticle-based drug carrier for photodynamic therapy is reported which can provide stable aqueous dispersion of hydrophobic photosensitizers, yet preserve the key step of photogeneration of singlet oxygen, necessary for photodynamic action. A multidisciplinary approach is utilized which involves (i) nanochemistry in micellar cavity to produce these carriers, (ii) spectroscopy to confirm singlet oxygen production, and (iii) in vitro studies using tumor cells to investigate drug-carrier uptake and destruction of cancer cells by photodynamic action. Ultrafine organically modified silica-based nanoparticles (diameter approximately 30 nm), entrapping water-insoluble photosensitizing anticancer drug 2-devinyl-2-(1-hexyloxyethyl) pyropheophorbide, have been synthesized in the nonpolar core of micelles by hydrolysis of triethoxyvinylsilane. The resulting drug-doped nanoparticles are spherical, highly monodispersed, and stable in aqueous system. The entrapped drug is more fluorescent in aqueous medium than the free drug, permitting use of fluorescence bioimaging studies. Irradiation of the photosensitizing drug entrapped in nanoparticles with light of suitable wavelength results in efficient generation of singlet oxygen, which is made possible by the inherent porosity of the nanoparticles. In vitro studies have demonstrated the active uptake of drug-doped nanoparticles into the cytosol of tumor cells. Significant damage to such impregnated tumor cells was observed upon irradiation with light of wavelength 650 nm. Thus, the potential of using ceramic-based nanoparticles as drug carriers for photodynamic therapy has been demonstrated.

Encapsulation of p-THPP into nanoparticles: cellular uptake, subcellular localization and effect of serum on photodynamic activity. Konan YN, Chevallier J, Gurny R, Allemann E. Photochem Photobiol. 2003 Jun;77(6):638-44.

The cellular uptake, localization and efflux of meso-tetra-(4-hydroxyphenyl)porphyrin (p-THPP)-loaded nanoparticles have been studied in EMT-6 tumor cells. The effect of blood serum on photocytotoxicity has also been evaluated. Sub-130 nm nanoparticles based on poly(D,L-lactide-co-glycolide) (PLGA) (50:50 PLGA and 75:25 PLGA) and poly(D,L-lactide) (PLA) have been examined in comparison with free p-THPP. For all formulations tested, uptake of photosensitizer into cells was dependent on concentration, time and temperature. All nanoparticulate formulations accumulated within the cells to a greater extent relative to free drug. Indeed, the fluorescence intensities measured on EMT-6 cells treated with p-THPP-loaded nanoparticulate formulations were at least two-fold higher than those obtained with free dye. Furthermore, the highest accumulation level was found with PLGA nanoparticles. Fluorescence microscopy revealed that endocytosis is a major intracellular sequestration mechanism of these p-THPP formulations and that these were localized into early and late endosomes. The efflux study performed on both nonirradiated and irradiated cells indicated that free and p-THPP-loaded nanoparticles gradually escaped from EMT-6 cells as a function of time. This was more pronounced when cells were treated with nanoparticles and irradiated, reflecting important photodamage. It was also found that regardless of the nanoparticulate formulations tested, p-THPP photocytotoxicity was influenced by the concentration of the serum.

Selective cell targeting with light-absorbing microparticles and nanoparticles. Pitsillides CM, Joe EK, Wei X, Anderson RR, Lin CP. Biophys J. 2003 Jun;84(6):4023-32.

We describe a new method for selective cell targeting based on the use of light-absorbing microparticles and nanoparticles that are heated by short laser pulses to create highly localized cell damage. The method is closely related to chromophore-assisted laser inactivation and photodynamic therapy, but is driven solely by light absorption, without the need for photochemical intermediates (particularly singlet oxygen). The mechanism of light-particle interaction was investigated by nanosecond time-resolved microscopy and by thermal modeling. The extent of light-induced damage was investigated by cell lethality, by cell membrane permeability, and by protein inactivation. Strong particle size dependence was found for these interactions. A technique based on light to target endogenous particles is already being exploited to treat pigmented cells in dermatology and ophthalmology. With exogenous particles, phamacokinetics and biodistribution studies are needed before the method can be evaluated against photodynamic therapy for cancer treatment. However, particles are unique, unlike photosensitizers, in that they can remain stable and inert in cells for extended periods. Thus they may be particularly useful for prelabeling cells in engineered tissue before implantation. Subsequent irradiation with laser pulses will allow control of the implanted cells (inactivation or modulation) in a noninvasive manner.

Magnetic drug targeting—biodistribution of the magnetic carrier and the chemotherapeutic agent mitoxantrone after locoregional cancer treatment. Alexiou C, Jurgons R, Schmid RJ, Bergemann C, Henke J, Erhardt W, Huenges E, Parak F. J Drug Target. 2003 Apr;11(3):139-49.

Magnetic Drug Targeting means the specific delivery of chemotherapeutic agents to their desired targets, e.g. tumors, by using magnetic nanoparticles (ferrofluids) bound to these agents and an external magnetic field which is focused on the tumor. This type of target directed drug injection attempts to concentrate a pharmacologic agent by enhancing its efficacy while simultaneously minimizing deleterious side effects. In previous studies, we have been able to demonstrate the efficacy of this type of localized intraarterial chemotherapy in VX2 squamous cell carcinoma among rabbits [Alexiou, C., Arnold, W., Klein, R.J., Parak, F.G., Hulin, P., Bergemann, C., Erhardt, W., Wagenpfeil, S. and Lubbe, A.S. "Locoregional cancer treatment with Magnetic Drug Targeting", Cancer Res. 60 (2000) 6641-6648]. In the present investigation, we have studied the biodistribution of ferrofluids and chemotherapeutic agent by measuring the amount in the tumor, peritumoral area, various organs and body fluids (e.g. blood and urine), with and without Magnetic Drug Targeting. We compared results to that of administering a chemotherapeutic agent soley. An external magnetic field was directed toward the tumor for 60 min. Biodistribution of ferrofluids in the tumor was investigated using histological cross sections and measured semi-quantitatively using 123I-labeled nanoparticles and quantitatively by the use of radioactive 59Fe-ferrofluids. Mitoxantrone was quantitatively measured using HPLC-analysis. The strength of the external magnetic field was 0.6 Tesla (permanent magnet) in the 123iodine study and 1.7 Tesla (electromagnet) in the 59Fe-study and HPLC-analysis. The concentration of the ferrofluids (FFs) in the tumor region i.e. the tumor tissue and the surrounding area, which was under the influence of an external magnetic field, was found to be much higher than in the absence of one. In contrast to systemic chemotherapy, a much higher concentration of mitoxantrone in the tumor and the peritumoral area (region surrounding the tumor

< or = 1 cm), by using only 50% and 20% of the normal dose was seen. Thus, the higher concentration of mitoxantrone could explain the therapeutic efficacy of Magnetic Drug Targeting in treatment of VX2 squamous cell carcinoma in rabbits in our previous studies with the advantage of no adverse clinical side effects.

A novel controlled release formulation for the anticancer drug paclitaxel (Taxol): PLGA nanoparticles containing vitamin E TPGS. Mu L, Feng SS. J Control Release. 2003 Jan 9;86(1):33-48.

Paclitaxel (Taxol) is one of the best antineoplastic drugs found from nature in the past decades. Like many other anticancer drugs, there are difficulties in its clinical administration due to its poor solubility. Therefore an adjuvant called Cremophor EL has to be employed, but this has been found to cause serious side-effects. However, nanoparticles of biodegradable polymers can provide an ideal solution to the adjuvant problem and realise a controlled and targeted delivery of the drug with better efficacy and fewer side-effects. The present research proposes a novel formulation for fabrication of nanoparticles of biodegradable polymers containing d-alpha-tocopheryl polyethylene glycol 1000 succinate (vitamin E TPGS or TPGS) to replace the current method of clinical administration and, with further modification, to provide an innovative solution for oral chemotherapy. In the modified solvent extraction/evaporation technique employed in this research, the emulsifier/stabiliser/additive and the matrix material can play a key role in determining the morphological, physicochemical and pharmaceutical properties of the produced nanoparticles. We found that vitamin E TPGS could be a novel surfactant as well as a matrix material when blended with other biodegradable polymers. The nanoparticles composed of various formulations and manufactured under various conditions were characterised by laser light scattering (LLS) for size and size distribution, scanning electron microscopy (SEM) and atomic force microscopy (AFM) for morphological properties, X-ray photoelectron spectroscopy (XPS) for surface chemistry and differential scanning calorimetry (DSC) for thermogram properties. The drug encapsulation efficiency (EE) and the drug release kinetics under in vitro conditions were measured by high performance liquid chromatography (HPLC). It was concluded that vitamin E TPGS has great advantages for the manufacture of polymeric nanoparticles for controlled release of paclitaxel and other anti-cancer drugs. Nanoparticles of nanometer size with narrow distribution can be obtained. A drug encapsulation efficiency as high as 100% can be achieved and the release kinetics can be controlled.

Integrin-mediated targeting of drug delivery to irradiated tumor blood vessels. Hallahan D, Geng L, Qu S, Scarfone C, Giorgio T, Donnelly E, Gao X, Clanton J. Cancer Cell. 2003 Jan;3(1):63-74.

The objective of this study was to target drug delivery to radiation-induced neoantigens, which include activated receptors within the tumor vasculature. These responses include posttranslational changes in pre-existing proteins, which can be discovered by phage-displayed peptide libraries administered to mice bearing irradiated tumors.

Phage-displayed peptides recovered from irradiated tumors included the amino acid sequence RGDGSSV. This peptide binds to integrins within the tumor microvasculature. Immunohistochemical staining of irradiated tumors showed accumulation of fibrinogen receptor alpha(2b)beta(3) integrin. We studied tumor targeting efficiency of ligands to radiation-induced alpha(2b)beta(3). Radiopharmaceuticals were localized to irradiated tumors by use of alpha(2b)beta(3) ligands conjugated to nanoparticles and liposomes. Fibrinogen-conjugated nanoparticles bind to the radiation-activated receptor, obliterate tumor blood flow, and significantly increase regression and growth delay in irradiated tumors. Radiation-guided drug delivery to tumor blood vessels is a novel paradigm for targeted drug delivery.

Antitumor properties of irinotecan-containing nanoparticles prepared using poly(DL-lactic acid) and poly(ethylene glycol)-block-poly(propylene glycol)-block-poly(ethylene glycol). Onishi H, Machida Y, Machida Y. Biol Pharm Bull. 2003 Jan;26(1):116-9.

Irinotecan-containing nanoparticles (NP) were prepared by coprecipitation with addition of water to acetone solution of poly(DL-lactic acid), poly(ethylene glycol)-block-poly(propylene glycol)-block-poly(ethylene glycol) and irinotecan, and subsequent evaporation of organic solvent. NP were purified by gel filtration and used for experiments after condensation by evaporation. The obtained NP showed the drug content of 4.5% (w/w) and the mean particle diameter of 118 nm with the particle diameter distribution between 80-210 nm. When the antitumor effect was examined at a repeated dose of 20 mg irinotecan eq/kg for 3 d (3 x 20 mg/kg) using mice bearing Sarcoma 180 subcutaneously, only NP suppressed tumor growth significantly. After i.v. injection in rats, NP maintained irinotecan plasma concentration longer than CPT-11 aqueous solution. The present nanoparticle formation is suggested as a possibly useful dosage form of irinotecan against solid tumor.

Nanoparticle-mediated gene delivery to tumour neovasculature. Reynolds AR, Moein Moghimi S, Hodivala-Dilke K. Trends Mol Med. 2003 Jan;9(1):2-4.

The alpha(v)beta(3) integrin is a potential pharmacological target for anti-angiogenic therapy. A recent report describes the use of alpha(v)beta(3)-targeted nanoparticles to deliver a gene to tumour vasculature selectively. This resulted in substantial tumour regression in several experimental mouse tumour models. Hence, this approach has great potential for the treatment of human cancer.

Cellular uptake and concentrations of tamoxifen upon administration in poly(epsilon-caprolactone) nanoparticles. Chawla JS, Amiji MM. AAPS PharmSci. 2003;5(1):E3.

PURPOSE: In an attempt to increase the local concentration of tamoxifen in estrogen receptor positive breast cancer cells, we have prepared and characterized poly(epsilon-caprolactone) (PCL) nanoparticle formulation. METHODS: PCL (mol wt 14,800 daltons) nanoparticles were prepared by the solvent displacement method in acetone-water system in the presence of Pluronic F- 68. PCL nanoparticles, labeled with rhodamine123, were incubated with MCF-7 estrogen receptor positive breast cancer cells to determine uptake, intracellular distribution, and localization as a function of time. Intracellular drug concentrations over a specified period of time using different initial doses were examined using tritiated [3H]-tamoxifen. RESULTS: A significant fraction of the administered rhodamine123-loaded PCL nanoparticles was found in the perinuclear region of the MCF-7 cells, where estrogen receptors are also localized, after 1 hour of incubation. Measurements of the intracellular concentrations revealed that most of the administered nanoparticle dose was internalized within the first 30 minutes of incubation, and the uptake followed saturable transport kinetics. CONCLUSION: Results of this study show that PCL nanoparticles were rapidly internalized in MCF-7 cells and intracellular tamoxifen concentrations followed a saturable process. This approach may provide better therapeutic benefit by delivering the drug locally, near the tumor cells, for a longer period of time.

Biodegradable nanoparticle mediated antigen delivery to human cord blood derived dendritic cells for induction of primary T cell responses. Diwan M, Elamanchili P, Lane H, Gainer A, Samuel J. J Drug Target. 2003;11(8-10):495-507.

Dendritic cells (DCs) in the peripheral tissues act as sentinels of the immune system. They detect and capture pathogens entering the body and present their antigens to T cells to trigger responses directed towards elimination of the pathogen. The induction of peripheral tolerance against self and certain foreign antigens is also believed to be mediated through DCs. The outcome of any immune response is largely controlled by the microenvironment of antigen capture, processing and presentation by DCs. The "context" of antigen delivery to DCs will directly influence the microenvironment of antigen presentation and hence the regulation of immune responses. We report here preliminary investigations describing the formulation of a pharmaceutically acceptable, biodegradable, and strategic nanoparticulate delivery system, and its application for efficient antigen loading of DCs to achieve antigen specific T cell activation. "Pathogen-mimicking" nanoparticles capable of interacting with DCs were fabricated by incorporating monophosphoryl lipid A (MPLA; toll-like receptor (TLR) 4 ligand) or CpG ODN (seq #2006; TLR9 ligand) in biodegradable copolymer, poly(D,L,-lactic-co-glycolic acid) (PLGA). The uptake of PLGA nanoparticles by human umbilical cord blood derived DCs (DCs propagated from CD34 progenitors) was conclusively demonstrated by scanning electron microscopy (SEM), fluorescence activated cell sorting (FACS) and confocal laser scanning microscopy (CLSM). Cell phenotype at day 12 of cultures was determined as immature DC using specific cell surface markers, i.e. CD11c (approximately 90%), MHC-II (approximately 70%), CD86 (approximately 20%), CD83 (approximately 5%), CD80 (approximately 40%), CD40 (approximately 40%), and CCR7 (approximately 5%). Tetanus toxoid (TT), a model antigen, was encapsulated in nanoparticles along with an immunomodulator, i.e. either MPLA or CpG ODN. DCs pulsed with various antigen formulations were co-cultured with autologous naive T cells at various cell ratios (DC: T cells were 1:5-20). The DCs pulsed with TT and

MPLA together in nanoparticles induced significantly higher T cell proliferation ($P<0.05$) as compared to when DCs pulsed with TT and MPLA in solution were employed. A similar trend was observed when CpG ODN was used instead of MPLA in the TT nanoparticles. This strategy of antigen delivery to DCs was then tested with a cancer vaccine candidate, a MUC1 lipopeptide. The T cell proliferation observed in the presence of nanoparticulate MUC1 and MPLA pulsed-DCs was much higher than DCs pulsed with soluble antigen ($P<0.0005$). These results indicate that PLGA nanoparticles mimicking certain features of pathogens are efficient delivery systems for targeting vaccine antigens to DCs and activation of potent T cell responses.

Biodegradable poly(epsilon -caprolactone) nanoparticles for tumor-targeted delivery of tamoxifen. Chawla JS, Amiji MM. Int J Pharm. 2002 Dec 5;249(1-2):127-38

To increase the local concentration of tamoxifen in estrogen receptor (ER) positive breast cancer, we have developed and characterized nanoparticle formulation using poly(epsilon -caprolactone) (PCL). The nanoparticles were prepared by solvent displacement method using acetone-water system. Particle size analysis, scanning electron microscopy, zeta potential measurements, and differential scanning calorimetry (DSC) were used for nanoparticle characterization. Biodegradation studies were performed in the presence and absence of Pseudomonas lipase in phosphate-buffered saline (PBS, pH 7.4) at 37 degrees C. Tamoxifen loading over different concentrations was analyzed by high-performance liquid chromatography (HPLC) and the optimum loading concentration was determined. In vitro release studies were performed in 0.5% (w/v) sodium lauryl sulfate (SLS) containing PBS at 37 degrees C. Cellular uptake and distribution of fluorescent-labeled nanoparticles was examined in MCF-7 breast cancer cells. SEM micrographs and Coulter analysis showed nanoparticles with spherical shape and uniform size distribution (250-300 nm), respectively. Zeta potential analysis revealed a positive surface charge of +25 mV on the tamoxifen-loaded formulation. Being hydrophobic crystalline polyester, PCL did not degrade in PBS alone, but the degradation was enhanced by the presence of lipase. The maximum tamoxifen loading efficiency was 64%. Initial burst release of tamoxifen was observed, probably due to significant surface presence of the drug on the nanoparticles. A large fraction of the administered nanoparticle dose was taken up by MCF-7 cells through non-specific endocytosis. The nanoparticles were found in the perinuclear region after 1 h. Results of the study suggest that nanoparticle formulations of selective ER modulators, like tamoxifen, would provide increased therapeutic benefit by delivering the drug in the vicinity of the ER.

Poly(ethylene glycol)-coated hexadecylcyanoacrylate nanospheres display a combined effect for brain tumor targeting. Brigger I, Morizet J, Aubert G, Chacun H, Terrier-Lacombe MJ, Couvreur P, Vassal G. J Pharmacol Exp Ther. 2002 Dec;303(3):928-36.

The aim of the present study was to evaluate the tumor accumulation of radiolabeled long-circulating poly(ethylene glycol) (PEG)-coated hexadecylcyanoacrylate nanospheres and non-PEG-coated hexadecylcyanoacrylate nanospheres (used as control), after intravenous injection in Fischer rats bearing intracerebrally well established 9L gliosarcoma. Both types of nanospheres showed an accumulation with a retention effect in the 9L tumor. However, long-circulating nanospheres concentrated 3.1 times higher in the gliosarcoma, compared with non-PEG-coated nanospheres. The tumor-to-brain ratio of pegylated nanospheres was found to be 11, which was in accordance with the ratios reported for other carriers tested for brain tumor targeting such as long-circulating liposomes or labels for magnetic resonance imaging. In addition, a 4- to 8-fold higher accumulation of the PEG-coated carriers was observed in normal brain regions, when compared with control nanospheres. Using a simplified pharmacokinetic model, two different mechanisms were proposed to explain this higher concentration of PEG-coated nanospheres in a tumoral brain. 1) in the 9L tumor, the preferential accumulation of pegylated nanospheres was attributable to their slower plasma clearance, relative to control nanospheres. Diffusion/convection was the proposed mechanism for extravasation of the nanospheres in the 9L interstitium, across the altered blood-brain barrier. 2) In addition, PEG-coated nanospheres displayed an affinity with the brain endothelial cells (normal brain region), which may not be considered as the result of a simple diffusion/convection process. The exact underlying mechanism of such affinity deserves further investigation, since it was observed to be as important as specific interactions described for immunoliposomes with the blood-brain barrier.

Paclitaxel-loaded PLGA nanoparticles: preparation, physicochemical characterization and in vitro anti-tumoral activity. Fonseca C, Simoes S, Gaspar R. J Control Release. 2002 Oct 4;83(2):273-286.

The main objective of this study was to develop a polymeric drug delivery system for paclitaxel, intended to be intravenously administered, capable of improving the therapeutic index of the drug and devoid of the adverse effects of Cremophor EL. To achieve this goal paclitaxel (Ptx)-loaded poly(lactic-co-glycolic acid) (PLGA) nanoparticles (Ptx-PLGA-Nps) were prepared by the interfacial deposition method. The influence of different experimental parameters on the incorporation efficiency of paclitaxel in the nanoparticles was evaluated. Our results demonstrate that the incorporation efficiency of paclitaxel in nanoparticles was mostly affected by the method of preparation of the organic phase and also by the organic phase/aqueous phase ratio. Our data indicate that the methodology of preparation allowed the formation of spherical nanometric (<200 nm), homogeneous and negatively charged particles which are suitable for intravenous administration. The release behaviour of paclitaxel from the developed Nps exhibited a biphasic pattern characterised by an initial fast release during the first 24 h, followed by a slower and continuous release. The in vitro anti-tumoral activity of Ptx-PLGA-Nps developed in this work was assessed using a human small cell lung cancer cell line (NCI-H69 SCLC) and compared to the in vitro anti-tumoral activity of the commercial formulation Taxol. The influence of Cremophor EL on cell viability was also investigated. Exposure of NCI-H69 cells to 25 microg/ml Taxol resulted in a steep decrease in cell viability. Our results demonstrate that incorporation of Ptx in nanoparticles strongly enhances the cytotoxic effect of the drug as compared to Taxol, this effect being more relevant for prolonged incubation times.

A study on the preparation and anti-tumor efficacy of bovine serum albumin nanospheres containing 5-fluorouracil. Santhi K, Dhanaraj SA, Joseph V, Ponnusankar S, Suresh B. Drug Dev Ind Pharm. 2002 Oct;28(9):1171-9.

The therapeutic profile of many anti-cancer drugs has been improved by their modified distribution through a colloidal carrier system. Hence, bovine serum albumin nanospheres containing 5-fluorouracil were prepared by pH-coacervation methods. To select the most suitable cryoprotector for the formulated nanosphere system, a study on the effect of cryoprotectors in the prevention of particle agglomeration was done. Using glucose and mannitol at various concentrations during freeze drying, glucose at a concentration of 5% was observed to be relatively more effective in the prevention of particle agglomeration than the other cryoprotectors. The carrier capacity was determined through the drug-to-albumin ratio. The particle size of all the drug-loaded batches was analyzed before and after freeze drying. The batch of nanospheres with uniform size distribution, and highest drug loading, was used for other subsequent studies. The effect of surfactant in drug loading was estimated through various concentrations of sodium lauryl sulfate, and it was observed that the surfactant has no influence on drug loading at the selected concentrations. The batch of nanospheres with highest drug loading was evaluated for its in-vitro release, and the drug release was found to be in a bi-phasic pattern. To evaluate the efficacy of 5-fluorouracil-loaded nanospheres against cancer cells, an in vitro cytotoxicity study was carried out using HEp-2 cell lines. The nanosphere-bound drug was observed to produce a better cytotoxic effect than the free drug. The anti-tumor efficacy of drug-loaded nanosphere was investigated in DLA tumor-induced mice models, and the percentage tumor inhibition was relatively higher in animals treated with nanosphere-bound drug than with free drug.

Tumor accumulation of gadolinium in lipid-nanoparticles intravenously injected for neutron-capture therapy of cancer. Watanabe T, Ichikawa H, Fukumori Y. Eur J Pharm Biopharm. 2002 Sep;54(2):119-24.

Gadolinium-incorporating lipid-nanoemulsions (Gd-nanoLE) for neutron-capture therapy were prepared. As a more convenient administration route than the intraperitoneal (i.p.) injection previously reported, the intravenous (i.v.) injections in tumor-bearing hamsters were carried out at an administration volume of 1 ml, which was the maximum tolerable injection volume of an i.v. injection and half that of an i.p. injection. When the standard-Gd-nanoLE of 1.5 mgGd/ml was administered, the absolute bioavailability in the i.p. injection was 57%, probably resulting from incomplete absorption from the peritoneal cavity into the blood stream. The biodistribution data revealed that the i.v. injection had three advantages over the i.p. injection, namely, a faster and higher accumulation of Gd-nano LE, and a more extended retention time in the tumor.

Two i.v. injections of the standard-Gd-nanoLE with a 24h interval doubled the tumor accumulation of Gd, resulting in 49.7 microg Gd/g wet tumor 12h after administration. By using a twofold Gd-enriched formulation (High-Gd-nanoLE) of 3.0 mgGd/ml in the repeated administration schedule, the accumulation was doubled again, reaching 101 microg Gd/g wet tumor. This level was comparable to the maximum level in the single i.p. injection previously reported. These results demonstrated that i.v. injection could be an alternative to i.p. injection as an administration route.

Antiangiogenics meet nanotechnology. Ruoslahti E. Cancer Cell. 2002 Aug;2(2):97-8.

A mutant Raf-1 gene loaded onto nanoparticles, delivered to tumor vasculature with an integrin binding compound, provides effective antiangiogenic gene therapy in mice.

Long-circulating poly(ethylene glycol)-modified gelatin nanoparticles for intracellular delivery. Kaul G, Amiji M. Pharm Res. 2002 Jul;19(7):1061-7.

PURPOSE: The objective of this study was to develop and characterize long-circulating, biodegradable, and biocompatible nanoparticulate formulation as an intracellular delivery vehicle. METHODS: Poly(ethylene glycol) (PEG)-modified gelatin was synthesized by reacting Type-B gelatin with PEG-epoxide. The nanoparticles, prepared by pH and temperature controlled ethanol-water solvent displacement technique, were characterized for mean size, size distribution, and surface morphology. Electron spectroscopy for chemical analysis (ESCA) was used to confirm the surface presence of PEG chains. In vitro release of tetramethylrhodamine-labeled dextran (TMR-dextran, Mol. wt. 10,000 daltons) from the nanoparticle formulations was examined in PBS, with and without 0.2-mg/ml protease, at 37 degrees C. Relative cytotoxicity profile of control and PEGylated gelatin was evaluated in BT-20 a human breast cancer cell line. The nanoparticles were incubated with BT-20 cells to determine uptake and cellular distribution using confocal microscopy. RESULTS: Gelatin and PEGylated gelatin nanoparticles were found to be spherical in shape with a smooth surface in a size range of 200-500 nm and a unimodal size distribution. ESCA results showed an increase in the ether carbon (-C-O-) peak in the PEGylated gelatin nanoparticles due to the presence of PEG chains. The presence of PEG chains decreased the percent release of TMR-dextran in the presence of proteolytic enzyme due to steric repulsion. Cytotoxicity assays indicated that both gelatin and PEGylated gelatin were completely non-toxic to the cells. A large fraction of the administered control gelatin and PEGylated gelatin nanoparticles were found to be concentrated in the perinuclear region of the BT-20 cells after 12 hours indicating possible vesicular transport through initial uptake by endocytosis and endosomal processing. CONCLUSION: The results of this study show that PEGylation of gelatin may prove beneficial as long-circulating delivery system in vivo. Additionally, the nanoparticles could encapsulate hydrophilic macromolecules and are internalized by tumor cells.

Tumor regression by targeted gene delivery to the neovasculature. Hood JD, Bednarski M, Frausto R, Guccione S, Reisfeld RA, Xiang R, Cheresh DA. Science. 2002 Jun 28;296(5577):2404-7.

Efforts to influence the biology of blood vessels by gene delivery have been hampered by a lack of targeting vectors specific for endothelial cells in diseased tissues. Here we show that a cationic nanoparticle (NP) coupled to an integrin alphavbeta3-targeting ligand can deliver genes selectively to angiogenic blood vessels in tumor-bearing mice. The therapeutic efficacy of this approach was tested by generating NPs conjugated to a mutant Raf gene, ATPmu-Raf, which blocks endothelial signaling and angiogenesis in response to multiple growth factors. Systemic injection of the NP into mice resulted in apoptosis of the tumor-associated endothelium, ultimately leading to tumor cell apoptosis and sustained regression of established primary and metastatic tumors.

Self-assembly of a virus-mimicking nanostructure system for efficient tumor-targeted gene delivery. Xu L, Frederik P, Pirollo KF, Tang WH, Rait A, Xiang LM, Huang W, Cruz I, Yin Y, Chang EH. Hum Gene Ther. 2002 Feb 10;13(3):469-81.

Molecular therapy, including gene therapy, is a promising strategy for the treatment of human disease. However, delivery of molecular therapeutics efficiently and specifically to the target tissue remains a significant challenge. A human transferrin (Tf)-targeted cationic liposome-DNA complex, Tf-lipoplex, has shown high gene transfer efficiency and efficacy with human head and neck cancer in vitro and in vivo (Xu, L., Pirollo, K.F., Tang, W.H., Rait, A., and Chang, E.H. Hum. Gene Ther. 1999;10:2941-2952). Here we explore the structure, size, formation process, and structure-function relationships of Tf-lipoplex. We have observed Tf-lipoplex to have a highly compact structure, with a relatively uniform size of 50-90 nm. This nanostructure is novel in that it resembles a virus particle with a dense core enveloped by a membrane coated with Tf molecules spiking the surface. More importantly, compared with unliganded lipoplex, Tf-lipoplex shows enhanced stability, improved in vivo gene transfer efficiency, and long-term efficacy for systemic p53 gene therapy of human prostate cancer when used in combination with conventional radiotherapy. On the basis of our observations, we propose a multistep self-assembly process and Tf-facilitated DNA cocondensation model that may provide an explanation for the resultant small size and effectiveness of our nanostructural Tf-lipoplex system.

Nanocapsules: lipid-coated aggregates of cisplatin with high cytotoxicity. Burger KN, Staffhorst RW, de Vijlder HC, Velinova MJ, Bomans PH, Frederik PM, de Kruijff B. Nat Med. 2002 Jan;8(1):81-4.

Cisplatin is one of the most widely used agents in the treatment of solid tumors, but its clinical utility is limited by toxicity. The development of less toxic, liposomal formulations of cisplatin has been hampered by the low water solubility and low lipophilicity of cisplatin, resulting in very low encapsulation efficiencies. We describe a novel method allowing the efficient encapsulation of cisplatin in a lipid formulation; it is based on repeated freezing and thawing of a concentrated solution of cisplatin in the presence of negatively charged phospholipids. The method is unique in that it generates nanocapsules, which are small aggregates of cisplatin covered by a single lipid bilayer. The nanocapsules have an unprecedented drug-to-lipid ratio and an in vitro cytotoxicity up to 1000-fold higher than the free drug. Analysis of the mechanism of nanocapsule formation suggests that the method may be generalized to other drugs showing low water solubility and lipophilicity.

Nanobiology and Cancer Nanotechnology

Single cell manipulation, analytics, and label-free protein detection in microfluidic devices for systems nanobiology. Hellmich W, Pelargus C, Leffhalm K, Ros A, Anselmetti D., Electrophoresis. 2005 Sep 9.

Single cell analytics for proteomic analysis is considered a key method in the framework of systems nanobiology which allows a novel proteomics without being subjected to ensemble-averaging, cell-cycle, or cell-population effects. We are currently developing a single cell analytical method for protein fingerprinting combining a structured microfluidic device with latest optical laser technology for single cell manipulation (trapping and steering), free-solution electrophoretical protein separation, and (label-free) protein detection. In this paper we report on first results of this novel analytical device focusing on three main issues. First, single biological cells were trapped, injected, steered, and deposited by means of optical tweezers in a poly(dimethylsiloxane) microfluidic device and consecutively lysed with SDS at a predefined position. Second, separation and detection of fluorescent dyes, amino acids, and proteins were achieved with LIF detection in the visible (VIS) (488 nm) as well as in the deep UV (266 nm) spectral range for label-free, native protein detection. Minute concentrations of 100 fM injected fluorescein could be detected in the VIS and a first protein separation and label-free detection could be achieved in the UV spectral range. Third, first analytical experiments with single Sf9 insect cells (Spodoptera frugiperda) in a tailored microfluidic device exhibiting distinct electropherograms of a green fluorescent protein-construct proved the validity of the concept. Thus, the presented microfluidic concept allows novel and fascinating single cell experiments for systems nanobiology in the future.

A bio-recognition device developed onto nano-crystals of carbonate apatite for cell-targeted gene delivery. Chowdhury EH, Akaike T. Biotechnol Bioeng. 2005 Apr 6; [Epub ahead of print].

The DNA delivery to mammalian cells is an essential tool for analyzing gene structure, regulation, and function. The approach holds great promise for the further development of gene therapy techniques and DNA vaccination strategies to treat and control diseases. Here, we report on the establishment of a cell-specific gene delivery and expression system by physical adsorption of a cell-recognition molecule on the

nano-crystal surface of carbonate apatite. As a model, DNA/nano-particles were successfully coated with asialofetuin to facilitate uptake by hepatocyte-derived cell lines through the asialoglycoprotein receptor (ASGPr) and albumin to prevent non-specific interactions of the particles with cell-surface. The resulting composite particles with dual surface properties could accelerate DNA uptake and enhance expression to a notable extent. Nano-particles coated with transferrin in the same manner dramatically enhanced transgene expression in the corresponding receptor-bearing cells and thus our newly developed strategy represents a universal phenomenon for anchoring a bio-recognition macromolecule on the apatite crystal surface for targeted gene delivery, having immediate applications in basic research laboratories and great promise for gene therapy.

Glossary

Actin filament (microfilament): Helical protein filament formed by the polymerization of globular actin molecules. A major constituent of the cytoskeleton of all eucaryotic cells and part of the con- tractile apparatus of skeletal muscle.

Activation energy: Extra energy that must be possessed by atoms or molecules in addition to their ground-state energy in order to undergo a particular chemical reaction. Distinct states correspond to minima of a potential energy surface in a configuration space. In this classical picture, the activation energy for transforming state A into state B is the maximum increase in energy (relative to the ground state of A) encountered on a minimum-energy path from A to B. Energy here refers to potential energy; an analogous definition based on free energy can be constructed. When tunneling is considered, lower energy paths become possible, but an activation energy can be associated with the reaction (at a given temperature) via the relationship between temperature and reaction rate.

Algorithm: A set of well-defined mathematical rules or operations for solving a problem in a finite number of steps.

Alkane: (1) A saturated, acyclic hydrocarbon structure; usually quite inert. (2) A hydrocarbon containing a double bond; often rather reactive.

Analyte: A chemical species targeted for qualitative or quantitative analysis.

Aromatic: A term used to describe cyclic pi-bonded structures of special stability.

Autoproductivity: The ability of a system, under external control, to automatically produce an identical copy of itself.

Bacteriophage (phage): Any virus that infects bacteria. They were the first organisms used for the study of molecular genetics and are now widely used as cloning vectors. (From Greek *phagein*, to eat.)

Bacteriorhodopsin: Pigmented protein found in the plasma membrane of a salt-loving bacterium, *Halobacterium halobium*; it pumps protons out of the cell in response to light.

Bacterium (plural bacteria): Common name for any member of the diverse group of procaryotic organisms. Most are single cells, but multicellular forms also exist.

Banoprism: Scientists at Northwestern University have created a nanoparticle with a new shape that could be a useful tool in the race to detect biological threats. The nanoprism, which resembles a tiny Dorito, exhibits unusual optical properties that could be used to improve biodetectors, allowing them to test for a far greater number of biological warfare agents or diseases at one time.

Banoscience: The study of phenomena and manipulation of materials at atomic, molecular and macromolecular scales, where properties differ significantly from those at a larger scale. Nanoscience is primarily the extension of existing sciences into the realms of the extremely small (*nanomaterials, nanochemistry, nanobio, nanophysics*, etc.) while *nanoengineering* represents the extension of the engineering fields into the nano- scale realm (*nanofabrication, nanodevices*, etc.). The exponential growth of nanoscience is largely due to the development of new instruments and related techniques that are used to "routinely" probe and manipulate material at the atomic and molecular level. *Scanning probe microscopies,* analytical electron-beam techniques, epitaxial growth facilities, and *synchrotron* radiation sources are all opening huge opportunities.

Banoshells: Procedures that target cancer cells while leaving normal cells untouched, patients controlling the release of medicine in their bodies with an infrared light, and medical test results produced in seconds rather than days – these are three new medical technologies currently being tested by nanotechnology researchers at Rice University. ... The research focuses on nanoshells, a new type of nanoparticle invented by Naomi Halas, Rice professor of electrical and computer engineering and chemistry. Nanoshells are layered nanoparticles whose ability to manipulate light and color can be designed into the nanoparticle by varying the thickness of the nanoparticle's layers.

Basal body: Short cylindrical array of microtubules plus their associated proteins found at the base of a eucaryotic cell cilium or flagellum. Serves as a nucleation site for the growth of the axoneme. Closely similar in structure to a centriole.

Basal lamina (plural basal laminae): Thin mat of extracellular matrix that separates epithelial sheets, and many types of cells such as muscle cells or fat cells, from connective tissue. Sometimes called a basement membrane.

Bionanotechnology: Includes molecular motors, biomaterials, single molecule manipulation technologies, biochip technologies, etc.

Binding energy: The reduction in the free energy of a system that occurs when a ligand binds to a receptor. Generally used to describe the total energy required to remove something, or to take a system apart into its constituent particles—for example, to separate two atoms from one another, or to separate an atom into electrons and nuclei.

Bio-assemblies or Biomolecular Assemblies: Containing several protein units, DNA loops, lipids, various ligands, etc.

Biochauvinism: The prejudice that biological systems have an intrinsic superiority that will always give them a monopoly on self-reproduction and intelligence.

Biocompatible coated materials: Biocompatible materials usually used in dental and bone implants that enhance biologic fixation, thereby increasing the bond strength between the coated material and bone, and minimize possible biological effects that may result from the implant itself. MeSH, 1999

Biocompatible materials: Synthetic or natural materials, other than drugs, that are used to replace or repair any body tissue or bodily function. MeSH, 1973

Biodefense: The need for improved national defenses against biological attacks are urgently needed. A real-time reporting system needs to be developed, whereby officials can be informed about an emerging threat before an agent has a chance to affect thousands of people. The development of integrated systems for detecting and monitoring biological agents is instrumental to this goal. The development of vaccines and novel detection platforms will enable preparedness, while also benefiting human health.

Biodynotics Biologically Inspired Multifunctional Dynamic Robotics: A multidisciplinary, multi- pronged approach with far reaching impact on robotic capabilities for national security applications. Biologically Inspired Multifunctional Dynamic Robotics: BIODYNOTICS will explore the following areas: DYNAMIC MOBILITY, BEHAVIOR, INTEGRATION Biological Sciences, Defense Sciences Office, DARPA.

Bioengineering: Is rooted in physics, mathematics, chemistry, biology, and the life sciences. It is the application of a systematic, quantitative, and integrative way of thinking about and approaching the solutions of problems important to biology, medical research, clinical proactive, and population studies. The NIH Bioengineering Consortium agreed on the following definition for bioengineering research on biology, medicine, behavior, or health recognizing that no definition could completely eliminate overlap with other research disciplines or preclude variations in interpretation by different individuals and organizations. Integrates physical, chemical, or mathematical sciences and engineering principles for the study of biology, medicine, behavior, or health. It advances fundamental concepts, creates knowledge for the molecular to the organ systems levels, and develops innovative biologics, materials, processes, implants, devices, and informatics approaches for the prevention, diagnosis, and treatment of disease, for patient rehabilitation, and for improving health.

Biofabrication: Includes nanoparticle delivery systems, biomaterials, tissue engineering, implants and prostheses. Using biological processes to synthesize and manufacture chemicals and materials of high value to the Department of Defense. Biological processes are characterized by: low energy barriers (~10Kcal, high temperatures/ pressures not required); high reaction-, regio- and stereo- specificity (protection/ deprotection wastes and catalyst poisoning avoided); spatio-temporal control of materials synthesis of defined composition and size with angstrom- level precision; and local control of the dielectric environment (largely eliminating the need for toxic solvents). Potential target materials and chemicals include composites with enhanced mechanical properties, ultra-low k dielectrics and thermoelectrics, optoelectronic materials, photonic devices (waveguides and logic elements), electronic materials (e.g., GaN, InGaN, AlGaN), elastomers, energetic materials, and adhesives. Biofabrication, DARPA.

Biomaterials: Synthetic or natural materials that can replace or augment tissues, organs or body functions.

Biomechanics: Mechanical structures of living organisms (especially muscles and bones).

Biomedical Nanotechnology: (1) the comprehensive monitoring, control, construction, repair, defense, and improvement of all human biological systems, working from the molecular level, using engineered nanodevices and nanostructures; (2) the science and technology of diagnosing, treating, and preventing disease and traumatic injury, of relieving pain, and of preserving and improving human health, using molecular tools and molecular knowledge of the human body; (3) the employment of molecular machine systems to address medical problems, using molecular knowledge to maintain and improve human health at the molecular scale. (4) A rapidly expanding field that includes many potential technologies and approaches. The key to this definition is that phenomena and materials at the nanometer scale are known to have properties that are uniquely attributable to that scale length. Nanomedicine could similarly be defined [as nanotechnology] as the design, synthesis, or application of materials, devices, or technologies in the nanometer-scale for the basic understanding, diagnosis, and / or treatment of disease. Canadian Institute of Health Research, Regenerative Medicine and Nanomedicine, RFA, 2003. The monitoring, repair, construction and control of human biological systems at the molecular level, using engineered nanodevices and nanostructures. (5) the comprehensive monitoring, control, construction, repair, defense, and improvement of all human biological systems, working from the molecular level, using engineered nanodevices and nanostructures; (6) the science and technology of diagnosing, treating, and preventing disease and traumatic injury, of relieving pain, and of preserving and improving human health, using molecular tools and molecular knowledge of the human body; (7) the employment of molecular machine systems to address medical problems, using molecular knowledge to maintain and improve human health at the molecular scale. The use of molecular-scale devices (nanotechnology) to repair damage and boost the immune system. Vapour grown carbon fibres are obtained as shown in this SEM image. The diameters of these fibres can vary from 100 nm to 500 nm. As part of the National Institutes of Health (NIH) Roadmap for Medical Research [nihroadmap.nih. gov], the NIH [nih.gov] will establish a handful of nanomedicine centers. These centers will be staffed by a highly interdisciplinary scientific crew including biologists, physicians, mathematicians, engineers and computer scientists. Research conducted over the first few years will be spent gathering extensive information about how molecular machines are built. A key activity during this time will be the development of a new kind of vocabulary, or lexicon, to define biological parts and processes in engineering terms. Once researchers have completely catalogued the interactions between and within molecules, they can begin to look for patterns and a higher order of connectedness than is possible to identify with current experimental methods. Mapping these networks and understanding how they change over time will be a crucial step toward helping scientists understand nature's rules of biological design. Understanding these rules will, in many years' time, enable researchers to use this information to address biological issues in unhealthy cells.

Biomedical polymers: More and more therapeutic problems are relevant to the use of polymer- based therapeutic aids for a limited period of time, namely the healing time related to the outstanding capacity of living systems to self- repair ... After healing the remaining prosthetic materials or devices become foreign residues or wastes that have to be eliminated from the body. Nowadays, biocompatible polymers that can degrade in the body are developed. The degradation and the elimination of degradation by-products depend on rather complex phenomena that are presently reflected inconsistently by terms issued from the tradition because each domain has developed its own terminology almost independently. This is a source of misunderstandings, confusions and misperceptions among scientists, surgeons, pharmacists, journalists and politicians, the situation being increased by the introduction of degradable polymers in plastic waste management and environmental protection. Therefore, it is urgent to reflect the various phenomena by specific terms, harmonize and enforce their use by the people active in the biomedical, pharmacological and environmental fields, and, last but not least by the publishing media and journalists. IUPAC, Terminology for biomedical (therapeutic) polymers.

BioMEMS Biological MicroElectro Mechanical Systems: Highlights the technical advances in the field that are leading a revolution in medicine, and creating a new generation of analytical devices for medical diagnostics. The meeting will encompass technology developments in micro & nano drug delivery, interface of nanotech and tissue engineering, microfluidics, and miniaturized total analysis systems (microTAS), biosensors, innovations in mass spec, and nanoscale imaging *BioMEMS and Nanotech World*, Aug. 16- 17, 2004, Washington DC

BioMEMS: MEMS used in medicine, that use microchips.

Biomimetic Chemistry: Knowledge of biochemistry, analytical chemistry, polymer science, and biomimetic chemistry is linked and applied to research in designing new molecules, molecular assemblies, and macromolecules having biomimetic functions. These new bio-related materials of high performance, including, for example, enzyme models, synthetic cell membranes, and biodegradable polymers, are prepared, tested, and constantly improved in this division for industrial scale production.

Biomimetic materials: Materials fabricated by BIOMIMETICS techniques, i.e., based on natural processes found in biological systems. [MeSH 2003] (2) Materials that imitate, copy, or learn from nature.

Biomimetic: Imitating, copying, or learning from biological systems.

Biomimetic: Imitating, copying, or learning from nature. Nanotechnology already exists in nature; thus, nanoscientists have a wide variety of components and tricks already available.

Biomimetics: An interdisciplinary field in materials science, ENGINEERING, and BIOLOGY, studying the use of biological principles for synthesis or fabrication of BIOMIMETIC MATERIALS. There is a need to develop the next generation of restorative materials and medical implants. New avenues of scientific inquiry may enable the development of biomaterials that are safe, reliable, "smart", long-lasting, and perform ideally in their respective biological environments.... Over the last few years biomimetics and tissue engineering have emerged as a new vision in the field of tissue and organ repair and restoration. Biomimetics and tissue engineering are interdisciplinary fields that combine information from the study of biological structures and their functions with physics, mathematics, chemistry and engineering for the generation of new materials, tissues and organs. In the area of craniofacial, oral and dental principles from biomimetics and tissue engineering are applied to developing dental and facial implants, new polymers for guided tissue regeneration used in treating periodontal disease and bone and connective tissue defects, coral- based hydroxyapatite replicas for reconstruction of alveolar ridges and other osseous defects, temporomandibular joint (TMJ) and other joint prostheses, formation of bone matrix substitutes, and artificial replicas of bone, skin, and mucosa. The term biomimetics was coined in 1972 in the context of artificial enzymes; it might be defined broadly as "the abstraction of good design from nature". It is a fact of everyday life that nature has managed to built materials and 'devices' with breathtaking functionality, heterogeneity and stability by using a comparatively limited number of building blocks (the whole range of synthetic materials is restricted to man- made engineering). The basic concepts of nature are often simple; it is the way in which building blocks and materials are arranged that results in functionality. Among the most simple and abundant themes of nature is self- assembly: lipids assemble in sheets to form cell membranes, proteins assemble into functional enzymes, cellular 'sensors', fibers, or virus coats, and DNA assembles in double strands to provide the very basis for live: replication Biomimetics Study of the structure and function of biological substances to make artificial products that mimic the natural ones.

BIOML Biopolymer Markup Language: Designed by the BIOML core team at Proteometrics, LLC and Proteometrics Canada Ltd. It is to be used "for the annotation of biopolymer sequence information. BIOML allows the full specification of all experimental information known about molecular entities composed of biopolymers, for example, proteins and genes."

Biomolecular materials: An emerging discipline, materials whose properties are abstracted from biology. They share many of the characteristics of biological materials but are not necessarily of biological origin. For example, they may be inorganic materials that are organized or processed in a biomimetic fashion. A key feature of biological and biomolecular materials is their ability to undergo *self- assembly*.

Biomolecular Nanotechnology: Nanotechnology existing in living systems and resulting from our ability to use biomolecules as components for molecular nanotechnology.

Biomotors: Driven by energy sources such as *adenosine triphosphate (ATP)* for chemical transduction and other processes. These biomotors are considered to be biomolecular and are discussed in the body of this report, but strictly speaking they do not conform to the panel's definition of self- assembly.

Biopolymeroptoelectromechanical Systems [BioPOEMS]: Combining optics and microelectromechanical systems, and used in biological applications.

Biopolymers: Macromolecules (including proteins, nucleic acids and polysaccharides) formed by living organisms.

Biorobotics: Our research focuses on the role of sensing and mechanical design in motor control, in both robots and humans. This work draws upon diverse disciplines, including biomechanics, systems analysis, and neurophysiology. The main approach is experimental, although analysis and simulation play important parts. In conjunction with industrial partners, we are developing applications of this research in biomedical instrumentation, teleoperated robots, and intelligent sensors.

Biosensor: The term "biosensor" is a general designation that denotes either a sensor to detect a biological substance or a sensor which incorporates the use of biological molecules such as antibodies or enzymes. Biosensors are a subcategory of chemical sensors.

Biostasis: A condition in which an organism's cell and tissue structure are preserved, allowing later restoration by cell repair machines. Applicable to cryonics.

Biovorous: From "biovore;" an organism capable of converting biological material into energy for sustenance.

Bipolar-junction transistor: Transistor with n-type and p-type semiconductors having base-emitter and collector-base junctions.

Bone substitutes: Synthetic or natural materials for the replacement of bones or bone tissue. They include hard tissue replacement polymers, natural coral, hydroxyapatite, beta- tricalcium phosphate, and various other biomaterials. The bone substitutes as inert materials can be incorporated into surrounding tissue or gradually replaced by original tissue.

Brownian Assembly: Brownian motion in a fluid brings molecules together in various position and orientations. If molecules have suitable complementary surfaces, they can bind, assembling to form a specific structure. Brownian assembly is a less paradoxical name for self-assembly (how can a structure assemble itself, or do anything, when it does not yet exist?).

Calorie: Unit of heat. One calorie (small "c") is the amount of heat needed to raise the temperature of 1 gram of water by 1 °C. A kilocalorie is the unit used to describe the energy content of foods.

CAP (catabolite gene activator protein): Gene regulatory protein in procaryotes that, when glucose is absent, activates genes responsible for the breakdown of alternative carbon sources.

Capacitor: Energy storage circuit element having two conductors separated by an insulator.

Cell pharmacology: Delivery of drugs by medical nanomachines to exact locations in the body.

Cell Repair Machine: A system including nanocomputers and molecular scale sensors and tools, programmed to repair damage to cells and tissues. Molecular and nanoscale machines with sensors, nanocomputers and tools, programmed to detect and repair damage to cells and tissues, which could even report back to and receive instructions from a human doctor if needed.

Cell surgery: In medical nanorobotics, modifying cellular structures using medical nanomachines. Modifying cellular structures using medical nanomachines.

Cell: A membrane-bound unit, typically microns in diameter. All plants and animals are made up of one or more cells (trillions, in the case of human beings). In general, each cell of a multicellular organism contains a nucleus holding all of the genetic information of the organism. (2) A small structural unit, surrounded by a membrane, making up living things.

Combinatorial materials design: Uses computing power (sometimes together with massive parallel experimentation) to screen many different materials possibilities to optimize properties for specific applications (e.g., catalysts, drugs, optical materials).

Configuration space: A mathematical space describing the three-dimensional configuration of a system of particles (e.g., atoms in a nanomechanical structure) as a single point; the configuration space for an N particle system has $3N$ dimensions.

Conformation: The shape of a molecule. A molecular geometry that differs from other geometries chiefly by rotation about single or triple bonds; distinct conformations (termed con formers) are associated with distinct potential wells. Typical biomolecules and products of organic synthesis can interconvert among many conformations; typical diamondoid structures are locked into a single potential well, and thus lack conformational flexibility.

Coulomb [C]: Measure of electrical charge: 1 C is an amount of charge equal to that of about 6.24×10^{18} electrons.

Coupling: A connection between more than one object or energy pathway so that together they function as a single unit.

Crosstalk: Electromagnetic noise transmitted between leads or circuits in close proximity to each other.

Crybiology: The science of biology at low temperatures; research in cryobiology has made possible the freezing and storing of sperm and blood for later use.

Crystal Lattice: The regular three-dimensional pattern of atoms in a crystal.

Crystal structure: For crystalline materials, the manner in which atoms or ions are arrayed in space. It is defined in terms of the unit cell geometry and the atom positions within the unit cell.

Crystallescence: In medical nanorobotics, the crystallization of solid solute that is offloaded by nanorobot sorting rotors at a concentration that exceeds the solvation capacity of the surrounding solvent.

Curie temperature (also Curie point) (T_c): The temperature above which a ferromagnetic or ferrimagnetic material becomes paramagnetic. For iron the Curie point is 760oC and for nickel 356°C.

Current [A]: Measure of rate of flow of electric charge: a one-ampere current is a flow of 1 C of charge per second.

Cutoff: Condition in a diode or bipolar junction transistor in which the potential across a p-n junction prevents current flow.

Cyclic: A structure is termed cyclic if its covalent bonds form one or more rings.

DNA (Deoxyriobonucleic Acid): DNA molecules are long chains consisting of four kinds of nucleotides; the order of these nucleotides encodes the information needed to construct protein molecules. These in turn make up much of the molecular machinery of the cell. DNA is the genetic material of cells.

DNA Chip: Also: Gene Chip and DNA Microchip. A purpose built microchip used to identify mutations or alterations in a gene's DNA.

DNA diagnostics - miniaturization of: In the areas of sample preparation and assay, it is clear that miniaturization is key. To reduce the size of samples by a factor of 10 or greater, barriers in microfluidics, micromachining, robotics, microchemistry, nucleic acid chemistry, and surface chemistry must be overcome. To implement miniaturized protocols accurately and efficiently, substantial automation of the process will be required. In the development of miniaturized systems, it is essential that the system can be adapted for high levels of parallelization. Miniaturization poses significant technological risks. Currently, there exists no universally accepted precedent for the handling, replication, amplification, or cloning of DNA in nanoliter volumes. Due to the size and charge of the DNA molecule, and the relative instability of many of the enzymes involved in the sample preparation processes, nanoliter and less volumes may pose substantial challenges. In addition, interactions of the biological molecules with the surfaces of the reaction chambers must be minimized. For some methodologies, it is not clear what the optimal sample will be, so substantial improvement in DNA fragmentation technologies or DNA cloning vectors may be required for the ultimate efficient application to diagnostics. Improvements in any of these areas are likely to be of value to other non- DNA based diagnostic applications such as antibody screening protocols and enzyme based diagnostics, because miniaturized robotic or micro- electro mechanical systems developed for DNA could be modified to be used for these purposes.

DNA Probes: A DNA or nucleic acid probe is a short strand of DNA that locates and binds to its complementary sequence in samples containing single strands of DNA or RNA enabling identification of specific sequences. Nucleic acid probe assays exploit the fundamental hybridization reaction that occurs spontaneously between two complementary DNA:DNA or DNA:RNA strands. As in immunoassays, detection of the hybrid requires that the probe be labeled. Various direct and indirect methods have been devised for the detection of the hybrid. Direct labeling involves attaching the label directly to the probe sequence; indirect labeling binds an antibody to the DNA:DNA or DNA:RNA hybrid. As in immunoassays, non-isotopically-labeled probes are preferred over radio-labeled probes primarily because of radiation hazards, disposal problems, and short reagent shelf life. In addition, the factors determining the detection limits of hybridization assays based on labeled probes are similar to those in immunoassays. Therefore, the development of a simple, inexpensive and sensitive direct detection system which eliminates the use of labels is highly desirable.

DNA Sequencing: There are two main classical methods for sequencing DNA: The first method, developed by Allan Maxam and Walter Gilbert, involves chemicals used to cleave the DNA at certain positions, generating a set of fragments that differ by one nucleotide. The second method, developed by Fred Sanger and Alan Coulson, involves enzymatic synthesis of DNA strands that terminate in a modified nucleotide. Analysis of fragments is similar for both methods and involves gel electrophoresis and autoradiography or fluorescence. The enzymatic method has largely replaced the chemical method as the technique of choice, although there are some situations where chemical sequencing can provide data more easily than the enzymatic method.

DNA: A molecule encoding genetic information, found in the cell's nucleus. Deoxyribose Nucleic Acid. DNA is a molecule used in biological systems to store genetic information. It has a long, double stranded structure with the two strands wrapped around each other to form a right handed helix. There are about 10 nucleotide pairs per helical turn. Each strand is composed of a sugar phosphate backbone and attached bases. It is connected to a complementary strand by hydrogen bonding (non- covalent) between paired bases, adenine (A) with thymine (T) and guanine (G) with cytosine (C). DNA is an abbreviation of DeoxyriboNucleic Acid. : It is the genetic 'blueprint' of all life on this planet. : DNA typing is often applied to Human Identification which has also been called DNA Fingerprinting. DNA Fingerprinting has been attributed to Alex Jeffries, who with his colleagues in 1985, discovered a sequence of nucleotides, subsequently termed short tandem repeats (STR) when comparing the DNA sequence of the human á-globin chromosomes isolated from different individuals.

Emergence: A complex whole created by simple parts, as in the brain where billions of neurons work individually, but collectively make up our consciousness and give us the ability to think, rationalize, and create.

Emulation: An absolutely precise simulation of something, so exact that it is equivalent to the original (for example, many computers emulate obsolete computers to run their programs). [AS] The Star Trek replicator is an example.

Enabling science and technologies: Areas of research relevant to a particular goal, such as nanotechnology. Also, technology that "enables" other technology to advance, such as the transistor enabled the computer chip revolution, as did photolithography.

Endocrine cell: Specialized animal cell that secretes a hormone into the blood; usually part of a gland, such as the thyroid or pituitary gland.

Endocytosis: Uptake of material into a cell by an invagination of the plasma membrane and its internalization in a membrane-bounded vesicle.

Endoplasmic reticulum (ER): Labyrinthine, membrane-bounded compartment in the cytoplasm of eucaryotic cells, where lipids are synthesized and membrane-bound proteins are made.

Endosome: Membrane-bounded organelle in animal cells that carries materials newly ingested by endocytosis and passes many of them on to lysosomes for degradation.

Endothelium: Single sheet of highly flattened cells (endothelial cells) that forms the lining of all blood vessels. Regulates exchanges between the bloodstream and surrounding tissues and is usually surrounded by a basal lamina.

Energy [J]: Capacity for performing work or to cause heat flow. Like work itself, it is measured in Joules.

EPROM: Electrically Programmable Read-Only Memory — nonvolatile memory device.

Eurisko: A computer program developed by Professor Douglas Lenat which is able to apply heuristic rules for performing various tasks, including the invention of new heuristic rules.

Face-centered cubic: A crystal structure found in common elemental metals also FCC. Within the cubic unit cell, atoms are located at all corner and face-centered positions.

Fail-stop: Describes a component or subsystem that, in the event of a failure, produces no output (e.g., of material or data) rather than producing a damaged or incorrect output.

Fermi energy: The energy level in a solid at which the probability of finding an electron is 1/2. For a metal, the energy corresponding to the highest filled electron state in the valence band at 0 K.

Field-Effect Transistor (FET): Field-Effect Transistor — semiconductor device whose insulated gate electrode controls current flow.

Ganglioside: Any glycolipid having one or more sialic acid residues in its structure. Found in the plasma membrane of eucaryotic cells and especially abundant in nerve cells.

GUI: Graphical User Interface — hardware, software, and firmware that produces the display on modern personal computers.

Guy Fawkes Scenario: If nanotechnology becomes widely available, it might become trivial for anyone to committ acts of terrorism (such as making nanomachines build a large amount of explosives under government buildings a la Guy Fawkes). This would either force strict control over nanotechnology (hard) or a decentralized mode of organization.

Hair cell: Specialized sensory epithelial cell in the ear with bundles of giant microvilli (stereocilia) protruding from its apical surface. Sound vibrations tilt the stereocilia, evoking an electrical change in the hair cell, which thus acts as a sound detector.

Hall effect: The phenomenon whereby a force is applied to a moving electron or hole by a magnetic field that is applied perpendicular to the direction of motion. The force direction is perpendicular to both the magnetic field and the particle motion directions.

Heme: An iron complex. Cyclic organic molecule containing an iron atom that carries oxygen in hemoglobin and carries an electron in cytochromes.

Imaging Contrast Agent: A molecule or molecular complex that increases the intensity of the signal detected by an imaging technique, including MRI and ultrasound. An MRI contrast agent, for example, might contain gadolinium attached to a targeting antibody. The antibody would bind to a specific target—a metastatic melanoma cell, for example – while the gadolinium would increase the magnetic signal detected by the MRI scanner.

Immortalization: Production of a cell line capable of an unlimited number of cell divisions. Can be the result of a chemical or viral transformation or of fusion with cells of a tumor line.

Insulator: A substance, object or material that does not conduct electricity. Material that conducts electricity very poorly.

Integrated Circuit (IC): An electronic circuit consisting of many interconnected devices on one piece of semiconductor, typically into 10 millimeters on a side. ICs are the major building blocks of today's computers. Semiconductor circuit, typically on a very small silicon chip, containing microfabricated transistors, diodes, resistors, capacitors, etc.

Jupiter-Brain: A posthuman being of extremely high computational power and size. This is the archetypal concentrated intelligence. The term originated due to an idea by Keith Henson that nanomachines could be used to turn the mass of Jupiter into computers running an upgraded version of himself.

Kilohertz (kHz): One thousand cycles per second.

Kilojoule: Standard unit of energy equal to 1000 joules, or 0.24 kilocalories.

Lectin: Protein that binds tightly to a specific sugar. Abundant lectins derived from plant seeds are often used as affinity reagents to purify glycoproteins or to detect them on the surface of cells.

Life (lifetime): The length of time the sensor can be used before its performance changes.

Life Science Applications of Deep Red T2-MP EviTag Quantum Dots: The new T2-MP EviTags are available in the deep red wavelengths used in life science research – Macoun Red (660nm) and Jonamac Red (680nm) – with carboxyl, amine, biotin and non-functionalized surfaces. Having a biologically compatible quantum dot label like the T2-MP EviTags allows further exploration and development of in vivo assays, following cells inside living organism to get a new understanding of life. In particular, it offers the possibility to create new classes of assays for use where current testing materials could produce cytotoxicity or cell death. Potential applications include live cell and even whole animal imaging, blood cancer assays, and numerous other applications.

Ligand: In protein chemistry, a small molecule that is (or can be) bound by a larger molecule is termed a ligand. In organometallic chemistry, a moiety bonded to a central metal atom is also termed a ligand; the latter definition is more common in general chemistry.

Light Emitting Diodes (LEDs): Light Emitting Diodes work on a completely different concept. Traditionally LEDs are created from two semiconductors. By running current in one direction across the semiconductor the LED emits light of a particular frequency (hence a particular color) depending on the physical characteristics of the semiconductor used. The semiconductor is covered with a piece of plastic that focuses the light and increases the brightness. These semiconductors are very durable, there is no filament, they donít require much power, theyíre brighter and they last a long time. By densely packing red, blue and green LEDs next to each other on a substrate one can create a display.

Liposome: A type of nanoparticle made of lipids, or fat molecules, surrounding a water core. Liposomes, several of which are widely used to treat infectious diseases and cancer, were the first type of nanoparticle to be used to create therapeutic agents with novel characteristics.

Liquid Crystal Display: LCD is the predominant technology used in flat panel displays. The principle that makes the display work is this: A crystalís alignment can be altered with an electric current. If the crystal is lined up one way ñ it will allow the light waves to pass through a polarized filter, but if the electric current alters the crystalís alignment, it will guide light so that the polarized filter blocks the light. By densely packing red, blue and green light emitting crystals next to each other on a sheet (ìcalled a substrateî), one can create a full color display. The great thing about LCD is that the crystals can be packed together closely, allowing for a higher-resolution, finer-detail display. The con is that LCDs are somewhat fragile, require a lot of power and are relatively less bright. Liquid Crystal Display — display device employing light source and electrically alterable optically active thin film.

Machine-phase chemistry: The chemistry of systems in which all potentially reactive moieties follow controlled trajectories (e.g., guided by molecular machines working in vacuum).

Macroscale: Larger than nanoscale; often implies a design that humans can directly interact with; too large to be built by a single assembler (one cubic micron of diamond contains 176 billion atoms).

Magnetization (M): The total magnetic moment per unit volume of material. Also, a measure of the contribution to the magnetic flux by some material within an H field. The magnitude of M is proportional to the applied field as: $M = \div_m \times H$, with $\div_m$ the magnetic susceptibility.

Magnetostrictive material: A material that changes dimension in the presence of a magnetic field or generates a magnetic field when mechanically deformed.

Manufacturing: The ability to make products, in this case ranging from clothing, to electronics, to medical devices, to books, to building materials, and much more.

Measurand: A physical quantity, condition, or property that is to be measured.

Meat Machine: AKA Cabinet Beast. A box containing assemblers and raw material, within which is formed meat [or whatever else it was programmed to make].

Mechanical: Pertaining to the positions and motions of atoms, as defined by the positions of their nuclei; see electronic. A purely mechanical device can be described in terms of atomic positions and motions without reference to electronic properties, save through their effect on the potential energy function.

Mechanochemistry: Chemistry accomplished by mechanical systems directly controlling the reactant molecules; the formation or breaking of chemical bonds under direct mechanical control. In this volume, the chemistry of processes in which mechanical systems operating with atomic-scale precision either guide, drive, or are driven by chemical transformations. In general usage, the chemistry of processes in which energy is converted from mechanical to chemical form, or vice versa.

Mechanosynthesis: Chemical synthesis controlled by mechanical systems operating with atomic-scale precision, enabling direct positional selection of reaction sites; synthetic applications of mechanochemistry. Suitable mechanical systems include AFM mechanisms, molecular manipulators, and molecular mill systems. Processes that fall outside the intended scope of this definition include reactions guided by the incorporation of reactive moieties into a shared covalent framework (i.e., conventional intramolecular reactions), or by the binding of reagents to enzymes or enzymelike catalysts. Molecular tools with chemically specific tip structures can be used, sequentially, to modify a work piece and build a wide range of molecular structures.

Mechatronics: The study of the melding of AI and electromechanical machines to make machines that are greater than the sum of their parts. The synergistic combination of precision mechanical engineering with electronic control.

Megabyte (MB): 2^{20} (= 1,048,576, or about one million) bytes of information.

Megahertz (Mhz): One million cycles per second

Meme: An idea that replicates through a society as it is propagated through person-to-person interaction, both direct and indirect. Memetics is a field of study that focuses on memes' role in the evolution of a culture. Self-reproducing idea or other information pattern which is propagated in ways similar to that of a gene. [Richard Dawkins, 1976]

Micellar nanocontainers: Block copolymer micelles are water- soluble biocompatible nanocontainers with great potential for delivering hydrophobic drugs. An understanding of their cellular distribution is essential to achieving selective delivery of drugs at the subcellular level. R. Savic, L. Luo, A. Eisenberg, D. Maysinger, Micellar nanocontainers distribute to defined cytoplasmic organelles Science 300 (5619): 615- 618, Apr. 25, 2003

Micelles: Micelles are small, spherical structures composed of molecules that attract one another to reduce surface tension. The head of the molecule is hydrophilic, meaning it likes water, while the interior portion is hydrophobic, meaning it avoids water.

Microelectronics: Narrower terms: MEMS, nanoelectronics, optoelectronics, SED Single Electron Devices. Related terms: molecular electronics, semiconductors

Microencapsulation: Individually encapsulated small particles.

Microspectrophotometry: Analytical technique for studying substances present at enzyme concentrations in single cells, in situ, by measuring light absorption. Light from a tungsten strip lamp or xenon arc dispersed by a grating monochromator illuminates the optical system of a microscope. The absorbance of light is measured (in nanometers) by comparing the difference between the image of the sample and a reference image. MeSH 1990

Microspheres: Microspheres are generally defined as small spheres made of any material and sized from about 0.5 μm to 100 μm. Similar, but smaller spheres sized 10 to 500 nm are called *nanospheres.* Ideally, microspheres are completely spherical and homogeneous in size, although less perfect particles are often termed microspheres as well. Depending on the preparation method and material used, microspheres show a typical size distribution which often deviates from the mono-sized ideal.

Microstructure: In material engineering the structural features of a material such as grain boundaries, grain size and structure, subject to observation under a microscope, selective etching etc. In MEMS microstructure unfortunately is also used to designate a micromachined feature. The last decade has seen rapid developments in the fabrication, characterization and conceptual understanding of synthetic microstructures in many different material systems including silicon, III-V and II-VI semiconductors, metals, ceramics and organics. The objective of this journal [Superlattices and Microstructures] is to provide a common interdisciplinary platform for the publication of the latest research results on all such "nanostructures" with dimensions in the range of 1 - 100 nm; the unifying theme here being the dimensions of these artificial structures rather than the material system in which they are fabricated.

Microsystem: A microscale machine that can sense information from the environment and act accordingly. Outside the U.S., it can also refer to microelectromechanical systems (MEMS). [smalltimes glossary, 2002]

MicroTas, micro Total Analysis Systems, uTAS: Although initial research dates back to the early 1970's, the field of micro- TAS formally started in 1990, when Manz et al described the possibility of creating microsystems that would take care of many or all the traditional analytical steps involved in a biochemical analysis (sample introduction, handling, extraction, purification, concentration, filtration, analysis, detection) .... Micro- TAS offer many advantages over traditional analysis systems. Low power consumption and small reaction volumes, faster analysis, ultrasensitive detection, and minimal human intervention are key parameters in the development of micro- TAS. Most biochemical reactions take place in liquid environments. Hence, the development of MicroTAS is intrinsically linked to the design of liquid handling micro- devices. [Biomedical Applications Group (GAB) Centro Nacional de Microelectronica (CNM- IMB) Bellaterra, Spain, 2000]

Microtubule: A hollow cylindrical tube used for the transport of materials inside cells.

Miller indices: A set of 3 integers (4 for hexagonal) that designate crystallographic planes, as determined from reciprocals of fractional axial intercepts.

Millimeter: One thousandth of a meter, or about 1/26 of an inch.

Molecular - Nanotechnology: The technology of precisely- constructed molecular-scale machines; from nanometer: a billionth of a meter. For a good introduction, see What is Nanotechnology? or Ralph C. Merkle's nanotechnology page.

Molecular assembler: A general-purpose device for molecular manufacturing, able to guide chemical reactions by positioning individual molecules to atomic accuracy (e.g. mechanosynthesis) and to construct a wide range of useful and stable molecular structures according to precise specifications. Also known as an assembler, a molecular assembler is a molecular machine that can build a molecular structure from its component building blocks. [ZY]

Molecular Beam Epitaxy: [MBE] Process used to make compound (multi-layer) semiconductors. Consists of depositing alternating layers of materials, layer by layer, one type after another (such as the semiconductors gallium arsenide and aluminum gallium arsenide).

Molecular Biology: [AKA: wet nano]

Molecular mechanics models: Many of the properties of molecular systems are determined by the molecular potential energy function. Molecular mechanics models approximate this function as a sum of 2-atom, 3-atom, and 4-atom terms, each determined by the geometries and bonds of the component atoms. The 2-atom and 3-atom terms describing bonded interactions roughly correspond to linear springs.

Molecular medicine: A variety of pharmaceutical techniques and gene therapies that address specific molecular diseases or molecular defects in biological systems. A variety of pharmaceutical techniques and therapies in use today. Studying molecules as they relate to health and disease, and manipulating those molecules to improve the diagnosis, prevention, and treatment of disease.

Molecular mill: A mechanochemical processing system characterized by limited motions and repetitive operations without programmable flexibility.

Molecular motors: Protein based machines that are involved in or cause movement such as the rotary devices (flagellar motor and the F1 ATPase) or the devices whose movement is directed along cytoskeletal filaments (myosin, kinesin and dynein motor families). [MeSH, 1999]

Molecular Nanoscience: An emerging interdisciplinary field that combines the study of molecular/ biomolecular systems with the science and technology of nanoscale structures and systems. The potential applications for this research are very broad and include such possibilities as 1) the use of biomolecules and cellular systems to self- assemble nanoelectronic circuitry and other nanoscale structures and 2) the use of lamellar host frameworks containing nano pores that can be tailored to include guest molecules for separation of chemicals for pharmaceutical and other applications. Molecular Nanoscience Alliance for Interdisciplinary Studies and Activities, Univ. of Minnesota, US,

Molecular Wire: A molecular wire - the simplest electronic component - is a quasi-one-dimensional molecule that can transport charge carriers (electrons or holes) between its ends. [Michael D Ward]

Molecularly imprinted polymers MIPs: A new class of materials that have artificially created receptor structures. Since their discovery in 1972, MIPs have attracted considerable interest from scientists and 3engineers involved with the development of chromatographic absorbents, membranes, sensors and enzyme and receptor mimics.

Nanobiology: Many fundamental biological functions are carried out by molecular machineries that have the sizes of 1-100 nm. You find many examples in molecular biology and cell biology: single enzymes, transcription complex, ribosome, transport complex, nuclear pore, and so on. To understand the functions of these machineries, one has to describe their movements, changes in their shapes, and their localization. This means the mechanistic study is equivalent to dynamic morphology at this level of size, making a new field that merges mechanistic biology and morphology. The emergence of nanobiology depended on the invention of nano- technology: scanning probe microscopy, modern optical techniques, and micro- manipulating techniques. This concept of nanobiology was first proposed in a group study named "Biological Nano- Mechanisms", which was supported by Japanese Agency of Science and Technology (1992-1997). [National Institute of Genetics, Japan]

Nanobiotechnology: An emerging area of scientific and technological opportunity. Nanobiotechnology applies the tools and processes of nano/ microfabrication to build devices for studying biosystems. Researchers also learn from biology how to create better micro- nanoscale devices. The Nanobiotechnology Center (NBTC), a National Science Foundation, Science and Technology Center is characterized by its highly interdisciplinary nature and features a close collaboration between life scientists, physical scientists, and engineers. [NBTC, NanoBiotechnology Center, Cornell Univ. US, 2002] Use of nanotechnology[ies] in the life sciences. These may include, but are not limited to, therapeutics, medical devices/implants, biosensors, and tools for the development of drugs. Applying the tools and processes of MNT to build devices for studying biosystems, in order to learn from biology how to create better nanoscale devices. Should hasten the creation of useful micro devices that mimic living biological systems.

Nanocentrifuge: In medical nanorobotics, a proposed nanodevice that can spin materials at very high speed, imparting rotational accelerations of up to one trillion gravities (g's), thus permitting rapid sortation.

Nanoscience: Nanoscience is the science of nanoscale phenomena. This is a very complicated area, since Quantum Mechanics is often needed to describe the physical properties of atoms and molecules properly. What makes it even more complicated is that nanostructured materials can still be quite a challenge to describe Quantum Mechanically, since the equations are feindishly difficult - this is where supercomputers come in to make the job approachable. The scientific discipline seeking to increase our knowledge and understanding of nanoscale phenomena, i.e. science on the scale of 0.1 nm to 100 nm.

Nanosensor: A chemical or physical sensor constructed using nanoscale components, usually microscopic or submicroscopic in size.

Nanosources: sources that emit light from nanometre-scale volumes.

Nanotech: Slang for nanotechnology.

Neophile: One who welcomes the future and who enjoys change and evolution.

Orbital: In the approximation that each electron in a molecule has a distinct, independent wave function, the spatial distribution of an electron wave function corresponds to a molecular orbital. These, in turn, can be approximated as sums of contributions from the orbitals characteristic of the isolated atoms. An electron added to a molecule—or, similarly, one excited to a higher-energy state within a molecule—would occupy a state with a different wave function from the rest; an unoccupied state of this kind corresponds to an unoccupied molecular orbital. Orbital-symmetry effects on reaction rates arise when a reaction requires overlap between two lobes of the orbitals on each of two reagents: if the algebraic signs of the wave functions in the facing lobes do not match, bond formation between those orbitals is prohibited.

Piezoelectric material: A ferroelectric material in which an electrical potential difference is created due to mechanical deformation, or conversely, in which the application of a voltage causes dimensional changes in the material.

Piezoelectric: A piezoelectric material expands or contracts when a voltage is applied across it. Piezoelectric tubes are often used to move the tip in an AFM or STM. If a piezoelectric crystal is compressed, a voltage appears across it. This effect is sometimes used to measure displacements. A commonly used piezoelectric is PZT $(Pb,Zr)TiO_3$.

Quantum Mechanics: A largely computational physical theory explaining the behavior of quantum phenomena, which incorporates the theory of special relativity. Despite dilignet attempts, general relativity has not been sucessfully incorporated into quantum mechanics. A physical model of chemical and optical phenomena, as well as the behaviour of matter in general on a small scale. Quantum mechanics describes a system of particles in terms of a wave function defined over the configuration space of the system. Although the concept of particles having distinct locations is implicit in the potential energy function that determines the wave function (e.g., of a ground-state system), the observable dynamics of the system cannot be described in terms of the motion of such particles from point to point. In describing the energies, distributions, and behaviors of electrons in nanometer-scale structures, quantum mechanical methods are necessary. Electron wave functions help determine the potential energy surface of a molecular system, which in turn is the basis for classical descriptions of molecular motion. Nanomechanical systems can almost always be described in terms of classical mechanics, with occasional quantum mechanical corrections applied within the framework of a classical model.

Quantum Well: A P-N-P junction in which the "N" layer is ~10 nm (where traditional physics leaves off and quantum effects take over) and an "electron trap" is created. "If one makes a heterostructure with sufficiently thin layers, quantum interference effects begin to appear prominently in the motion of the electrons. The simplest structure in which these may be observed is a quantum well, which simply consists of a thin layer of a narrower-gap semiconductor between thicker layers of a wider-gap material."

Recessive: Refers to the member of a pair of alleles that fails to be expressed in the phenotype of the organism when the dominant member is present. Also refers to the phenotype of an individual that has only the recessive allele.

Scanning Near Field Optical Microscopy: A method for observing local optical properties of a surface that can be smaller than the wavelength of the light used. A type of scanning probe microscopy that maps the near-field optical properties of an interface.

Scanning Probe Microscopy (SPM): A method for imaging nanoscale features of surfaces by scanning a sensor (probe) over a surface. Near-field effects such as tunneling, van der Waals forces, local fields and more are serially detected at localized points on the surface and used to create an SPM image.

Scanning Thermal Microscopy (SThM): A type of scanning probe microscopy that maps the local temperature and thermal conductivity of an interface. A method for observing local temperatures and temperature gradients on a surface. [NTN]

Scanning Tunneling Microscope (STM): An instrument able to image conducting surfaces to atomic accuracy; has been used to pin: molecules to a surface. A device in which a sharp conductive tip is moved across a conductive surface close enough to permit a substantial tunneling current (typically a nanometer or less). In a common mode of operation, the voltage is kept constant and the current is monitored and kept constant by controlling the height of the tip above the surface; the result, under favorable conditions, is an atomic-resolution map of the surface reflecting a combination of topography and electronic properties. The STM has been used to manipulate atoms and molecules on surfaces. A type of scanning probe microscopy that maps the local electrical properties of an interface.

Schwann cell: Glial cell responsible for forming myelin sheaths in the peripheral nervous system.

Science: The process of developing a systematized knowledge of the world through the variation and testing of hypotheses. The pursuit of knowledge and understanding, from the Latin term *scientia,* which means 'knowledge'.

Tesla [T]: Unit of magnetic induction: 1T = 1 weber/m^2 (also, 1T = 10^4 gauss).

Thermistor: A temperature-measuring device, that contains a resistor or semiconductor whose resistance varies with temperature.

Thermocouple: A temperature-measuring device, which contains a pair of end-joined dissimilar conductors in which an electromotive force is developed by thermoelectric effects when the joined ends and the free ends of the conductors are a different temperature.

Thiol: An SH group, or a molecule containing one. Also known as a sulfhydryl or mercapto group.

Threshold: The smallest input signal that will cause a readable change in the output signal.

Thrombin : The enzyme that converts *fibrinogen* to *fibrin* to form a blood clot. Thrombin circulates in the inactive form, prothrombin.

Thylakoid: Flattened sac of membrane in a chloroplast that contains pigment and carries out the light-gathering reactions of photosynthesis. Stacks of thylakoids form the grana of chloroplasts.

Upload: (a) To transfer the consciousness and mental structure of a person from a biological matrix to an electronic or informational matrix (this assumes that the strong AI postulate holds). The term "Downloading" is also sometimes used, mainly to denote transferring the mind to a slower or less spacious matrix. (b) The resulting infomorph person. [The origin of the term is uncertain, but obviously based on the computer technology term 'uploading' (loading data into a mainframe computer).] [AS]

URL: Universal Resource Locator — address of a World Wide Web site.

Vasculocyte: In medical nanorobotics, a theorized (nanorobotic) device capable of performing repairs of an injured vascular luminal surface.

Vasculography: A physical description (and mapping) of the human vascular system.

Vasculoid: Is a single, complex, multisegmented nanotechnological medical robotic system capable of duplicating all essential thermal and biochemical transport functions of the blood, including circulation of respiratory gases, glucose, hormones, cytokines, waste products, and cellular components. The vasculoid [concept] is a single, complex, multisegmented nanotechnological medical robotic system capable of duplicating all essential thermal and biochemical transport functions of the blood, including circulation of respiratory gases, glucose, hormones, cytokines, waste products, and cellular components. Vasculoid is a single, complex, multisegmented nanotechnological medical robotic system capable of duplicating all essential thermal and biochemical transport functions of the blood, including circulation of respiratory gases, glucose, hormones, cytokines, waste products, and cellular components.

VC (Venture capitalist): A person or organisation that invests in promising companies during early development stages.

Vector: In cell biology, an agent (virus or plasmid) used to transmit genetic material to a cell or organism.

Vegetal pole: The end at which most of the yolk is located in an animal egg. The end opposite the animal pole.

Virtual Nanomedicine: Using VR to perform surgery and other functions inside the body.

Virtual reality system: A combination of computer and interface devices (goggles, gloves, etc.) that presents a user with the illusion of being in a three dimensional world of computer-generated objects.

Virus: A parasite (consisting primarily of genetic material) that invades cells and takes over their molecular machinery in order to copy itself. A small replicator consisting of little but a package of DNA or RNA which, when injected into a host cell, can direct the cell's molecular machinery to make more viruses. Particle consisting of nucleic acid (RNA or DNA) enclosed in a protein coat and capable of replicating within a host cell and spreading from cell to cell. Often the cause of disease.

Viscoelasticity: A type of deformation exhibiting the mechanical characteristics of viscous flow and elastic deformation.

Viscosity: The ratio of the magnitude of an applied shear stress to the velocity gradient that it produces; in other words: a measure of a noncrystalline material's resistance to permanent deformation.

Wet Nanotechnology: The study of biological systems that exist primarily in a water environment. The functional nanometer-scale structures of interest here are genetic material, membranes, enzymes and other cellular components. The success of this nanotechnology is amply demonstrated by the existence of living organisms whose form, function, and evolution are governed by the interactions of nanometer-scale structures. [Rice University]

Whisker: A very thin, single crystal of high perfection which has an extremely large length-to-diameter ratio. Whiskers are used as the reinforcing phase in some composites.

White biotechnology: An emerging field within modern biotechnology that serves industry. It uses living cells like moulds, yeasts or bacteria, as well as enzymes to produce goods and services. Living cells can be used as they are or improved to work as "cell factories" to produce enzymes for industry.

White blood cell (leucocyte): Nucleated blood cell lacking hemoglobin; includes lymphocytes, neutrophils, eosinophils, basophils, and monocytes.

Index

J

L

M

N

O

P

R

S

T

V

W